CYTOKINES IN VETERINARY MEDICINE

Cytokines in Veterinary Medicine

Edited by

Virgil E.C.J. Schijns, PhD*

and

Marian C. Horzinek, DVM, PhD
Virology Unit
Department of Infectious Diseases and Immunology
Faculty of Veterinary Medicine
Utrecht University
The Netherlands

**Present address*
Immunology R&D
Intervet International BV
Boxmeer
The Netherlands

CAB INTERNATIONAL

CAB INTERNATIONAL
Wallingford
Oxon OX10 8DE
UK

Tel: +44 (0)1491 832111
Fax: +44 (0)1491 833508
E-mail: cabi@cabi.org

CAB INTERNATIONAL
198 Madison Avenue
New York, NY 10016-4314
USA

Tel: +1 212 726 6490
Fax: +1 212 686 7993
E-mail: cabi-nao@cabi.org

A catalogue record for this book is available from the British Library, London, UK

ISBN 0 85199 209 9

Typeset in Sabon by AMA Graphics Ltd, UK
Printed and bound in Great Britain by
Biddles Ltd, Guildford and King's Lynn

Contents

Contributors

Adair, B.M., *Veterinary Sciences Division, Stormont, Belfast BT4 3SD, UK*
Andrews, A., *Centre for Animal Biotechnology, The University of Melbourne, Parkville, Victoria 3052, Australia*
Asthana, D., *Department of Surgery, Microbiology and Immunology at the University of Miami School of Medicine and the Miami Veterans Administration Hospital, PO Box 016310, Miami, Florida 33101, USA*
Babiuk, L.A., *Veterinary Infectious Disease Organization, 120 Veterinary Road, University of Saskatchewan, Saskatoon, Saskatchewan, Canada S7N 5E3*
Baca-Estrada, M.E., *Veterinary Infectious Disease Organization, 120 Veterinary Road, University of Saskatchewan, Saskatoon, Saskatchewan, Canada S7N 5E3*
Barcham, G., *Centre for Animal Biotechnology, The University of Melbourne, Parkville, Victoria 3052, Australia*
Billiau, A., *Laboratory of Immunobiology, Rega Institute for Medical Research, Minderbroedersstraat 10, B-3000 Leuven, Belgium*
Blecha, F., *Department of Anatomy and Physiology, College of Veterinary Medicine, 1600 Denison Avenue, Kansas State University, Manhattan, Kansas 66506-5602, USA*
Bradley, W.G., *Eckerd College, 4200 54th Avenue South, St Petersburg, Florida 33711, USA*
Brown, J., *Division of Immunobiology, Moredun Research Institute, 408 Gilmerton Road, Edinburgh EH17 7JH, UK*

Brown, W.C., *Department of Veterinary Microbiology and Pathology, College of Veterinary Medicine, Washington State University, Pullman, Washington 99164, USA*

Cameron, A., *CSL Ltd, 45 Poplar Road, Parkville, Victoria 3052, Australia*

Carreno, M., *Department of Surgery, Microbiology and Immunology at the University of Miami School of Medicine and the Miami Veterans Administration Hospital, PO Box 016310, Miami, Florida 33101, USA*

Coran, A.G., *CS Mott Children's Hospital, University of Michigan Medical Center, Ann Arbor, Michigan 48109-0245, USA*

Drongowski, R.A., *CS Mott Children's Hospital, University of Michigan Medical Center, Ann Arbor, Michigan 48109-0245, USA*

Elhay, M., *TB Research Unit, Statens Seruminstitut, Artillerivej, 5, Copenhagen, 2300 S, Denmark*

Entrican, G., *Division of Immunobiology, Moredun Research Institute, 408 Gilmerton Road, Edinburgh EH17 7JH, UK*

Esquenazi, V., *Department of Surgery, Microbiology and Immunology at the University of Miami School of Medicine and the Miami Veterans Administration Hospital, PO Box 016310 Miami, Florida 33101, USA*

Estes, D.M., *Department of Veterinary Pathobiology, College of Veterinary Medicine, University of Missouri, Columbia, Missouri 65211, USA*

Flint, A.P.F., *Department of Physiology and Environmental Science, Faculty of Agricultural and Food Sciences, University of Nottingham, Sutton Bonington Campus, Loughborough, Leicestershire LE12 5RA, UK*

Fuller, L., *Department of Surgery, Microbiology and Immunology at the University of Miami School of Medicine and the Miami Veterans Administration Hospital, PO Box 016310, Miami, Florida 33101, USA*

Godson, D.L., *Veterinary Infectious Disease Organization, 120 Veterinary Road, University of Saskatchewan, Saskatoon, Saskatchewan, Canada S7N 5E3*

Grünig, G., *DNAX, 901 California Avenue, Palo Alto, California 93404, USA*

Haagmans, B.L., *Virology Unit, Department of Infectious Diseases and Immunology, Veterinary Faculty, Utrecht University, Yalelaan 1, de Uithof, 3584 CL Utrecht, The Netherlands*

Hamers, C., *Department of Immunology/Vaccinology, Faculty of Veterinary Medicine, University of Liège, B43 bis, B-4000 Liège, Belgium*

Hanon, E., *Department of Immunology/Vaccinology, Faculty of Veterinary Medicine, University of Liège, B43 bis, B-4000 Liège, Belgium*

Hill, F.W.G., *Department of Clinical Studies, Faculty of Veterinary Science, University of Zimbabwe, Mount Pleasant, Harare, Zimbabwe*

Himmler, A., *Boehringer Ingelheim R&D Vienna, Bender + Co GmbH, Dr Boehringer-Gasse 5–11, A-1121 Vienna, Austria*

Horohov, D.W., *Department of Veterinary Microbiology and Parasitology, School of Veterinary Medicine, Louisiana State University, Baton Rouge, Louisiana 70803, USA*

Horzinek, M.C., *Virology Unit, Department of Infectious Diseases and Immunology, Veterinary Faculty, Utrecht University, Yalelaan 1, de Uithof, 3584 CL Utrecht, The Netherlands*

Innes, E.A., *Division of Immunobiology, Moredun Research Institute, 408 Gilmerton Road, Edinburgh EH17 7JH, UK*

Kaiser, P., *Institute for Animal Health, Compton, Berkshire RG20 7NN, UK*

Klei, T.R., *Department of Veterinary Microbiology and Parasitology, School of Veterinary Medicine, Louisiana State University, Baton Rouge, Louisiana 70803, USA*

Klein, W.R., *Department of General and Large Animal Surgery, Veterinary Faculty, Utrecht University, Yalelaan 1, de Uithof, 3584 CL Utrecht, The Netherlands*

McInnes, C.J., *Division of Immunobiology, Moredun Research Institute, 408 Gilmerton Road, Edinburgh EH17 7JH, UK*

McNulty, M.S., *Veterinary Sciences Division, Stormont, Belfast BT4 3SD, UK*

May, S.A., *Department of Farm Animal and Equine Medicine and Surgery, Royal Veterinary College, University of London, Hawkshead Lane, North Mymms, Hatfield, Hertfordshire AL9 7TA, UK*

Miller, J., *Department of Surgery, Microbiology and Immunology at the University of Miami School of Medicine and the Miami Veterans Administration Hospital, PO Box 016310, Miami, Florida 33101, USA*

Nash, A., *Centre for Animal Biotechnology, The University of Melbourne, Parkville, Victoria 3052, Australia*

den Otter, W., *Department of Functional Morphology, Veterinary Faculty, Utrecht University, Yalelaan 1, de Uithof, 3584 CL Utrecht, The Netherlands*

Pastoret, P-P., *Department of Immunology/Vaccinology, Faculty of Veterinary Medicine, University of Liège, B43 bis, B-4000 Liège, Belgium*

Pinelli, E., *Department of Immunology, Faculty of Veterinary Medicine, Yalelaan 1, 3584 CL Utrecht, The Netherlands*

Robinson, J.A., *Department of Veterinary Clinical Sciences, School of Veterinary Medicine, Louisiana State University, Baton Rouge, Louisiana 70803, USA*

Rothel, J.S., *CSIRO, Division of Animal Health, Private Bag No.1, Parkville, Victoria, Australia 3052*

Ruitenberg, E.J., *Central Laboratory of The Netherlands Blood Transfusion Service, Amsterdam, The Netherlands*

Rutten, V.P.M.G., *Immunology Unit, Department of Infectious Diseases and Immunology, Veterinary Faculty, Utrecht University, Yalelaan 1, de Uithof, 3584 CL Utrecht, The Netherlands*

Schat, K.A., *Unit of Avian Medicine, Department of Microbiology and Immunology, College of Veterinary Medicine, Cornell University, Ithaca, New York 14853, USA*

Schijns, V.E.C.J., *Virology Unit, Department of Infectious Diseases and Immunology, Veterinary Faculty, Utrecht University, Yalelaan 1, de Uithof, 3584 CL Utrecht, The Netherlands (present address: Immunology R&D, Intervet International BV, PO Box 31, 5830 AA Boxmeer, The Netherlands)*

Swiderski, C.E., *Department of Veterinary Microbiology and Parasitology, School of Veterinary Medicine, Louisiana State University, Baton Rouge, Louisiana 70803, USA*

Vandenbroeck, K., *Laboratory of Immunobiology, Rega Institute for Medical Research, Minderbroedersstraat 10, B-3000 Leuven, Belgium*

Wood, P.R., *CSIRO, Division of Animal Health, Private Bag No.1, Parkville, Victoria, Australia 3052*

Zhen, S., *Department of Surgery, Microbiology and Immunology at the University of Miami School of Medicine and the Miami Veterans Administration Hospital, PO Box 016310, Miami, Florida 33101, USA*

Zucker, K., *Department of Surgery, Microbiology and Immunology at the University of Miami School of Medicine and the Miami Veterans Administration Hospital, PO Box 016310, Miami, Florida 33101, USA*

List of Abbreviations

2′5′OAS	2′-5′-oligo(A) synthetase-RNase
aa	amino acid(s)
AC	accessory cells
ADCC	antibody-dependent cytotoxicity
AIDS	acquired immunodeficiency syndrome
AM	alveolar macrophage(s)
AMCF	alveolar macrophage chemotactic factor
APC	antigen presenting cell(s)
BAL	bronchoalveolar lavage
BCG	Bacillus Calmette-Guérin
BHV	bovine herpesvirus
BOSCC	bovine ocular squamous cell carcinoma
bp	base pair
BVDV	bovine viral diarrhoea virus
BVPC	bovine vulval papilloma carcinoma
CAA	chicken anaemia agent
CAT	chloramphenicol acetyl transferase
CAV	chicken anaemia virus
CEF	chick embryo fibroblast
CHO	chinese hamster ovary
CMI	cell-mediated immunity
Con A	concanavalin A
CrFK	Crandell feline kidney
CNS	central nervous system

CSF	colony-stimulating factor(s)
CTL	cytotoxic T lymphocyte(s)
DK	dog kidney 6247
DTH	delayed-type hypersensitivity
ELISA	enzyme-linked immunosorbent assay
FAIDS	feline acquired immunodeficiency syndrome
FIP(V)	feline infectious peritonitis (virus)
FIV	feline immunodeficiency virus
GAS	gamma activation site
GBP	guanylate-binding protein
G-CSF	granulocyte-colony stimulating factor
GM-CSF	granulocyte-macrophage colony stimulating factor
HIV	human immunodeficiency virus
IBD(V)	infectious bursal disease (virus)
IBR(V)	infectious bovine rhinotracheitis (virus)
ICS	interferon consensus sequence
i.d.	intradermal
IDO	indoleamine 2,3-dioxygenase
IFN	interferon
Ig(s)	immunoglobulin(s)
IL	interleukin
IL-1R	IL-1 receptor
i.m.	intramuscular(ly)
i.n.	intranasal(ly)
IRF	IFN regulatory factor
ISRE	interferon-stimulated response elements
i.v.	intravenous
kDa	kilodaltons
KT1	kidney cell associated antigen
LAK	lymphokine activated killer
LPS	lipopolysaccharide
mAb(s)	monoclonal antibody (antibodies)
MCF	mononuclear cell factor
MCP	monocyte chemotactic protein(s)
MDBK	Madin-Darby bovine kidney
MDCK	Madin-Darby canine kidney
MD(V)	Marek's disease (virus)
mgF	myelomonocytic growth factor
MHC	major histocompatibility complex
MLIC	primary autolymphoproliferative response of T lymphocytes to islets of Langerhans
MLKC	primary autolymphoproliferative response of T lymphocytes to kidney cells
n	native

NAP	neutrophil-activating peptide
NDV	Newcastle disease virus
ng	nanogram
NK	natural killer
NO	nitric oxide
NRAMP	natural resistance-associated macrophage protein
NSAID	non-steroidal anti-inflammatory drugs
OV	orf virus
PBMC	peripheral blood mononuclear cell(s)
PBP	platelet basic protein
PCR	polymerase chain reaction
pg	picogram
PGE2	prostaglandin E2
p.i.	post (after)-infection/-inoculation/-injection
PMN	polymorphonuclear neutrophils
poly (I : C)	polyinosinic : polycytidylic acid
PPD	purified protein derivative
PRE	pig repetitive element
PWM	pokeweed mitogen
r	recombinant
rBo	recombinant bovine
rCa	recombinant canine
rEq	recombinant equine
rFe	recombinant feline
rHu	recombinant human
rOv	recombinant ovine
rPo	recombinant porcine
RAP-1	rhoptry-associated protein-1
RLN	regional lymph node
RT-PCR	reverse transcription-polymerase chain reaction
SCF	stem cell factor
SIDT	single intradermal tuberculin test
SPF	specified pathogen-free
TCR	T cell receptor
TGEV	transmissible gastroenteritis virus
TGF	transforming growth factor
Th	T helper
TIL	tumour infiltrating lymphocytes
TNF	tumour necrosis factor
TPO	thrombopoietin
VL	visceral leishmaniasis
VSV	vesicular stomatitis virus

Preface

This book was compiled at the same time that one of us was engaged in founding and organizing the Graduate School of Animal Health at the Faculty of Veterinary Medicine, Utrecht University. In this country, gathering veterinary research topics around a few common themes is a political and financial necessity, and after a lengthy incubation period the first progeny hatched – the cluster 'Intercellular communication in proliferation and differentiation'. It is part of a core programme focused on disturbances of central homeostatic mechanisms caused by humoral, nutritional and/or genetic factors that lead to aberrations in the growth and differentiation of animal cell and tissue types, to disrupted embryonic and fetal development, and to metabolic disorders. Intercellular communication is paramount for these mechanisms, and cytokines (amongst others) are communicators *par excellence*.

And so are the contributors to this volume. It gives us great pleasure to acknowledge the speed and efficiency of the authors, their helpfulness, patience and above all their willingness to share unpublished data. In contrast to similar exercises, this one was a pleasure. We should also like to thank Mr Tim Hardwick from CAB INTERNATIONAL for making quite clear what the publisher wanted and for providing guidance. Our secretary Dr Jolanda van Luipen did much of the desk editing and cross-checking, often in her own time.

Dr Virgil E.C.J. Schijns
Prof. Dr Dr h.c. Marian C. Horzinek

Utrecht, December 1996

Introduction

V.E.C.J. Schijns and M.C. Horzinek

Cytokines form a group of secreted regulatory proteins that act as intercellular communication signals in a variety of cellular responses such as embryonic development, haematopoiesis, stress, inflammation, immunity and tissue repair. They are heterogeneous with respect to molecular mass which ranges between 6 and 70 kDa, and often contain protective carbohydrates and intramolecular disulphide bonds. Most cytokines are secreted in minute amounts, at pg to ng ml^{-1} concentrations that are active only locally, in the immediate vicinity of the producer cell. Ligand binding to only a few receptors on the cell membrane is sufficient to activate intracellular signalling pathways.

Cytokines are produced by a variety of cell types. A given cytokine may both stimulate and inhibit the production of several others, leading to cascades of synergistic or cross-inhibitory interactions. Most cytokines exert multiple biological activities (pleiotropy) that may overlap with those of others (redundancy). For example, IFN-γ displays antiviral activity, as does tumour necrosis factor (TNF), but it also stimulates the expression of major histocompatibility complex (MHC) molecules, an activity shared with other interferons. This suggests that the loss of one cytokine activity is compensated by that of others, and it stresses the vital importance of these interactions.

Since the early 1980s, an increasing number of human and murine cytokine genes has been cloned, sequenced and expressed. During the last decade we have also seen more and more studies using recombinant cytokines and antibodies neutralizing their activities, as well as investigations involving gene-deleted ('knock-out') mice lacking cytokines or their receptors. These studies established the vital *in vivo* role of cytokines for the immune, endocrine and neuronal system. Of particular importance was the discovery by Mosmann and Coffman of a differential development of T helper (Th) cell phenotypes and their polar cytokine profiles. A concept emerged which contributed to the understanding of the antagonism between cell-mediated immunity (CMI) and antibody-mediated immunity. Numerous studies have revealed the tremendous importance of the local cytokine environment on the quality and magnitude of starting and ongoing immune responses *in vitro* and *in vivo*. Also knowledge of cytokines in man has grown remarkably during the past 5 years; many observations made in rodents have been found to apply also for humans.

In comparison with the mouse and man only little is known about cytokines and their activities in animals of veterinary importance. This is due to the species specificity of many cytokines and to the small number of groups working in this area. However, knowledge has considerably expanded during recent years, predominantly of cytokine homologues in food production animals – and lately also in companion animals – thanks mainly to recombinant DNA and polymerase chain reaction (PCR) technology. Many ruminant cytokines have now been cloned and most of them expressed as recombinant proteins. Several cytokine-specific enzyme-linked immunoassays (ELISA) and bioassays are available, and more are in the development phase. However, application of these reagents in veterinary medicine is still in its infancy. This book addresses basic research on cytokines and its clinical potential for veterinary medicine.

CYTOKINES AS CORRELATES OF DISEASE

As in man and mice most studies of veterinary cytokines have focused on their immunoregulatory properties. The detection of cytokines as immunological markers may be useful for monitoring disease progression, facilitating prognosis, increasing vaccine efficacy and improving meat quality in food-production animals. A good example is the use of an IFN-γ assay as an alternative to diagnose tuberculosis in cattle through measurement of cellular immune responses (Chapter 3).

Joint disease, in particular osteoarthritis, is very important in the equestrian industry. Elevated IL-1-like activity has been identified in the synovial fluids of horses with a variety of arthritides, including osteoarthritides, osteochondrosis, traumatic and infectious arthritis. Levels of IL-1, but also of TNF and IL-6, tended to be highest in animals with severe arthritis. Furthermore, TNF-α activity has been detected in adult horses and foals after experimental induction of endotoxaemia, in horses with colic attributable to gastrointestinal tract disease and in foals with presumed sepsis (Chapter 11).

In cattle, correlations between circulating TNF-α and the severity of *Escherichia coli*-induced mastitis were observed (Chapter 1). Taken together, these studies indicate that the association between high or inappropriate expression of cytokines with disease states in animals may become important for diagnostic tests with prognostic value.

COMPARATIVE CYTOKINOLOGY

Information collected for animal species of veterinary importance, especially through *in vivo* studies, may be applicable to man. The recent cloning of duck IFNs opens the possibility for treatment studies of chronic duck hepatitis B virus infection in White Pekin ducks – one of the few practicable models of human hepatitis B (Chapter 20). Most human cytokines are more homologous to those of ruminants and carnivores than to those of rodents. Measurement of the *in vivo* dynamics of local cytokine production is easily achieved in ruminants (e.g. sheep) after cannulation of lymphatic vessels around single lymph nodes at the site of immunization or infection (Chapter 7). Also, studies of infectious bovine rhinotracheitis (IBR) in cattle may help to define beneficial and clinico-pathological features of cytokines in viral and bacterial respiratory disease in humans (Chapter 6). The study of cytokine responses during *Leishmania* infections in the dog, which is the main parasite reservoir, may benefit strategies to prevent visceral leishmaniasis in both dog and man (Chapter 14). The remarkable similarities between the feline immunodeficiency syndrome and human AIDS support the contention that cats infected with feline immunodeficiency virus (FIV) may serve a model function for the human condition. It will be especially relevant for studies concerning antiviral drugs and vaccines, AIDS-associated dementia, and the role of the cytokine response in disease progression (Chapter 18).

THE TYPE I–TYPE II PARADIGM

To further our understanding of appropriate and inappropriate immune responses in livestock and companion animals it will be important to follow the kinetics of cytokine expression in resolving and exacerbating disease conditions. However, the existence of a dichotomy in type 1 and type 2 cytokine responses – a paradigm in immunology – has yet to be demonstrated in many animal species.

An inverse relationship between IFN-γ responses and antibody production has been observed in cattle infected with *Mycobacterium tuberculosis* (Chapter 3). The dichotomy in cross-regulation of type 1 and type 2 responses in cattle and humans appears less distinct than that observed in the mouse (Chapter 2). In *Leishmania infantum*-infected dogs, resistance is associated with high levels of type I cytokines, as compared to low levels in sick dogs (Chapter 14). Little information is available regarding type I and type II cytokine responses in horses. However, increased mRNA levels of IL-4 have been noted in colonic lymph nodes from a pony undergoing spontaneous reduction in faecal egg counts of *Strongylus vulgaris*. Furthermore, different cytokine profiles have been noted in horses infected with or vaccinated against influenza (Chapter 11). Analysis of mRNA levels during the acute phase of FIV infection in cats revealed a rise in IL-4, IL-6, IL-10 and TNF-α gene expression, indicative for a Th1 > Th2 shift (Chapter 18).

IMMUNE ENHANCEMENT

Research has gained momentum after the observation in mice that application of cytokines or their antagonists is especially suitable for prophylactic or therapeutic modulation of immune responses. Conceivably, cytokine-based immune stimulation during non-responsiveness, and reconstitution treatment in immunodeficiencies are practicable in vaccination or immunotherapy. Being valuable in the veterinary field, these studies may also serve as models for applications in man; apart from IFNs and IL-2 little is known about the side effects of cytokine treatment in humans.

Promising results consisting of potentiated antitumour responses and increased tumour regression were obtained after low-dose locoregional therapy with IL-2 in cattle suffering from ocular squamous cell carcinoma and the vulval papilloma/carcinoma complex (Chapter 5). Recombinant bovine (rBo) IFN-α and IFN-γ given before bovine herpesvirus (BHV)-1 infection significantly reduced the clinical signs and mortality associated with subsequent bacterial infection (Chapters 1 and 6). Intramammary administration of IL-1β, IL-2 or IFN-γ augmented local mammary gland defence mechanisms against *Staphylococcus aureus* and *E. coli* (Chapter 1).

In pigs, cross-reacting (human IL-2, bovine IL-1) and homologous recombinant cytokines (porcine TNF and IFN-γ) have been shown to influence non-specific and specific immune functions. IFN-α administration was beneficial in transmissible gastroenteritis virus (TGEV) and *Actinobacillus pleuropneumoniae* infections. Recombinant porcine (rPo) IFN-γ reversed dexamethasone-induced immunosuppression, while cross-reacting rBoIL-1 reduced the severity of *Streptococcus suis* infection (Chapter 9). Further studies must show the usefulness of these immunoregulatory cytokines in strategies to prevent periparturient immunosuppression and unresponsiveness observed in weaning pigs.

ADJUVANTS

Cytokines are attractive as adjuvants in rational vaccine design. Cytokine-directed immunization aiming at the induction of desired Th cell responses should lead to potent vaccines against particular intracellular or extracellular pathogens. In cattle, rBoIL-1β and rBoIL-2 act as effective and safe adjuvants in bovine vaccines containing BHV-1, BHV-1/parainfluenza-3 combinations and killed bovine viral diarrhoea virus (BVDV; Chapters 1 and 8). In sheep, recombinant ovine (rOv) IL-1β acts as an effective, safe and stable adjuvant in a commercial multivalent clostridial vaccine and an experimental vaccine against *Lucilia cuprina* (Chapter 8).

Cytokines can be applied as immunostimulatory adjuvants also in antimicrobial and antitumour therapy, in combination with chemotherapy, and especially in immunocompromised hosts. We may expect adjuvant studies using recently cloned cytokines from various species – cytokines that influence B cells, cytolytic T cells and Th cells (e.g. IL-12, IFN-γ and IL-4; Chapter 19).

IMMUNE DEVIATION

Murine models have demonstrated the feasibility of immune deviation during aberrant immune responses using inhibitory cytokines or cytokine antagonists. In dogs, the administration of antibodies against canine IFN-γ prolonged the survival of renal allografts (Chapter 15). Neutralization of TNF-α activity by infusion of anti-TNF-α antibodies has been shown to improve the clinical course of endotoxaemia and Gram-negative shock in horses (Chapter 11). Natural or engineered cytokine antagonists like soluble receptors, cytokine mutants or peptide antagonists are likely to become useful in veterinary medicine for the treatment of pathological conditions, including autoimmune and inflammatory diseases.

DISCREPANCIES

The identification of types I and II chicken IFNs and of the avian homologue to the recently recognized mammalian IL-15 demonstrates the biological importance of the cytokine system and its conservation: birds and mammals diverged some 350 million years ago (Chapter 20). The comparison of cytokine activities in different species is of interest from an evolutionary point of view and may even open new therapeutic avenues. However, some discrepancies have emerged from cytokine research in the veterinary environment. Thus porcine IFN-γ is expressed in tropho-ectodermal cells, aside from lymphocytes and natural killer (NK) cells. Porcine IL-1β exists in two forms (Chapter 10). The structure and function of most ovine cytokines are similar to their human counterparts, but IL-3 stimulates the production of mast cell colonies, an activity it shares with rodent, but not with human IL-3 (Chapter 7). In ruminants, an unequivocal example of endocrine regulation by cytokines has been found: during early gestation IFN is expressed at high levels in the trophoblast in a narrow time window – it prevents blocking of luteolysis in pregnancy (Chapter 4).

PROSPECTS

In this book the role of cytokines in the pathogenesis, diagnosis and therapy of animal diseases is reviewed, as is the use of animal models to investigate diseases analogous to those in humans.

For the veterinary scene, several fields of application can be anticipated:

- The detection and measurement of cytokines can be used as an immunological yardstick to monitor disease progression and prognosis, vaccine efficacy and meat quality in food-production animals.
- In production animals the health status can be improved by cytokine application, leading to increased economic and reproductive performance, for example during periparturient immunosuppression.
- Cytokines have great potential as vaccine adjuvants or as adjuncts to specific antimicrobial and antitumour chemotherapy.

The present volume also demonstrates that cytokine research in veterinary medicine is yet in its adolescence. Cytokine genes known to exist in mice and humans need to be identified in other species, expressed and studied. In the forthcoming years we may expect an avalanche of data from cytokine research in fundamental and applied veterinary medicine. New applications may become available well before they are approved in humans. Future studies will reveal whether the promise of cytokine research held by rodent models can be fulfilled, to the benefit of animal patients. In addition to providing basic information about cytokine functions *in vivo*, such studies will undoubtedly increase our understanding of human disease conditions.

I

Bovine Cytokines

Edited by L.A. Babiuk

Applications of Bovine Cytokines 1

D.L. Godson, M.E. Baca-Estrada, and L.A. Babiuk

Veterinary Infectious Disease Organization, 120 Veterinary Road, University of Saskatchewan, Saskatoon, Saskatchewan, Canada S7N 5E3

INTRODUCTION

Research during the past decade has clearly demonstrated that cytokines play a central role in host immune regulation. As such, these molecules are of prime interest to researchers concerned with modulating immune responses and reproductive performance. As integral components involved in immune regulation, they are essential ingredients in the study of pathogenesis of disease. As we better understand how cytokines choreograph host responses, we can use that information as follows:

1. To characterize the presence of cytokines which may be of diagnostic significance, whether that be by detection of cytokines in blood or secretions from the animal, or after *in vitro* stimulation of leukocyte populations to indicate the presence of specific immunity, or as a marker of genetic innate resistance.
2. To develop strategies for cytokine immunomodulation to improve the host response. This would include stimulation of innate immune mechanisms such as the expansion and maturation of leukocyte populations, or enhancement of effector functions such as phagocytosis. Cytokines can also be used as adjuvants to improve the quantity or quality (Th1 versus Th2) of the response to vaccination.
3. To enhance reproductive efficiency of animals.

In addition to their pivotal role in immune regulation, cytokines are also part of the more complicated web of immune, neural and endocrine

 Cytokines in Veterinary Medicine (eds V.E.C.J. Schijns and M.C. Horzinek)

interactions responsible for growth, development and maintenance of homeostasis in the host. As such, cytokines exert some of their pleiotrophic effects in these other systems. Thus, understanding of the roles of cytokines in normal body function may see non-immune applications for cytokine therapy in addition to those presently proposed.

The present series of chapters on bovine cytokines describes the production of cytokines by different cells and their application in disease models as well as in reproduction. In this chapter, applications of cytokines as adjuncts for the prevention and treatment of disease and as potential diagnostic aids will be described.

DETECTION OF CYTOKINES – DISEASE DIAGNOSIS AND PROGNOSIS

Detection of cytokines allows us to tap into the messages of the immune system. Unfortunately, most cytokines act over short distances, necessitating precise sampling (e.g. lymph cannulation, or collection of local secretions) for detection. For instance, in calves suffering from BHV-1 infection, IFN activity was readily detected in nasal secretions, but was only marginally elevated in serum (Bielefeldt Ohmann and Babiuk, 1985). However, some cytokines are released into systemic circulation to amplify and expand the response, or in some cases, the intensity or extensiveness of the local response may allow spill-over of detectable levels of cytokines into the circulation. Thus, the level and extent of immune activation may be determined from measurement of levels of cytokines in serum or secretions. For example, in *E. coli* mastitis, TNF-α could be detected in both serum and milk (Sordillo and Peel, 1992). The concentration of TNF-α was especially elevated in cows that developed more severe clinical symptoms and eventually died (Sordillo and Peel, 1992). Thus, detection of cytokines may be used not only to detect disease but also to determine disease severity and predict outcome.

Since cytokines are often not detectable systemically, an alternative to direct detection is measurement of certain biological effects induced by cytokines. In many cases, these biological effects are more easily detected and of longer duration than the presence of the cytokine in circulation. For instance, 2′,5′-oligoadenylate synthetase activity can be measured in peripheral blood after IFN administration (Bielefeldt Ohmann *et al.*, 1989c) and thus may be a sensitive way of evaluating IFN production during viral infection (Bielefeldt Ohmann *et al.*, 1989d). The level of serum acute phase proteins, such as haptoglobin, rises markedly after bacterial infection (Godson *et al.*, 1996) and may serve to indicate the production of proinflammatory cytokines such as IL-1 (Godson *et al.*, 1995).

As well as detecting *in vivo* cytokine production, *in vitro* stimulation of appropriate leukocyte populations, and measurement of the cytokines produced, can be used to detect antigen-specific responses, such as the diagnosis of tuberculosis by measurement of IFN-γ production (see Chapter 4). Alternatively, the quality and quantity of cytokine response to certain antigens or pathogens may be used as a marker of genetic resistance. For instance, the constitutive production of nine of ten cytokines tested is higher in macrophages from cattle which have been selected for resistance to brucellosis infection (Adams *et al.*, 1995). Thus, our ability to detect cytokines, or the effect of cytokines, should provide us with an opportunity to incorporate this knowledge into future management tools in veterinary medicine.

IMMUNOMODULATION

Adjuvant Activity

The ability of cytokines to influence not only the magnitude but also the type of immune response when co-administered with a vaccine antigen has been demonstrated in several animal models (reviewed in Heath and Playfair, 1992), including cattle (Reddy *et al.*, 1990; Hughes *et al.*, 1991). This is especially the case with those cytokines capable of exerting their immunoregulatory activity by direct action upon antigen presenting cells (APC) and/or T cells. It now appears that protection against infectious agents is determined, in part, by the pattern of cytokines secreted by different subsets of T lymphocytes (Mosmann *et al.*, 1986; Mosmann and Coffman, 1989a, b). For example, protection against intracellular pathogens is considered to be mediated by cellular immunity provided by Th1 cells, whereas humoral immune responses are preferentially regulated by the cytokines secreted by the Th2 subset. In this context, cytokines capable of influencing the development of Th1 or Th2 responses offer the potential of selectively tailoring vaccines to be capable of inducing more effective immune responses.

Only a few adjuvants are licensed for veterinary use, because most adjuvants induce strong tissue reactions or have residue problems. Therefore, the use of recombinant cytokines as vaccine adjuvants holds great promise with regard to the production of new types of highly effective and safe vaccines.

The use of cytokines in the modulation of immune responses to vaccines in cattle has been limited to few molecules. The adjuvant effects of IL-1β have been demonstrated using modified live and subunit vaccines against BHV-1 (Reddy *et al.*, 1990; Gao *et al.*, 1995). Cattle immunized with modified live BHV-1 vaccine and IL-1β, at different sites, had significant increases in BHV-1-specific cytotoxic responses and virus-neutralizing antibody titres (Reddy *et al.*, 1990, 1993). Similarly, subcutaneous administration of a subunit vaccine against BHV-1 and IL-1β at the base of the ear primed the nasal mucosal surface for an IgA response following challenge and protected animals against disease (Gao *et al.*, 1995). The authors concluded that in this model, induction of mucosal immune responses can be achieved by immunization at sites that share lymphatic drainage with mucosal surfaces. Interestingly, the doses of IL-1β used to potentiate mucosal immune responses by subcutaneous immunization (Gao *et al.*, 1995) were approximately 200 times less than the doses used for systemic immunization (Reddy *et al.*, 1990, 1993). These observations may indicate the importance of immunization site and formulation to achieve more effective adjuvant activity.

Several studies have demonstrated the ability of IL-2 to enhance responses to a number of antigens (Nunberg *et al.*, 1989; Reddy *et al.*, 1989, 1993; Hughes *et al.*, 1991, 1992b; Pighetti and Sordillo, 1995) and to augment resistance to infection (Reddy *et al.*, 1989). Potentiation of antigen-specific responses by IL-2 requires multiple administration of the cytokine (Reddy *et al.*, 1989, 1993; Hughes *et al.*, 1991, 1992b; Pighetti and Sordillo, 1995) and is influenced by the vaccine formulation (Hughes *et al.*, 1992b). In one study, IL-2 enhanced humoral and cellular responses to BHV-1 only when the gD subunit vaccine was formulated with Avridine (Hughes *et al.*, 1992b), a known IFN-γ and IL-1 inducer (Jansen, 1986), but there was no effect when mineral or vegetable oils were used as vaccine adjuvants (Hughes *et al.*, 1992b). This information suggested that induction of other cytokines may be necessary for IL-2 to exert its adjuvant activity. This seems logical since cytokines do act in a cascade with IL-1 acting to initiate the response and IL-2 expanding the response. However, no additive effects were observed when IL-1β (Reddy *et al.*, 1993), IFN-γ or IFN-α (Hughes *et al.*, 1992b) were administered together with IL-2. These results demonstrate a significant point in the use of cytokines as adjuvants: the importance of delivery of the cytokine for the appropriate duration and in the appropriate sequence if given in combination with other cytokines.

Delivery of the cytokine to the appropriate site may also be important. Administration of IL-2 enhanced antigen-specific responses in the bovine mammary gland (Pighetti and Sordillo, 1995). In this study, i.m. injection with antigen and IL-2 near the supramammary lymph node significantly increased serum and lacteal antibody titres and antigen-specific proliferative responses in peripheral blood mononuclear cells (PBMC).

Conjugates of antigen and cytokines have been used as an alternative way to enhance immunogenicity of vaccines (Heath and Playfair, 1990; Hughes *et al.*, 1992a). When a synthetic peptide fragment from IL-1, which contained the immunostimulatory region, was conjugated to an antigen, an adjuvant effect on antibody responses was observed in mice (Rao and Nayak, 1990). Similar results have been reported using other cytokine–antigen chimeric molecules. In cattle, a chimeric molecule containing antigen and IL-2 was also tested (Hughes *et al.*, 1992a). Although no increase in antigen-specific antibody response was observed in immunized cattle, there was an increase in the antigen-specific cell responses as assessed by limiting-dilution analysis. Chimerization of an antigen to a cytokine will not only be effective in enhancing immunogenicity but will also provide economic advantages in production and purification of vaccines. Administration of IFN-γ and vesicular stomatitis virus (VSV) glycoprotein G to cattle enhanced the magnitude of the secondary virus-neutralizing antibody response after immunization with antigen alone (Anderson *et al.*, 1988). In addition, cattle treated with IFN-γ at the time of a single primary immunization were more resistant to VSV challenge than those immunized without IFN-γ.

The use of cytokines *in vivo* is hampered by their rapid clearance or inactivation and the toxicity associated with systemic administration. One alternative to this problem is the use of delivery systems to entrap both cytokines and antigen. This approach will not only protect the cytokine but will also deliver the two components simultaneously to the same site. Encapsulation in a delivery system, such as liposomes, can extend the circulation time, modify biodistribution and reduce dosage and systemic toxicity (Lasic and Papahadjopoulos, 1995). In a number of models, it has been shown that liposome formulations of this nature are more efficient at inducing immunity than simple mixtures of antigen and cytokine alone (Abraham and Shah, 1992; Duits *et al.*, 1993).

A novel approach to vaccination has been developed in recent years by the direct immunization with genes encoding for protective antigens (Tang *et al.*, 1992; Cox *et al.*, 1993; Robinson *et al.*, 1993; Ulmer *et al.*, 1993). Several animal models have demonstrated that the protein encoded in the transferred genes can induce protective long-lasting immune responses (Robinson *et al.*, 1993; Ulmer *et al.*, 1993). Using this approach, it has been demonstrated that immunization with plasmid containing cytokine genes enhances the immune response to the vaccine antigen and can shift the type of immune response from a Th1 to a Th2 response or vice versa depending on the form of the antigen and the cytokine (Raz *et al.*, 1993). Perhaps as understanding of cytokine biology increases, together with the development of more efficacious delivery systems, we will be able to create vaccine formulations with a specific combination of cytokines to promote the particular response required in a given circumstance.

Enhancement of Immune Function for the Prophylaxis and Treatment of Diseases

Major advances in the study of bovine cytokines were initially spurred by the interest of pharmaceutical firms in those cytokines to which they had commercial rights. These companies provided purified recombinant bovine cytokines in sufficient quantity to conduct large-scale *in vivo* trials. The cytokines studied were IFN-α and IFN-γ, IL-1β and IL-2, and colony-stimulating factors (CSF). The outcome of these studies provided considerable advancement in the understanding of the *in vivo* and *in vitro* biological effects of the cytokines studied. This information was used to design therapeutic strategies that were applied to various disease models. Two of the disease models studied extensively in cattle include bovine respiratory disease and mastitis.

Bovine Respiratory Disease

The effects of cytokines in bovine respiratory disease have been evaluated in several experimental trials (reviewed in Babiuk *et al.*, 1991; Campos *et al.*, 1994). Treatment of cattle with IFN-α or IFN-γ before exposure to BHV-1 dramatically reduced the clinical signs and mortality associated with the subsequent bacterial infection (Babiuk *et al.*, 1985, 1987). These beneficial effects of IFN-α were not correlated with its antiviral property since no reduction in viral shedding from the respiratory tract was observed (Babiuk *et al.*, 1987), and it was concluded that immunomodulatory effects of IFN treatment were responsible for improved disease resistance. When a field trial was conducted in which IFN-α was administered to animals entering the feedlot, the results were not as dramatic as those observed in experimental trials. Interestingly, the beneficial effects of IFN-α were not as pronounced if cytokine treatment occurred after infection (Babiuk *et al.*, 1985, 1987). These results may explain the disparity between the outcomes seen in experimental trials as compared to field trials, where animals may have been infected at different times relative to IFN-α treatment.

The effectiveness of cytokine treatment has been shown to be valuable in instances where the host immune system is compromised. For example, IFN-γ had greater immunomodulatory activity on neutrophil function in immunosuppressed cattle compared to normal cattle (Roth and Dagmar, 1989). As well, administration of IFN-γ reduced the severity of *Haemophilus somnus* induced pneumonia in dexamethasone-treated cattle (Chiang *et al.*, 1990). Although IFN-γ treatment did not influence the course of the disease in normal infected animals, there is evidence for increased resistance to bacterial infections by IFN-γ treatment in other animal models (Campos *et al.*, 1989).

IL-1 plays an important role in host defence mechanisms against infection (Durum *et al.*, 1985). Considerable interest has been generated in using IL-1β as a prophylactic and therapeutic molecule in various disease situations including bovine respiratory disease (Baca-Estrada *et al.*, 1995; Van Kessel *et al.*, 1996). A single administration of IL-1β was able to induce a dose-dependent increase in polymorphonuclear cells and monocytes in peripheral blood (A.G. Van Kessel, 1996 unpublished results). In experimental trials, treatment of cattle with IL-1β alone was unable to impact significantly on the establishment or progression of BHV-1 or BHV-1/*P. haemolytica* induced diseases (Baca-Estrada *et al.*, 1995; Van Kessel *et al.*, 1996). Thus, these results demonstrated that IL-1β was capable of modulating diverse immune parameters, and although not therapeutically effective alone, could be administered safely as a vaccine adjuvant or adjunct therapy in cattle.

The outcome of these experiments was an emphasis on the appropriate delivery of the cytokine to achieve the desired effects. The timing of administration and the duration of effect are important factors in determining therapeutic efficacy.

Mastitis

Subclinical and clinical mastitis are major causes of loss in the dairy industry. One of the more prevalent aetiological agents contributing to this loss is *Staphylococcus aureus*. Antibiotic therapy is only moderately effective in the control of *S. aureus* infections; thus offering the opportunity for cytokine therapy to be a viable alternative to improve bacterial clearance. Mastitis is a localized infection; thus cytokine therapy could be delivered by the intramammary route directly to the site of infection, possibly reducing the dose required and the development of systemic side-effects. The development of the chronic nature of *S. aureus* mastitis depends on the interaction between invading bacteria and the cells of the defence system, primarily polymorphonuclear neutrophils (PMN). Thus, cytokines which modulate neutrophil recruitment and function were primary candidates for mastitis immunotherapy.

Intramammary infusion of rBoIL-1β (200 μg; 6×10^6 IU), rBoIL-2 (2 mg; 40×10^6 IU) and rBoGM-CSF (1 mg; 38×10^6 IU) elicited quantitative and qualitative changes in mammary gland neutrophils (Daley *et al.*, 1991, 1993). Both IL-1β and IL-2 but not GM-CSF increased the somatic cell count in milk. The minimal effective dose and maximal tolerable intramammary dose for IL-1β and IL-2 was 0.5–1000 μg and 0.5–30 mg, respectively, while intramammary infusion of up to 5 mg of GM-CSF failed to elicit a significant influx of cells into the gland (Daley *et al.*, 1993). All three cytokines enhanced inducible oxygen radical formation by milk neutrophils. However, phagocytosis was increased only in the cells from GM-CSF or IL-2 treated glands.

All three cytokines (IL-1β, 1 μg; IL-2, 1 mg; GM-CSF, 1 mg) were effective in preventing *S. aureus* infection (75–100%) if given 24 h before infection (Daley *et al.*, 1993). However, when IL-1β or IL-2 was given 48 h before, or at the time of infection, they had lower efficacy.

When existing *S. aureus* infections were treated with cytokines administered three times at 24 h intervals, 83% of IL-1β treated quarters and 54% of IL-2 treated quarters responded by cessation of *S. aureus* shedding for

1 day or more. However, the number of treated quarters free of *S. aureus* for a 14 day consecutive milk sample period after the last treatment was similar for both groups at about 40% (Daley *et al.*, 1991). Thus, the high relapse rate suggests that although host cells recruited into the mammary gland may ingest *S. aureus*, these cells are inefficient at killing the bacteria. There were no cures in untreated control quarters, but the therapeutic efficiency was no better than standard antibiotic treatment. Both cytokines showed dose–response therapeutic efficacy and were most efficacious if administered over three consecutive milkings, rather than in a single bolus (Daley *et al.*, 1993). This observation was corroborated in a separate study in which a single treatment with rBoIL-2 (500 μg; 7.5×10^5 IU) induced less dramatic increases in somatic cell counts and reductions in bacterial shedding (Quiroga *et al.*, 1993). Also observed in this study was that treatment with IFN-γ (20 μg; 10^5 IU) did not have any effect on somatic cell count or bacterial shedding (Quiroga *et al.*, 1993).

Systemic administration of cytokines has also been evaluated as a method to combat the periparturient immunosuppression thought to be important in the establishment of new intramammary infections. Daily subcutaneous injection of rHuG-CSF (3 μg kg^{-1} for 15 days) significantly increased the circulating pool of leukocytes, especially the neutrophil population (Nickerson *et al.*, 1989). Despite the marked increase in circulating neutrophils, there was only a trend for increased somatic cell counts. There was also a tendency for cows to be somewhat protected from experimental challenge with *S. aureus* after 10 days of treatment with rHuG-CSF.

The use of the homologous cytokine, rBoG-CSF, resulted in even higher neutrophil numbers in the blood, primarily through its effect on increasing the bone marrow output of mature neutrophils (Kehrli *et al.*, 1991; Cullor *et al.*, 1992). Fc receptor-mediated bacterial ingestion and cytotoxic activities of the neutrophils were markedly enhanced. However, the increase in milk somatic cell count was not significant and there was no effect on the number of *S. aureus* shed from chronically infected glands of treated cows (Kehrli *et al.*, 1991).

These results demonstrated that recombinant cytokines are effective immunomodulators that can augment various mammary gland defence mechanisms. Although the level of efficacy of cytokine therapy for *S. aureus* mastitis was not commercially viable, there remained the potential for cytokines to be used as an adjunct to antibiotic therapy. Combining IL-2 (3.3 or 10 mg) with sodium cephapirin treatment resulted in approximately 20% improvement in therapeutic efficacy, although this was not a statistically significant improvement above antibiotic alone (Daley *et al.*, 1992). In a similar study, co-administration of 2 mg of rBoIL-2 and sodium cephapirin in cows with established *S. aureus* mastitis decreased somatic cell counts and shedding of *S. aureus* more effectively than either treatment alone (Reddy *et al.*, 1992).

Administration of 100 μg of IL-2 (daily for 21 days) accelerated involution and resulted in increased infiltration by plasma cells, suggesting a role for this cytokine as a general immunostimulant at drying off (Nickerson *et al.*, 1992). However, when a single administration of IL-2 (1 mg) was given immediately preceding intramammary infusion of antibiotic dry cow therapy, cure rates for intramammary infections present at drying off did not improve. Also, IL-2 did not affect the rate of new intramammary infections during the dry period (Hogan *et al.*, 1995). The duration of IL-2 treatment may have been an important difference in these two studies.

A second mastitic model has also been studied with regards to cytokine therapy. Bovine coliform mastitis is regarded as a serious problem because it is severe, sometimes fatal, and difficult to control with antibiotics formulated for intramammary use. The acute symptoms most often associated with coliform mastitis are due to the rapid and unrestricted growth of the organism, the release of endotoxin, and the subsequent development of an unlimited inflammatory reaction. The ability of rBoIFN-γ to ameliorate experimentally induced *Escherichia coli* mastitis during the parturient period was investigated. Dairy cows intramammarily treated with IFN-γ 24 h before *E. coli* challenge had fewer infected quarters, lower clinical scores, and infections of shorter duration compared to placebo-treated animals. All IFN-γ treated cows survived the challenge, while placebo-treated cows had a 42% mortality rate (Sordillo and Babiuk, 1991). Elevated TNF concentrations were especially evident in cows that developed severe clinical symptoms and eventually died from endotoxaemia. Pretreatment of cows with recombinant IFN-γ possibly may downregulate the generation of TNF-α within infected quarters and thus reduce the systemic effects of TNF-α which may lead to death (Sordillo and Peel, 1992).

The mammary gland response to endotoxin has been investigated. High concentrations of IL-1 and IL-6 were detected in milk of infused glands within 2.5 to 4 h after administration of 10 μg of endotoxin (Shuster *et al.*, 1993), while higher levels of endotoxin (25 μg) were required to induce detectable levels of TNF (Shuster and Kehrli, 1995). Mammary gland permeability changes preceded the increase in IL-1 and IL-6 activity indicating that these clinical signs were not mediated by these cytokines. However, systemic responses and leukocytic influx developed after or concurrent with increases in IL-1 and IL-6 suggesting that these cytokines are important in the development of these responses.

The results described above indicate that anticytokine therapy could be considered as an approach to lessen the pathologic consequences of *E. coli* infection. However, the intramammary administration of IL-1 receptor antagonist had no effect on local inflammation including neutrophil influx, systemic reaction, or impairment in milk production. Thus, despite IL-1 being present in the milk, it does not mediate its effect within the milk compartment (where it was effectively blocked by the IL-1 receptor antagonist), and must move to the mammary tissue or systemic circulation to interact with cellular receptors. Alternatively, IL-1 may not be critical to the mastitic response, and other cytokines can orchestrate the response to endotoxin. This is an illustration of the complexity of cytokine regulation of the host response, and explains why cytokine therapy has not always achieved the anticipated results.

Outside of these two disease models, there have been reports of cytokine therapy in a variety of conditions. Calves with salmonellosis treated with IFN-α had reduced septicaemia and, in some cases, reduced fever, but there were no improvements in the incidence of diarrhoea or mortality (Peel *et al.*, 1990). Systemic administration of IFN-α did not have a beneficial therapeutic effect on experimentally induced warts (Lassauzet and Salamin, 1993). Interleukin-2 has been used to treat vulvar papilloma, and a neoplastic rather than infectious disease, squamous cell carcinoma (see Chapter 2).

OTHER APPLICATIONS

Roles for cytokines in non-immune functions have been discovered, and with these functions exist the possibility for cytokine application in other physiologic processes. Cell growth and development, reproductive function, and various physiologic processes can be influenced by cytokines. For instance, IFN-like molecules play a role in the recognition of pregnancy, which will be discussed in Chapter 5. Further work on the role of cytokines such as IL-1 and TGF-β (in articular cartilage metabolism) may lead to strategies for more effective repair of joint injury and disease (Morris *et al.*, 1992; Morales, 1994)

EPILOGUE

The recent availability of purified cytokines, biological assay systems, as well as specific antibodies for many cytokines in cattle, has allowed us to begin to unravel the complex interactions within the immune system. Furthermore, since the immune system does not function in a vacuum, the availability of cytokines and assays has enhanced our understanding of the intimate interactions between the immune system and various other body systems such as the nervous, endocrine and reproductive systems. From the limited studies conducted to date, it is clear that these complex interactions must be maintained in a very critical balance if proper physiological homeostasis and health of the animal are to be maintained. Although in some instances a large bolus of a specific cytokine demonstrated a beneficial effect, it appears that a sustained, slow release of cytokines or directed delivery of cytokines will be more beneficial in many situations.

Although many of the cytokines are becoming available in large quantities for use in therapy, diagnosis and characterization of immune responses, considerable effort is still required to dissect the complex interactions between various cells before we can unequivocally state which cytokine profile and consequent type of immune response is best for recovery from a specific pathogen, or maintenance of a specific physiological state. Combining the application of cytokines for diagnosis with our understanding of the pathogenesis of different infectious agents should make it possible to test vaccine quality without resorting to large challenge studies. This will be especially important as society becomes more concerned about the use of animals for challenge studies. However, we must ensure that the diagnostic tests are clearly correlated to protection before we embark on registering vaccines that may stimulate immune responses which are not protective. This will require considerable background information before these assays become commonplace.

In addition to using cytokines in modulating immune responses to vaccination or infectious diseases, Chapter 3 clearly indicates the importance of cytokines in disease diagnosis. It is envisaged that a battery of diagnostic kits could be developed to diagnose a variety of different infectious agents, in addition to tuberculosis and brucellosis. Furthermore, the testing of cytokines, or the effects of cytokines, has the opportunity to revolutionize our veterinary inspection service at abattoirs to ensure a safe meat supply for the consumer. Whether it will be possible to use a simple immunological assay for cytokines to measure conception is presently unknown, but if it were possible it would greatly alleviate the time-consuming present-day practice of pregnancy testing in cattle. We feel that we are on the verge of a new era where a better understanding of the cytokine network and its role in modulating homeostasis will rapidly advance to application in clinical practice not only for prophylaxis and treatment, but also for the diagnosis of specific conditions. Thus in summary, we are confident that cytokine applications will become an essential component in the repertoire of every clinician, diagnostic laboratory and innovative producer in the future.

Type 1 and Type 2 Responses in Cattle and their Regulation

2

W.C. Brown[1] and D.M. Estes[2]

[1]Department of Veterinary Microbiology and Pathology, College of Veterinary Medicine, Washington State University, Pullman, Washington 99164, USA; [2]Department of Veterinary Pathobiology, College of Veterinary Medicine, University of Missouri, Columbia, Missouri 65211, USA

INTRODUCTION

The past decade has produced a wealth of information on the characterization of type 1 helper (Th1) and type 2 helper (Th2) T cell subsets based on differential cytokine expression patterns, the cross-regulation of functionally distinct types of T cells by cytokines, and the role of cytokines in regulating the differentiation of Th1 and Th2 cells from naive T cells. Mosmann and Coffman (Mosmann *et al.*, 1986) originally characterized long-term murine $CD4^+$ T cell clones as either IL-2, IFN-γ and TNF-β producing cells (Th1) or IL-4, IL-5 and IL-10 producing cells (Th2), and as more cytokines were discovered, the Th2 profile was expanded to include IL-6, IL-9 and IL-13 (Mosmann and Sad, 1996). This Th1/Th2 paradigm has provided a useful framework for understanding the observed bias in immune responses that are often dominated by either cell-mediated or humoral responses, and for devising therapeutic strategies to stimulate T cell- or antibody-mediated immunity. However, as recently argued by Kelso (1995), this Th1/Th2 paradigm may be an oversimplification of a more complex immunoregulatory network. Many studies with cloned $CD4^+$ T cell lines from mice and humans revealed the presence of additional phenotypes of T cells, referred to as Th0 cells, that expressed any or all combinations of cytokines (Del Prete *et al.*, 1988; Maggi *et al.*, 1988; Paliard *et al.*, 1988; Umetsu *et al.*, 1988; Firestein *et al.*, 1989; Mosmann and Coffman, 1989; Salmon *et al.*, 1989; Street *et al.*, 1990). Thus, it can be envisioned that Th1 and Th2 cytokine profiles represent ends of a continuing spectrum of cytokine profiles (Clerici and Shearer, 1994; Kelso, 1995), and that immune responses are not simply regulated by a balance between two discrete subsets (Kelso, 1995). Furthermore, the contribution of Th1 or Th2 cytokines by other cell types, including $CD8^+$ T cells, γ/δ T cell receptor (TCR)-bearing T cells, NK cells and APC further blurs the definition of a Th1 vs. Th2 response, and Bloom and colleagues (1992b) introduced the nomenclature of type 1 and type 2 responses to reflect the overall immune response to an antigen or complex pathogen. Thus, IFN-γ-dominated responses are referred to as type 1 responses, whereas IL-4-dominated responses are referred to as type 2 responses (Mosmann and Sad, 1996). In general, type 1 responses involve cell-mediated inflammatory reactions, whereas type 2 responses involve strong antibody and allergic reactions.

This chapter will attempt to highlight the progress that has been made in defining type 1 and type 2 responses in cattle. We review evidence for the presence of different subtypes of antigen-specific Th cell clones of cattle and cytokine-mediated regulation of both Th cell and B cell responses, with a comparative view towards current knowledge of T and B cell function in mouse and man. Identification of Th0, Th1 and Th2 clones based on cytokine mRNA expression, regulation of bovine T cell growth and cytokine expression by immunomodulatory cytokines IL-10, IL-12 and IL-4, and regulation of B lymphocyte growth and differentiation by IFN-γ and IL-4 are specifically discussed.

SPECTRUM OF CYTOKINE PROFILES EXPRESSED BY ANTIGEN-SPECIFIC $CD4^+$ T CELL CLONES

Several studies were performed to characterize cytokine mRNA profiles of bovine MHC class II-restricted, $CD4^+$ Th clones specific for a variety of antigens derived from bovine pathogens. IFN-γ production was also determined either by bioassay, relying on neutralization of VSV in comparison with a rHuIFN-α2 reference standard, or by ELISA, using the IFN-γ kit from IDEXX Corporation (Westbrook, Maine) and a bovine IFN-γ standard determined by bioassay. In all experiments, T cell lines were derived by stimulating PBMC with specific antigen for several weeks in the absence of exogenous growth factor, and T cell clones were obtained by limiting dilution cloning in the presence of specific antigen and 10% bovine T cell growth factor (TCGF), prepared as conditioned medium from concanavalin A (Con A)-stimulated PBMC (Brown and Grab, 1985). To analyse cytokine expression, clones were stimulated with either Con A for 8 h or specific antigen and irradiated PBMC as a source of APC for 14 h. There were no qualitative differences in cytokine profiles expressed by T cells stimulated with either mitogen or antigen (Brown *et al.*, 1994b), although mitogen stimulation generally resulted in quantitatively higher levels of IFN-γ, the only cytokine product that was definitively measured.

Evidence for Type 1 and Type 2 Responses Against Protozoan or Helminth Parasite Antigens

Based on evidence from human and murine systems that intracellular pathogens and helminth parasites tended to induce type 1 and type 2 responses, respectively, when measured *ex vivo* (Mosmann and Coffman, 1989a, b; Romagnani, 1994), a similar strategy was used to define these responses in cattle. In the first set of experiments, antigen-specific Th cell clones derived from cattle infected with either the protozoan parasite *Babesia bovis* (Brown *et al.*, 1992, 1993a–c, 1994c), or the helminth liver fluke, *Fasciola hepatica* (Brown *et al.*, 1994b,c) were compared for cytokine mRNA expression following stimulation of the cells with either Con A or specific antigen. When mRNA was analysed by Northern blotting using bovine IL-2 (Reeves *et al.*, 1986), IL-4 (Heussler *et al.*, 1992) and IFN-γ (Cerretti *et al.*, 1986) specific cDNA probes, several Th1 clones specific for *B. bovis* and several Th2-like clones specific for *F. hepatica* were detected. The Th1 clones expressed IL-2 and IFN-γ, but no detectable IL-4 mRNA, whereas the Th2-like clones expressed high levels of IL-4 and low levels of IFN-γ message. Importantly, we identified only a single Th clone that lacked expression of IFN-γ transcript, and this clone co-expressed IL-2 and IL-4 (Brown *et al.*, 1994b). In both parasite systems, the majority of over 40 $CD4^+$ T cell clones expressed an unrestricted or Th0 cytokine profile, defined by the expression of IL-2, IL-4 and IFN-γ transcripts, or in some cases IL-4 and IFN-γ transcripts.

Despite the prevalence of Th0 clones based on mRNA analysis, measurement of IFN-γ production by individual Th clones indicated a bias towards a type 1 response in the *B. bovis*-specific clones. A comparison of IFN-γ titres in the supernatants of Th cell clones specific for *B. bovis* (Brown *et al.*, 1992, 1993b), with those of Th cell clones specific for *F. hepatica* (Brown *et al.*, 1994b) revealed that the latter set of clones secreted relatively low levels of IFN-γ. Only one (6%) of the 17 *F. hepatica*-specific clones produced more than 128 IU ml^{-1} IFN-γ, whereas 12 (57%) of the 21 *B. bovis*-specific Th cell clones, which were generated and tested under the same culture conditions, secreted between 200 and 800 IU ml^{-1} IFN-γ. In fact, six of eight *B. bovis*-specific Th cell clones that secreted > 200 IU ml^{-1} IFN-γ were derived from cattle G3 and G6, the genetically identical siblings of *F. hepatica*-infected cattle G1 and G8 (Fig. 2.1). Since the T cell clones were derived from genetically identical cattle, and cultured and cloned in the same manner, these collective results indicated that the predominant type of response (type 1 or type 2) is not solely explained by host genetic differences, and may be related to the parasite infection itself.

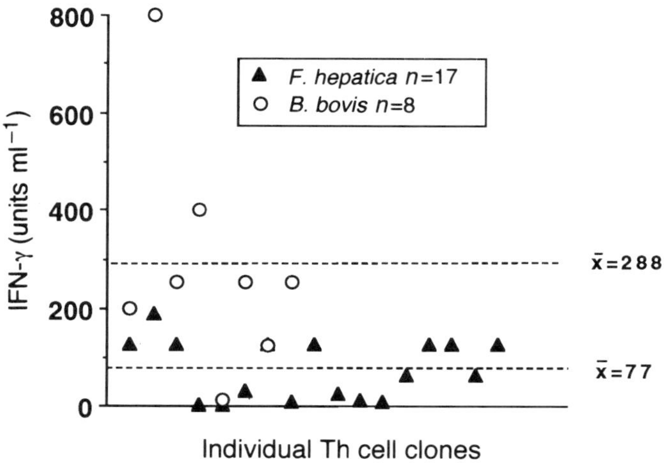

Fig. 2.1. Production of IFN-γ by individual Th cell clones derived from two sets of genetically identical cattle infected with either *Fasciola hepatica* or *Babesia bovis. F. hepatica-* (closed triangles) and *B. bovis-* (open circles) specific Th cell clones were stimulated at a density of 2×10^6 cells ml^{-1} with 2.5 μg ml^{-1} Con A without APC for 24 h. Serially diluted culture supernatants were assayed for IFN-γ by neutralization of VSV (Brown *et al.*, 1993b, 1994b) in comparison with an IFN-α2 reference reagent. The mean titres of eight *B. bovis-* and 17 *F. hepatica*-clones are indicated by dashed lines.

Additional experiments were performed with Th cell clones specific for proteins localized to the babesial rhoptry organelle. $CD4^+$ clones were derived from either *B. bovis*-infected cattle or calves immunized with a *B. bigemina* rhoptry protein antigen, designated rhoptry-associated protein-1 (RAP-1), in RIBI adjuvant (RIBI Immunochem Research Inc., Hamilton, Montana). Six *B. bovis* RAP-1-specific clones and eight *B. bigemina* RAP-1-specific clones were stimulated with Con A or specific antigen in the presence of APC, and reverse transcription-polymerase chain reaction (RT-PCR) was employed with bovine cytokine-specific primers to analyse cytokine mRNA expression. This assay was optimized with regard to input RNA concentration and cycle numbers to ensure amplification of specific cytokines within the linear portion of the plotted curve (Chitko-McKown *et al.*, 1995). The *B. bovis* RAP-1-specific clones from both sets of cattle expressed either Th1 (seven clones) or Th0 (six clones) cytokine mRNA patterns, based on low levels of IL-2 and IL-4 (where present), and relatively high levels of IFN-γ mRNA and product (Rodriguez *et al.*, 1996; Brown *et al.*, 1996b).

T Cell Clones Specific for PPD Express Type 1 Cytokine Profiles

In an attempt to obtain additional Th1 clones for our experiments, purified protein derivative (PPD) of *Mycobacterium tuberculosis* was used to generate T cell lines from previously non-TB-exposed, skin-test-negative cattle G3 and C97, which were subsequently infected with *B. bovis* (Brown *et al.*, 1991, 1993b). This complex antigen was selected since both murine (Street *et al.*, 1990) and human (Del Prete *et al.*, 1991) Th cell clones specific for PPD preferentially expressed type 1 cytokine profiles. Bovine PPD-induced T cell lines were cloned by limiting dilution, and the clones responded to *M. bovis* and *M. tuberculosis* PPD, but did not respond to *B. bovis* antigen (W.C. Brown, unpublished observations). Of 12 clones examined, nine expressed Th1 cytokine mRNA profiles, whereas three clones expressed low levels of IL-4 mRNA in addition to IL-2 and IFN-γ, and were classified as Th0 cells. IFN-γ titres were also high for antigen-stimulated cells, ranging from 20 to 440 IU ml^{-1}, with a mean titre of 170 IU ml^{-1}, in supernatants of antigen-stimulated clones.

INTERLEUKIN-10 EXPRESSION IS NOT RESTRICTED TO Th2-LIKE CELLS

The immunoregulatory cytokine IL-10 is, by definition, produced selectively by Th2 cells, and not Th 1 cells in the mouse (Fiorentino *et al.*, 1989b). Production of IL-10 by T cells differs in humans, where both Th1 and Th2 clones secrete the cytokine (Yssel *et al.*, 1992; Del Prete *et al.*, 1993). Furthermore, in studies with human leprosy, it was shown that lepromatous lesions which contained predominantly $CD8^+$ cells, expressed IL-4, IL-5 and IL-10, cytokines associated with the type 2 phenotype in mice (Bloom *et al.*, 1992a). We were interested to determine whether IL-10 was selectively expressed by bovine Th2 cells, as in mice, or if IL-10 expression was less restricted. The bovine IL-10 cDNA was cloned (Hash *et al.*, 1994), and used to probe RNA from selected clones expressing Th0, Th1, or Th2-like cytokine profiles. In this initial study, four Th0, two Th1, and four Th2-like antigen-specific clones were examined, and IL-10 was expressed by all three subsets and all clones within a subset of $CD4^+$ T cells (Brown *et al.*, 1994c). Further studies demonstrated that *B. bovis* RAP-1-specific clones expressed weak or undetectable levels of IL-10 mRNA (Brown *et al.*, 1996b), whereas all *B. bigemina* RAP-1-specific clones expressed IL-10 mRNA (Rodriguez *et al.*, 1996). Only two PPD-specific clones (one Th0 and one Th1) expressed detectable IL-10 transcripts. Thus, although clones of all phenotypes are capable of expressing IL-10, not every $CD4^+$ T cell clone expressed detectable IL-10 mRNA. These findings, which are similar to those with IL-10 expression by human Th cells, provide additional evidence against the existence of a strict Th1/Th2 dichotomy in cattle. The immunoregulatory properties of IL-10 will be discussed below.

Table 2.1 summarizes the frequencies of clones specific for *Babesia*, *Fasciola*, or PPD that expressed one or a combination of the type 1 and type 2 cytokines IL-2, IL-4 and IFN-γ. Of over 60 antigen-specific Th cell clones derived from eight cattle that were experimentally infected with bovine parasites or immunized with parasite antigen, cytokine mRNA profiles indicated that all but one clone expressed detectable levels of IFN-γ mRNA. With the exception of PPD-specific clones, the majority of Th clones co-expressed IL-4 and IFN-γ. When considering the variable expression of IL-10 by these clones, additional phenotypes were expressed by this panel of clones that are not represented in Table 2.1. It is clear from this analysis that, unlike PPD-specific clones which were biased towards a Th1 profile, the majority of clones specific for either *Babesia* or *Fasciola* antigens expressed unrestricted or Th0 cytokine profiles.

Although the majority of clones did not fall into distinct Th1 or Th2 subclasses, infection with different parasites tended to induce different types of cytokine responses in memory cells stimulated *ex vivo*. When the relative levels of cytokine mRNA expression were compared, all of the *B. bigemina* RAP-1-specific clones and many of the *B. bovis*-specific clones expressed a type 1 cytokine profile, based on relatively high levels of IFN-γ mRNA and product and correspondingly low levels of IL-4 mRNA (Brown *et al.*, 1993b, 1996a,b; Rodriguez *et al.*, 1996; and unpublished observations). No type 2 *Babesia*-specific clones were detected. In contrast, several clones specific for

Table 2.1. Co-expression of type 1 and type 2 cytokines among bovine CD4$^+$ T cell clones.

Antigen specificity of Th cell clones[b]	Frequency (%) of positive clones expressing one or a combination of cytokines[a]						
	IFN-γ	IL-2	IL-4	IFN-γ + IL-2	IL-2 + IL-4	IFN-γ + IL-4	IFN-γ + IL-2 + IL-4
Babesia bovis (*n* = 25)	8.0	0	0	24.0	0	8.0	60.0
B. bigemina RAP-1 (*n* = 8)	12.5	0	0	25.0	0	12.5	50.0
Fasciola hepatica (*n* = 17)	0	0	0	0	5.9	29.4	64.7
PPD (*n* = 12)	0	0	0	75.0	0	0	25.0

[a]The table shows frequencies observed for the unique expression of interferon γ (IFN-γ), interleukin 2 (IL-2), interleukin 4 (IL-4) mRNA or various combinations of cytokine mRNA.
[b]Cytokine mRNA expression was detected by Northern blot analysis for the majority of *B. bovis*- and *F. hepatica*-specific T cell clones, and by reverse transcription-polymerase chain reaction (RT-PCR) for PPD- and *B. bigemina* RAP-1-specific Th cells, following *in vitro* stimulation with either Con A or specific antigen and APC. The number of clones in each group (*n*) is indicated in parentheses.

the trematode parasite, *F. hepatica*, expressed high levels of IL-4 and relatively low levels of IFN-γ and IL-2 transcripts, consistent with a type 2 profile, and no *F. hepatica*-specific type 1 clones were identified (Brown *et al.*, 1994b). These *in vitro* findings parallel those described for human Th clones specific for different antigens, and support the concept of Clerici and Shearer (1994) and Kelso (1995) that during an immune response, individual cytokine-producing T cells likely constitute a spectrum of response patterns, with Th1 and Th2 responses at extreme ends. The capacity to mount a protective immune response to a pathogen or defined antigen likely reflects the kinetics of cytokine induction (i.e. IL-10 usually appears later than other cytokines to dampen ongoing immune responses) and the ratio of different cytokines produced, rather than the absolute presence or absence of a given cytokine (Del Prete *et al.*, 1988; Yssel *et al.*, 1992; Clerici and Shearer, 1994; Kelso, 1995).

REGULATION OF T CELL ACTIVATION BY CYTOKINES

The major immunoregulatory cytokines important for differentiation of naive T cells to type 1 or type 2 cells are IL-12 and IL-4, respectively (reviewed by Romagnani, 1994; Seder and Paul, 1994; Trinchieri, 1995). Experiments in our laboratories designed to compare the effects of these cytokines on the differentiation of Th cell responses in cattle immunized with parasite antigen are at the initial stages, and will not be considered here. However, type 1 and type 2 cytokines also exert important immunoregulatory effects on differentiated T cells, and the effects of IL-10, IL-12 and IL-4 on bovine T cells will be presented. Notable differences in the actions of these cytokines on murine versus human T cells have been described. IL-10, a murine type 2 cytokine, selectively inhibited cytokine production by murine Th1 cells, with no effect on Th2 cells (Fiorentino *et al.*, 1989a,b). In contrast, human IL-10 inhibited cytokine synthesis by both Th1 and Th2 cells (Del Prete *et al.*, 1993), which underscores the differences between species and the value of performing similar studies in ruminants. Murine IL-12, a type 1 cytokine produced mainly by macrophages and dendritic cells, selectively upregulated proliferation and IFN-γ production by murine Th1 clones, with no effect on either Th0 or Th2 clones (Germann *et al.*, 1993; Kennedy *et al.*, 1994). In contrast, human Th1 and Th0 clones produced increased levels of IFN-γ when cultured with antigen and IL-12, and Th2 clones exhibited transient IFN-γ expression as well (Manetti *et al.*, 1994). In contrast to the murine system where IFN-γ selectively inhibited proliferation of Th2 and not Th1 clones (Gajewsky and Fitch, 1988), human IFN-γ inhibited and IL-4 enhanced antigen-driven proliferation of both Th1 and Th2 clones, although the effects were more marked on the Th2 cells (Del Prete *et al.*, 1993). Human IL-4 also enhanced both IL-4 and IL-5 production by Th2 clones, but contrary to dogma, failed to inhibit IFN-γ production by Th1 clones. Thus, the mutually exclusive effects of type 1 and type 2 cytokines apparent in the mouse system were less obvious when human T cells were studied. To determine the immunoregulatory potential for these cytokines in cattle, we have similarly examined the effects of human IL-10, human IL-12 and bovine IL-4 on proliferation and limited cytokine expression of selected Th cell clones characterized as Th0, Th1, or Th2-like cells.

Interleukin-10 is Downregulatory for all Subtypes of Bovine Th Cells

Interleukin-10 has been shown to suppress T cells both directly (Taga *et al.*, 1993) and indirectly by impairing macrophage function (Fiorentino *et al.*, 1989b; reviewed by Howard and O'Garra, 1992). Effects on macrophages included downregulation of: (i) IL-12 production (D'Andrea *et al.*, 1993); (ii) the induction of B7 (CD80) expression (Ding *et al.*, 1993); and (iii) MHC class II expression (de Waal Malefyt *et al.*, 1991). At the T cell level, the indirect effects of IL-10 resulted in reduced levels of: (i) IFN-γ production (Fiorentino *et al.*, 1989a,b); (ii) IL-2Rα expression (Ding and Shevach, 1992); and (iii) nuclear factor-κB/Rel nuclear activity, an important step in T cell activation (Romano *et al.*, 1996). It is believed that IL-10 functions to deactivate macrophages, thereby preventing otherwise potentially pathologic inflammatory reactions (Spits and de Waal Malefyt, 1992). The kinetics of bovine IL-10 mRNA expression were determined by Northern blot analysis (Hash *et al.*, 1994). In comparison with human IL-10, bovine IL-10 was expressed earlier in both mitogen-stimulated bovine PBMC (reaching maximal levels by 8–24 h) and Th cell clones.

To evaluate the immunosuppressive potential of IL-10 for bovine Th cells, human recombinant IL-10 was added to antigen- or mitogen-stimulated T cells. Antigen-induced proliferative responses of Th0, Th1, and Th2-like cells were inhibited by 75–94% by rHuIL-10, which could be restored by low levels of exogenous human IL-2 (Brown *et al.*, 1994c). However, using the same dose range (5–80 ng ml^{-1}), proliferation of one Th1 clone was never inhibited, but was frequently enhanced by IL-10, indicating differential responses to this cytokine by individual Th cells that was independent of cytokine expression. In the absence of accessory cells (AC), the responses of Th clones to Con A or IL-2 were not inhibited by IL-10, whereas antigen-specific responses of Th1 and Th2 cells were reduced when IL-10-pretreated macrophages were used as APC, showing that the effects of IL-10 on these T cell clones were indirect and likely macrophage dependent.

To begin to elucidate the mechanism of IL-10-mediated suppressed Th cell proliferation, experiments were performed to determine whether the effect of IL-10 on T cell proliferation corresponded with downregulated expression of cytokines, or their receptors, important for T cell growth. The effects of IL-10 on antigen-driven cellular proliferation and expression of IL-2, IL-4, IL-2 receptor (IL-2Rα; p55), and IFN-γ by *B. bovis*- or *F. hepatica*-specific Th cell clones were simultaneously evaluated (Chitko-McKown *et al.*, 1995). Suppressed proliferation of all types of Th cell clones did not correspond with reduced expression of either IL-2 or IL-4 mRNA. Where measured, IL-2 activity in culture supernatants was also not affected, although levels of IL-2 were typically very low. In contrast, expression of IL-2Rα mRNA was reduced by 41–91% in five of six IL-10-treated clones. Flow cytometric analysis of IL-2Rα (CD25) expression on T cells treated with IL-10 indicated a slight reduction in the level of IL-2Rα expressed by one of two clones (C.G. Chitko-McKown and W.C. Brown, unpublished observations), but in general the levels of CD25 expressed by our Th clones tend to be very low (Brown *et al.*, 1994a), so the data were not interpretable using the available mAbs. These results indicated that IL-10 does not inhibit AC-dependent proliferation of bovine Th cell clones by downregulating T cell cytokines required for autocrine growth, and suggested that one way IL-10 acts is by downregulating IL-2Rα expression and subsequent signal transduction leading to decreased cellular proliferation. This effect could involve IL-10-mediated downregulation of IL-12 (D'Andrea *et al.*, 1993), since IL-12 has been shown to augment IL-2-induced IL-2Rα expression on murine Th1 clones (Yanagida *et al.*, 1994) as well as purified splenic T cells (Guler *et al.*, 1996). The ability of IL-2 to override the suppressive effects of IL-10 in our system is also consistent with this possibility. A second potential role of IL-12 in antigen-driven proliferation was recently described by Maruo and co-workers, whereby IL-12 production by APC, rather than IL-2 production by Th1 cells, was required for the proliferation of antigen-driven murine Th1 clones (Maruo *et al.*, 1996). The study utilized Th1 clones that produced very low levels of IL-2 mRNA and no detectable IL-2 product (similar to what we have observed with terminally differentiated bovine Th clones that have been maintained in culture for several months) and showed that antigen and APC-induced cellular proliferation could not be blocked by neutralizing antibodies against IL-2, but was blocked by anti-IL-12 antibody. This was observed even with a clone that did not proliferate in response to exogenous IL-12, which partially reflected the low level of IL-12R expression at the time of stimulation. Thus it is possible that in our system, IL-10-mediated suppression of APC-produced IL-12 could negatively affect IL-2Rα expression and/or IL-12-mediated proliferation by the Th clones. Once anti-bovine IL-12 antibodies become available, these possibilities can be tested. The effects of IL-10 on bovine AC costimulatory molecules, such as CD40 or CD80, have also not been explored and await development of specific reagents.

The effects of IL-10 on IFN-γ production by over 20 antigen-stimulated Th0 and Th1 clones were also evaluated by ELISA. IFN-γ production was consistently downregulated by IL-10, resulting in 42–69% inhibition in one study with *B. bovis-* and *F. hepatica*-specific T cell clones (Chitko-McKown *et al.*, 1995), and in 62–96% inhibition in a second study with *Babesia* RAP-1-specific Th cell clones (Brown *et al.*, 1996; W.C. Brown, unpublished observations). In summary, our results with bovine T cells support the concept that IL-10 is an immunoregulatory cytokine for outbred species, acting on antigen-stimulated Th cell clones through the macrophage to affect AC function. We additionally showed that the downregulatory effects of IL-10 were not selectively directed at Th1 populations, as they are in mice. The dichotomy between cross-regulation of type 1 and type 2 responses in humans and bovines is less distinct than that observed in the mouse, and our results with IL-10 underscored the need to characterize additional immunoregulatory cytokines in cattle.

Effects of Human Interleukin-12 on Proliferation and Interferon-γ Production by PBMC and Th Cells

Interleukin-12 is an important initiation factor for the differentiation of type 1 responses and is an immunoregulatory cytokine for differentiated T cells. IL-12 stimulates cytotoxic T lymphocytes (CTL) and NK cell activity, proliferation of activated and resting α/β and γ/δ TCR^+ T cells, and IFN-γ expression by NK cells and differentiated T cells of mice and humans (reviewed by Trinchieri, 1995). To evaluate the potential for IL-12 to cause naive T cells to differentiate into type 1, IFN-γ secreting cells, preliminary experiments were performed with bovine PBMC stimulated either non-specifically with T cell mitogens or specifically with antigen. Stimulation with mitogen could potentially activate both naive and memory T cells, whereas stimulation with specific antigen should selectively activate memory T cells. Recombinant human IL-12 (purchased from R&D Systems, Minneapolis, Minnesota, or provided by Genetics Institute, Andover, Massachusetts) was added to PBMC and evaluated for costimulation of proliferation. Over a range of concentrations from 0.25 to 50 ng ml^{-1}, IL-12 induced a modest (i.e. 2.5-fold) increase in proliferation of phytohaemagglutinin (PHA)-stimulated PBMC, but had no costimulatory effects on Con-A-activated cells (Brown *et al.*, 1996). In contrast, IL-12 dramatically increased IFN-γ production by mitogen or parasite antigen-stimulated PBMC (Table 2.2), consistent with its effects in other species. The relative contributions of IFN-γ by T cells and NK cells present in PBMC were not determined.

The effects of IL-12 on proliferation and IFN-γ production by different subsets of bovine Th cell clones were then examined (Brown *et al.*, 1996). When a panel of five Th0 or Th1 clones specific for PPD or *B. bovis* were stimulated with IL-12, only marginal (i.e. 1.3–2.7-fold) increases in baseline proliferation were observed. When assayed as a costimulator with specific

Table 2.2. Human IL-12 enhances IFN-γ production by *Babesia bigemina* RAP-1- or *Fasciola hepatica*-immune PBMC stimulated *in vitro* with mitogen or specific parasite antigen.

	IFN-γ produced (IU ml^{-1}) following stimulation of PBMC[b] with:					
	Phytohaemagglutinin		Concanavalin A		Parasite antigen	
Donor of PBMC[a]	–IL-12	+IL-12	–IL-12	+IL-12	–IL-12	+IL-12
Bovine 2216	2.0	589.0	ND[c]	ND	2.4	371.3
Bovine 2234	0.3	316.0	ND	ND	0	320.0
Bovine G1	1.9	539.0	ND	ND	1.0	252.5
Bovine G1	ND	ND	78	281	ND	ND

[a]Cattle 2216 and 2234 were immunized with *Babesia bigemina* RAP-1 in RIBI adjuvant (Rodriguez *et al.*, 1996). Animal G1 was infected with *Fasciola hepatica* (Brown *et al.*, 1994b).
[b]PBMC were cultured with either 1 μg PHA ml^{-1}, 2.5 Con A μg ml^{-1}, 25 μg *B. bigemina* merozoite membrane antigen ml^{-1} (animals 2216 and 2234) or 50 μg soluble *F. hepatica* worm antigen (animal G1) ml^{-1} without or with 1 ng ml^{-1} recombinant human IL-12. Supernatants were harvested at 48 h and tested for IFN-γ by ELISA using a bovine IFN-γ standard that contained 440 IU IFN-γ ml^{-1}.
[c]Not determined.

antigen, IL-12 stimulated less than twofold increases in proliferation of these clones. Marginal effects of IL-12 on proliferation of two *B. bovis*-specific Th1 and two *F. hepatica*-specific Th2-like clones stimulated with 10 IU ml^{-1} rHuIL-2 were similarly observed. IL-12 augmented IFN-γ production by all subtypes of Th clones examined (ranging from 17 to 279% above baseline values; Brown *et al.*, 1996). Results from selected clones are presented in Table 2.3. No apparent differences in the immunoregulatory effects of IL-12 on Th0 or Th1 cells were observed; however IL-12 had less of an effect on the single Th2-like clone examined (Table 2.3). These results are consistent with the reported effects of IL-12 on differentiated T cells of humans, where IFN-γ was upregulated in Th0, Th1 and Th2 clones (Manetti *et al.*, 1994), and differ from the selective effects of IL-12 for murine Th1 cells (Germann *et al.*, 1993; Kennedy *et al.*, 1994).

Effects of Recombinant Bovine Interleukin-4 on Th Cell Proliferation and Interferon-γ Production

Interleukin-4 is a pleiotropic cytokine which is key in controlling the immune response. IL-4 exhibits diverse effects on a variety of cell types including T cells, B cells, monocytes, neutrophils, fibroblasts, endothelial cells and haematopoietic progenitor cells (reviewed in Callard, 1991; Paul, 1991). In cattle, the activities of IL-4 are beginning to be described for a variety of different cell lineages. The bovine IL-4 homologue was originally

Table 2.3. Effect of recombinant bovine IL-4, human IL-13 and human IL-12 on IFN-γ production by antigen-stimulated Th cell clones.

		Percentage variation of baseline IFN-γ production after culture with the following cytokines:[a]		
Th clone	Subtype[b]	IL-4	IL-13	IL-12
G1.1H5	Th0	−91	−54	+279
G1.3G4	Th2	0	+371	+17
G3.3F2	Th1	+39	+20	+223
G3.3A12	Th1	−75	ND[c]	+190
G3.2D7	Th1	+5	ND[c]	+100

[a]T cell clones were cultured at a density of 1×10^6 cells ml^{-1} with 1×10^6 APC ml^{-1} and intermediate concentrations of specific antigen which consisted of 5 μg ml^{-1} soluble *F. hepatica* worm antigen (G1 clones) or 0.2 or 1 μg ml^{-1} *M. bovis* PPD (G3 clones). Cytokines were added at the beginning of culture at a final concentration of 20 ng ml^{-1} IL-4, 20 ng ml^{-1} IL-13, or 1 ng ml^{-1} IL-12. Supernatants were harvested after 24 h (IL-4 and IL-12) or 36 h (IL-13). Results are expressed as the percentage variation of the IU ml^{-1} IFN-γ produced by individual clones cultured with antigen in the absence of cytokine.
[b]The Th cell subtype was determined by relative expression profiles of IL-2, IL-4 and IFN-γ mRNA. The Th1 cells expressed IL-2 and IFN-γ, but no IL-4 mRNA, the Th0 cells expressed both IFN-γ and IL-4 mRNA, and the Th2 cells expressed high levels of IL-4 mRNA but little IFN-γ or IL-2 mRNA.
[c]Not determined.

cloned by Heussler and colleagues (1992) from Con A activated lymphocyte cDNA using PCR. Sequence analysis has shown that the bovine molecule is truncated in comparison with both the mouse and human homologues due to a 51 nucleotide deletion in the 5′ proximal end of the molecule, yielding a secreted protein from a precursor molecule of 136 amino acids (aa) with a mature molecular mass of approximately 14 kDa in non-glycosylated form; glycosylated forms from COS-7 cell transfectants range in size from 15 to 19 kDa (Estes *et al.*, 1995). The exon–intron structure of the bovine IL-4 homologue is unknown at present as genomic sequences have not yet been reported. Defining exon–intron structure for the bovine IL-4 locus may have important consequences for the interpretation of biological activities of various constructs, as alternatively spliced versions of human IL-4 lacking exon 2 have been reported (Sorg *et al.*, 1993). These exon D2 variant molecules have been shown to inhibit human T cell proliferation (Atamas *et al.*, 1996), a finding that may impact on immunoregulation in the bovine host. A comparative analysis of the simple homology of the human sequence for exon 2 (nucleotides 1578–1625) with the bovine cDNA sequence indicates a region of homology (73%) constituting nucleotides 202–251, and suggests that the truncated form reported for bovine IL-4 does not involve exon 2 (Arai *et al.*, 1989). Two *N*-linked glycosylation sites on the bovine

IL-4 molecule, which would appear to be utilized in eukaryotic expression systems, were predicted based on the primary aa sequence (Estes *et al.*, 1995). Non-glycosylated forms of mouse or human IL-4 retain their biological activity, but the effects of post-translational modification on activity have not been established for the bovine homologue.

A variety of expression systems have been used to express rBoIL-4, including bacterial, viral and yeast expression systems. To date, only mammalian or eukaryotic expression systems have resulted in biologically active rBoIL-4, suggesting that glycosylation or protein folding or both are important to retention of biological activity (D.M. Estes, R. Collins, and H. F. Seow, personal communication). Chou and Fasman protein predictions of the secondary structure of the bovine molecule indicate that the four helix bundle up-up-down-down helix topology of the mouse and human IL-4 molecules is conserved in the bovine protein with the potential for disulphide bonding due to the presence of eight cysteine residues.

While the effects of rBoIL-4 on B cells appear to be conserved with those described for the human and mouse homologues (see below), the activities of rBoIL-4 on T cell proliferative responses have been variable, presumably due to the cellular source of recombinant cytokine (Fig. 2.2). T cells isolated from PBMC by panning and magnetic bead depletion of B cells were cultured in the presence of medium alone, pokeweed mitogen (PWM), or Con A in combination with baculovirus-derived rBoIL-4 (viral IL-4, a gift from Robert Collins, Compton Laboratories, UK), *Pichia pastoris*-derived IL-4 (yIL-4), or COS-7 cell derived IL-4 (control IL-4) as previously described (Estes *et al.*, 1995). Costimulators were added at a submitogenic concentration (1 $\mu g\ ml^{-1}$) so that the growth costimulatory activity of the different IL-4 preparations could be evaluated. *Staphylococcus aureus* Cowan strain I (0.005%) was utilized as an additional negative control for T cell proliferation. Despite the comparable effects of rBoIL-4 on proliferation of bovine B cells from each expression system (data not shown), IL-4 from either yeast or baculoviral sources did not significantly enhance the growth of peripheral blood-derived T cells (Fig.2.2). This is in contrast to results obtained with COS cell-derived rBoIL-4, which in agreement with previously described studies, inhibited T cell growth in mitogen costimulation assays in the presence of AC (Estes *et al.*, 1995). As both yeast and COS cell material were purified by similar methodologies, it is unclear at present whether variable degrees of glycosylation or inhibitory co-purifying molecules are responsible for the largely negative effects of COS cell-derived IL-4 on T cell growth. Hyperglycosylation of secreted recombinant proteins has been reported in COS-7 cells (Aruffo and Seed, 1987), which may be a contributing factor to the previously observed suppressive activity of COS-7-derived IL-4 (Estes *et al.*, 1995) when compared with the marginally upregulatory activity of baculovirus-expressed rBoIL-4 on bovine T cell clones (see Fig. 2.3). Co-purification of suppressive factors or molecules

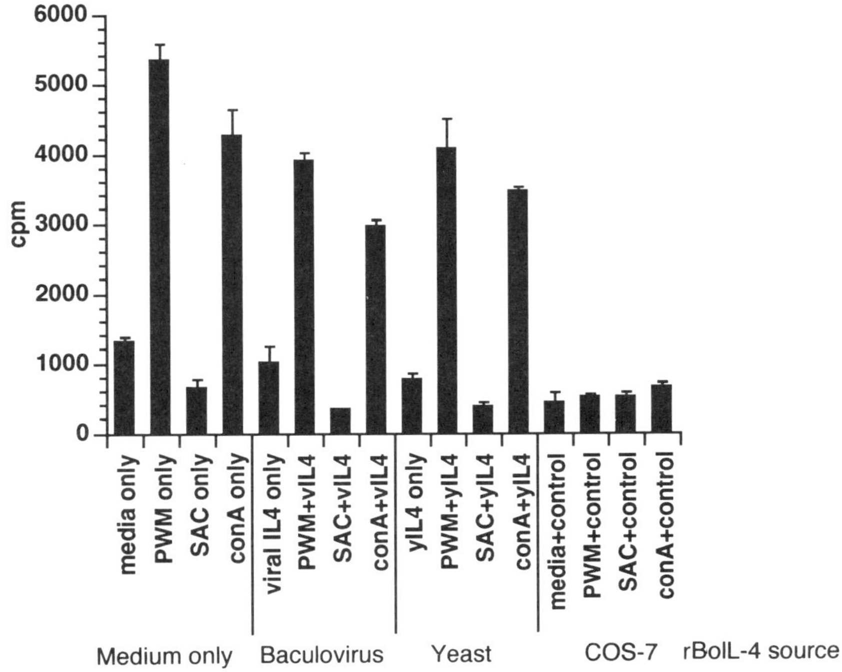

Fig. 2.2. Comparative analysis of T cell proliferation following mitogen co-stimulation with rBoIL-4 from different expression systems. T cells and monocytes from B-depleted PBMC (< 95% surface immunoglobulin negative (sIg)) were cultured at 1×10^5 cells per well in 200 μl total culture volume in the presence of medium alone, baculovirus-derived rBoIL-4 (1 ng ml^{-1}), *Pichia pastoris*-derived rBoIL-4 (1 ng ml^{-1}), or COS-7-derived rBoIL-4 (1 ng ml^{-1}) as shown. Yeast and COS-7-derived rBoIL-4 were partially purified by ultrafiltration and filter-sterilized before addition to cultures. Con A (1 μg ml^{-1}), PWM (1 μg ml^{-1}), or SAC (0.005% final) were added as costimulators as indicated. T cell growth was assessed by [^{3}H]-thymidine uptake at 72 h after stimulation when cells were harvested 18 h later. Thymidine incorporation was assessed by scintillation counting, and results are presented as the arithmetic mean and standard error determined from triplicate cultures for each culture variable.

would appear unlikely, as control COS cell supernatants (vector only transfectants) did not exhibit inhibitory activity for PBMC-derived T cells or antigen-stimulated Th cell clones (Estes *et al.*, 1995).

As IL-4 and IFN-γ are differentially expressed by Th1 and Th2 cells, it was of interest to determine the effects of rBoIL-4 on IFN-γ production by antigen-specific T cell clones derived from genetically identical cattle and stimulated by their cognate antigen in the presence of APC

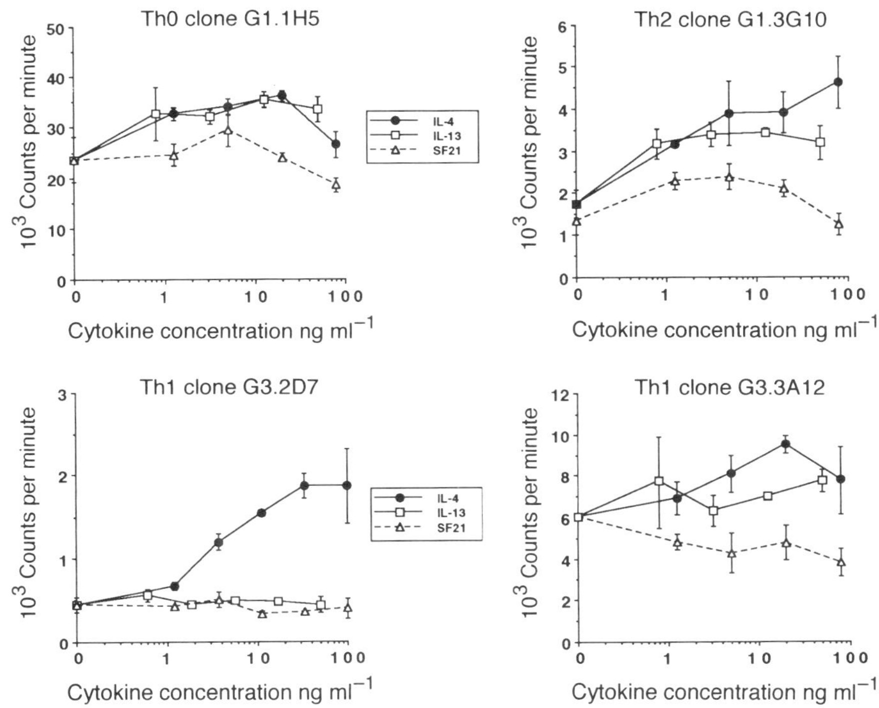

Fig. 2.3. Effects of rBoIL-4 and human IL-13 on antigen-induced proliferation of bovine Th cell clones. *F. hepatica*-specific Th0 and Th2-like clones G1.1H5 and G1.3G10 (upper panels) and *M. bovis* PPD-specific Th1 clones G3.2D7 and G3.3A12 (bottom panels), which were derived from genetically identical cattle G1 and G3, were stimulated with 5 μg ml^{-1} *F. hepatica*-soluble worm extract (clones G1.1H5 and G1.3G10) or 0.2 (clone G3.3A12) or 1.0 (clone G3.2D7) μg ml^{-1} *M. bovis* PPD for 72 h in the presence of syngeneic APC. Bovine baculovirus-derived IL-4 (solid circles), or human IL-13 (open squares) were added to the assays at concentrations ranging from 0.8 to 100 ng ml^{-1}. An equivalent amount of supernatant from *Spodoptera frugiperda* cells (open triangles) was used as a negative control for IL-4. Cells were radiolabelled with [^{3}H]-thymidine added during the final 6 h of culture, and results are presented as the mean and standard error of duplicate cultures.

(Table 2.3). Addition of intermediate concentrations of specific antigen and rBoIL-4 (baculovirus-derived) added at an optimal concentration of 20 ng ml^{-1} in the presence of APC either down-modulated IFN-γ production (*F. hepatica*-specific Th0 clone G1.1H5 and PPD-specific Th1 clone

G3.3A12), had no effect (*F. hepatica*-specific Th2-like clone G1.3G4), or moderately upregulated IFN-γ production (PPD-reactive Th1 clones G3.3F2 and G3.2D7). These results contrast with those using human IL-12, which reproducibly enhanced IFN-γ production (see above). These findings are consistent with those described for the effects of rBoIL-4 on T cell growth, in that regulatory effects of IL-4 for bovine T cells do not fit the classical Th1/Th2 paradigm, but rather are variable for individual Th clones in a manner independent of antigen specificity, donor animal genetic background, or cytokine expression profile.

Interleukin-4 and IL-4-like cytokines (i.e. IL-13, originally designated p600) play pivotal roles in the development of antibody responses regulated by Th2 cells (reviewed by Zurawski and de Vries, 1994). Human IL-13 and IL-4 are produced by activated T cells, albeit with different kinetics (Kuhn *et al.*, 1991; Paul, 1991). Unlike human IL-4, IL-13 appears to be produced by $CD4^+$ and $CD8^+$ T cells regardless of cytokine expression profile (Zurawski *et al.*, 1993). Analysis of bovine Th0, Th1 and Th2-like $CD4^+$ T cell clones as well as $CD8^+$ T cell clones for cytokine mRNA expression by RT-PCR indicated that transcripts for the bovine IL-13 homologue are induced upon activation with Con A or antigen in the presence of APC at time points ranging from 8 to 22 h post-stimulation (D.M. Estes, A. Hirano, and W.C. Brown, unpublished observations). Partial nucleotide sequence analysis of the bovine IL-13 molecule based on homology with human IL-13 (comparing 255 of 399 bases) indicated a relatively high degree of sequence conservation (81%) between species at the cDNA level. Human IL-13 has IL-4-like effects on human B lymphocytes, many of which appear to be conserved in the bovine (A. Hirano and D.M. Estes, unpublished observations). These include enhancement of the proliferative response of bovine B cells stimulated with anti-IgM coated beads or through ligation of the bovine CD40 homologue. Based on the relatively high degree of sequence conservation and the observed cross-reactivity of human IL-13 on bovine B cells, comparative studies were performed to assess the effects of human IL-13 and bovine IL-4 on bovine T cell growth and IFN-γ production (Fig. 2.3 and Table 2.3). In preliminary studies, representative Th0, Th1, or Th2-like $CD4^+$ T cell clones were evaluated for proliferation following coculture with APC and an intermediate concentration of antigen. RBoIL-4 (baculovirus-derived) and rHuIL-13 (R&D Systems, Minneapolis, Minnesota) were titrated by log dilution ranging from 0 to 100 ng ml^{-1}. RBoIL-4 stimulated moderate growth-enhancing activity for several clones (Th2-like clone G1.3G10, and Th1 clones G3.2D7 and G3.3A12). Th0 clone G1.1H5 responded only marginally to IL-4 costimulation over the concentration range tested. In the human system, IL-13 differs from IL-4 in its inability to stimulate growth of activated T cells or T cell clones (Zurawski *et al.*, 1993; Zurawski and de Vries, 1994), although IL-13 may regulate T cell responses indirectly through its effects on AC. Bovine T cell proliferation induced by human IL-13 was similarly modest or lacking and, with one exception (Th1 clone G3.2D7), did not significantly differ from that induced by IL-4. As with IL-4, the effects of IL-13 on IFN-γ production by selected Th clones were variable, but where tested, tended to parallel those for IL-4. The specific mechanism by which IFN-γ production is enhanced is unclear, but may be attributable to upregulation of class II MHC molecules on AC. Surface expression of class II MHC molecules on bovine B cells has been shown to be upregulated by rBoIL-4 (Estes *et al.*, 1995), but this biological activity has not been examined for IL-13.

CYTOKINE-MEDIATED REGULATION OF B CELL PROLIFERATION AND DIFFERENTIATION

Interleukin-4 is a key regulatory cytokine for B lymphocytes. Recombinant bovine IL-4 upregulated the expression of surface antigens, including IL-2Rα, CD23, and IgM, on B lymphocytes, and thereby enhanced the capacity of B cells to act as AC (Estes *et al.*, 1995). As has been observed in the mouse and human, bovine IL-4 upregulated expression of IgG1 and IgE from $sIgM^+$ B cells in the presence of anti-IgM, *S. aureus* Cowan Strain I, or PWM (Estes *et al.*, 1995). Under these conditions IL-4 also acted to enhance B cell proliferation alone and in a synergistic manner with IL-2. The production of polyclonal IgG1 and IgE induced by rBoIL-4 was inhibitable in a dose-responsive manner by addition of rboIFN-γ (CIBA-GEIGY, St Aubin, Switzerland) to *in vitro* cultures.

Similar *in vitro* studies have been conducted with rboIFN-γ using T-independent costimulation systems (Estes *et al.*, 1994). It was demonstrated that bovine peripheral blood-derived B cells can exclusively upregulate production of IgG2 and IgM but not IgG1 or IgA in the presence of anti-bovine IgM-coated beads or PWM. These results and those described for IL-4 suggest that antibody responses *in vivo* can be biased in the production of specific immunoglobulin (Ig) isotypes by T cells which express mutually exclusive or skewed cytokine production profiles. Although direct evidence is lacking from *in vivo* studies, these results may be correlated with the hyper IgG1 production associated with susceptibility to *Brucella abortus* infection in cattle versus a more predominant IgG2 response in resistant cattle (Templeton *et al.*, 1990). *Brucella abortus* infection of mice induces a predominant IgG2a response via the production of IFN-γ (reviewed by Finkelman *et al.*, 1988). Studies are presently underway to dissect signalling for T-cell-dependent antibody responses via the bovine CD40–CD40 ligand interaction in the presence or absence of recombinant bovine and/or human cytokines. From these studies, the contribution of type 1 or type 2 cytokines induced via CD40-dependent signalling can be elucidated and correlated with the results described previously for T-independent type 2 antigens.

EPILOGUE

The immune system of ruminants differs in many ways from that of rodents and humans, and assumptions on immune function in these outbred species that are based solely on information from murine systems may not always be valid. For example, striking differences occur in the composition of T cells in these different species. In young cattle and sheep, γ/δ T cells may comprise up to 50% of the total T cell population in peripheral blood (Hein and Mackay, 1991), whereas in healthy mice and humans, γ/δ T cells constitute approximately 1–10% of the population of T cells in peripheral blood and organized lymphoid tissues, as well as gut- and skin-associated lymphoid tissues (Kabelitz *et al.*, 1991; Roberts and Shevach, 1991). At the B cell level, cattle express a unique distribution of Ig subclasses in colostrum, where IgG1, and not IgA, predominates (Butler *et al.*, 1972). Furthermore, recent studies by Griebel and Hein (1996) suggest that in ruminants, Peyer's patches are a major site for the production of B cells and the generation of primary antibody responses. The analysis of type 1 and type 2 responses and their regulation in cattle provides additional evidence for basic differences in the immune response of mice and ruminants.

In this chapter, we have highlighted variations observed in patterns of cytokine expression by bovine Th cell clones and cytokine-mediated regulation of antigen-stimulated Th cells and polyclonally activated B cells, when compared to the murine, and in some cases, human systems. Although such *in vitro* analyses are limited in scope and possibly reflect a bias inherent in culture systems that favour those clones with good growth potential, results that differ from the murine Th1/Th2 paradigm were nevertheless obtained when similar systems were originally used to derive long-term murine clones. In contrast to the prevailing dogma that views Th1 and Th2 as polarized and distinct immune responses, examination of cytokine profiles of over 60 bovine antigen-specific $CD4^+$ T cell clones revealed that the primary type 1 and type 2 cytokines, IFN-γ and IL-4, were co-expressed by the majority of clones. The T cell clones derived from *Babesia*- or *Fasciola*-infected cattle are believed to constitute memory cells arising from the primary or secondary *in vivo* challenge. The memory phenotype of all subtypes of Th clones was verified by their proliferation to specific recall antigen, and by the expression of the CD45 isoform (CD45RO) present on memory cells (Brown *et al.*, 1993b). *Babesia* and *Fasciola* parasites are complex organisms that do not reside within macrophages, and the nature of innate immune responses induced by these pathogens important for influencing the differentiation of type 1 or type 2 cells in cattle is currently unknown. However, the observed bias towards type 1 cytokines expressed by Th clones derived from *Babesia*-immune cattle versus a bias towards type 2 cytokine expression by Th clones derived from *Fasciola*-immune cattle would predict the preferential induction of IL-12 or IL-4 by these different parasites.

In contrast to results derived from restimulation of parasite-specific T cells, PPD-stimulated clones, which expressed little or no IL-4 and high titres of IFN-γ, were the most restricted. Mycobacteria are known to induce IL-12 in macrophages (D'Andrea *et al.*, 1992), so the predominance of type 1 T cell clones specific for PPD may be explained by enhanced IL-12 production by the APC in culture. Since the cattle used to derive the PPD-specific clones were not experimentally or naturally infected with *M. bovis*, and were skin-test negative for intradermal (i.d.) challenge with *M. bovis* PPD, it is not known if these clones arose from naive precursor T cells or memory T cells responsive to cross-reactive epitopes shared with common microbes.

Immunization with a purified native parasite antigen, *B. bigemina* RAP-1, also produced a panel of type 1 Th clones. This *in vitro* finding correlated well with the analysis of lymph node cytokine mRNA obtained 3 days post-antigen inoculation. Elevated levels of IFN-γ, but not IL-4, mRNA were observed in the lymph node cells draining the site of antigen inoculation compared with the contralateral lymph node which drained the site of control adjuvant inoculation (Ruef *et al.*, 1997). Restimulation of antigen-draining lymph node cells *ex vivo* with specific antigen induced higher titres of IFN-γ over a 4 day period when compared with control lymph node cells. We are finally well poised to examine the *in vivo* cytokine response to specific antigens or infection with a complex pathogen, as many bovine cytokine reagents are now accessible (Zarlenga *et al.*, 1995). As more reagents become available to dissect the immunoregulatory network of the ruminant immune system at the molecular level, similar as well as unique characteristics will undoubtedly be revealed.

ACKNOWLEDGEMENTS

We would like to thank Robert Collins for providing baculovirus-derived bovine IL-4 for these studies, and William Davis for critically reviewing the manuscript. The research described in this review was partially supported by National Institutes of Health grant AI30136, United States Department of Agriculture NRICGP grants 92-37204-8180, 93-37206-9075 and 93-37204-9601, and the University of Missouri Agricultural Experiment Station Project VMHB0442.

3

The IFN-γ Assay as a Diagnostic Test for Bovine Tuberculosis

P.R. Wood and J.S. Rothel

CSIRO, Division of Animal Health, Private Bag No. 1, Parkville, Victoria, Australia 3052

INTRODUCTION

Since the tuberculin reaction was first described in 1891 by Koch, this assay has remained the backbone for the immunodiagnosis of tuberculosis in animals and humans until the present day. In cattle the i.d. tuberculin test, using bovine tuberculin PPD, has been used extensively for the diagnosis of bovine tuberculosis. However, the bovine tuberculin test suffers from problems such as moderate sensitivity (72%: Francis *et al.*, 1978; 65.6%: Wood *et al.*, 1991) and in some geographical areas, low specificity due to antigenic cross-reactivity with atypical mycobacteria. Additional problems relate to practical aspects of the tuberculin test and include the need to wait 3 days to read the test and the requirement that cattle cannot be retested for 60–90 days after initial injection of PPD.

With the above problems apparent, there have been many attempts to develop *in vitro* diagnostic assays for bovine tuberculosis (reviewed by Wood and Rothel, 1994). A large amount of this research has been directed towards the development of serological assays. However, the majority of serological assays have used crude antigen preparations and demonstrated poor sensitivity and specificity. Assays employing *M. bovis*-specific antigens, such as MPB-70 (Fifis *et al.*, 1989), have also been reported to have low sensitivity and, surprisingly, unacceptable specificity (Wood *et al.*, 1992).

The failure of serological tests to achieve high levels of sensitivity can be attributed to the nature of the immune response generated in *M. bovis* infected cattle. It has been demonstrated that most *M. bovis*-infected cattle mount an effective cellular immune response, have a low antibody response, and contain the infection within localized foci for long periods (Thorns and Morris, 1983). High levels of circulatory antibody are thought to be associated with a failure of the immune system to contain mycobacterial growth and the progression of disease in both man (Lenzini *et al.*, 1977) and cattle (Plackett *et al.*, 1989; Harboe *et al.*, 1990; Ritacco *et al.*, 1990). Thus, there is a reciprocal relationship between T cell reactivity and localized infection on the one hand, and antibody levels and progressive disease on the other. Taking the above findings into consideration, it is not surprising that assays based on the detection of cellular immune responses have been more successful for the diagnosis of tuberculosis.

Cellular assays based on the measurement of *in vitro* proliferation of PPD-stimulated T cells from *M. bovis*-infected cattle (Outteridge and Lepper, 1973; Muscoplat *et al.*, 1977; Thoen *et al.*, 1980) were useful for research purposes but were unsuitable as a practical alternative to tuberculin testing in the field. With the availability of recombinant bovine cytokines and sensitive biological assays it became possible to investigate alternative strategies for the measurement of cellular immune responses *in vitro*.

This chapter reviews the development and application of a novel diagnostic test for bovine tuberculosis based on the *in vitro* production of IFN-γ in whole blood cultures.

DEVELOPMENT OF THE INTERFERON-γ ASSAY

In attempting to develop an *in vitro* T cell assay for the diagnosis of bovine tuberculosis we decided to look at the release of a cytokine. IFN-γ was finally chosen for several reasons. Firstly IFN-γ was known to be predominantly released by T cells in response to antigen stimulation. Data existed that suggested IFN-γ was involved in immunity to mycobacterial infections, it could be released *in vitro* in quantities measurable by an ELISA and finally, unlike IL-2 (Baroja and Ceuppens, 1987), IFN-γ did not appear to be consumed during short-term culture of lymphocytes. In hindsight it now appears that our choice of IFN-γ as an indicator of the T cell response to mycobacterial antigens (PPD) was appropriate.

Our initial studies investigated the production of IFN-γ by purified lymphocytes from *M. bovis* infected cattle cultured *in vitro* with PPD. A bioassay was developed to detect IFN-γ in these culture supernates and the results showed a good correlation with those obtained from a lymphocyte proliferation assay (Wood *et al.*, 1990a). Although this assay system was able to diagnose *M. bovis* infection, a major barrier was that it (and other *in vitro* T cell assays) was both time consuming and expensive to perform.

The problem of isolating peripheral blood lymphocytes was solved by using a simple whole blood culture that only requires overnight incubation at 37°C of 1 ml of heparinized blood with an appropriate antigen. The following day plasma samples are removed and assayed for IFN-γ. The antigens used in the whole blood culture are: no antigen (negative control); bovine PPD; and avian PPD. Avian PPD is included to remove the problem of false-positive reactions due to the cross-reactive nature of the mycobacterial antigen preparations used. To overcome the requirement for the bioassay for IFN-γ, we produced mAbs to rBoIFN-γ (Wood *et al.*, 1990b) and used these to develop a sensitive enzyme immunoassay (Rothel *et al.*, 1990). The final assay system involves the overnight culture of whole blood with antigen followed the next day by quantification of IFN-γ in separated plasma samples using the IFN-γ specific assay; it can be completed in 24 h and facilitates the testing of large numbers of animals.

FIELD TRIALS OF THE INTERFERON-γ ASSAY

In large-scale field trials in Australia the IFN-γ assay proved to be significantly more sensitive than the single intradermal tuberculin test (SIDT) for the diagnosis of bovine tuberculosis (Wood *et al.*, 1991). Over 6000 cattle from *M. bovis* infected herds were tested with both the SIDT and the IFN-γ assay and of the 125 animals confirmed by culture to be infected with *M. bovis*, 93.6% were detected by the IFN-γ assay as compared to 65.6% by the SIDT. Interestingly, although the IFN-γ assay detected many infected cattle missed by the SIDT, there were some infected animals that were SIDT positive and IFN-γ negative. Therefore the use of the two diagnostic tests in parallel led to an overall sensitivity of 95.2%. The specificity of the IFN-γ assay was investigated by testing more than 6000 cattle from herds known to be free of *M. bovis* and was found to be 96.3%.

Field trials of the IFN-γ assay have now been conducted in several countries. In studies in Ireland, 81% of infected cattle had a positive result in the IFN-γ assay and 90% were identified by a combination of the tuberculin test (severe interpretation) and the IFN-γ assay (Monaghan *et al.*, 1995). It was also found that while the i.d. injection of tuberculin significantly reduced the capacity of cattle to respond to a second tuberculin test 7 days later, the *in vitro* production of IFN-γ was unaffected (Doherty *et al.*, 1995). The injection of tuberculin does not induce a primary immune response in cattle as detected by an IFN-γ response but in animals that were already sensitized, an anamnestic response was seen (Rothel *et al.*, 1992). In New Zealand, the sensitivity of the SIDT and the IFN-γ assay were found to be similar (Ryan, 1994). It has recently been recommended in New Zealand that the IFN-γ assay be used in cattle as an ancillary test in tuberculosis-free herds either 10 days or more after a positive caudal fold test or in parallel with the SIDT when a high sensitivity test is required. A very large trial involving approximately 100,000 cattle has been conducted in Northern

Ireland and the sensitivity of the IFN-γ assay was found to be equal to that of the standard skin test with little loss in test specificity (Neill *et al.*, 1995). In addition, when the IFN-γ assay was considered as a herd test, significantly better sensitivity than the standard tuberculin skin test was observed. Field trials in Spain found that the IFN-γ assay had an average sensitivity of 88% compared to 75% for the tuberculin skin test (Domingo *et al.*, 1995). They also concluded, that in herds with recent outbreaks of tuberculosis, the IFN-γ assay detects infected animals earlier than the skin test.

A recently completed trial in the USA indicated that the sensitivity of the SIDT ranged from 80.4 to 84.4% and for the IFN-γ assay it ranged from 55.4 to 97.1% depending on the method of interpretation (Whipple *et al.*, 1995). One difference between the USA trial and the trials conducted in the other countries mentioned above was that the test kit used to detect bovine IFN-γ in the USA was manufactured by the IDEXX Labs (USA), whereas all other trials used an EIA from CSL Ltd (Australia; Jones *et al.*, 1992). The significance of this difference is unknown, but most likely reflects variations in the recommended methods of interpretation for these two EIAs for bovine IFN-γ. An additional source of variation between the protocols used in the various trials was the source and relative concentrations of PPDs used. In the Australian trials we used the same source of PPDs (CSL Ltd) that were used for the SIDT and the same relative concentrations of bovine and avian PPD as used in the comparative tuberculin test. One of the advantages of the IFN-γ assay is that it is an *in vitro* comparative test. However, the relative concentration of bovine and avian PPD used needs to be appropriate to the conditions that exist in various countries. This was exemplified by our initial studies in Ireland where the use of bovine and avian PPDs from CSL Ltd resulted in an unacceptable specificity but replacement with the PPDs used for tuberculin testing in Ireland resulted in a specificity commensurate with that obtained in the Australian field trials (M.L. Monaghan and J.S. Rothel, 1991, personal communication).

THE USE OF THE INTERFERON-γ ASSAY IN COMBINATION WITH OTHER TESTS

While the majority of *M. bovis* infected cattle are positive in both the SIDT and the IFN-γ assay there are groups of animals that are only positive in one of these tests. This observation can be explained by the findings that while IFN-γ may be involved in the delayed-type hypersensitivity (DTH) reaction measured in the SIDT, there is not an absolute requirement for IFN-γ. Additionally, there has been an active debate as to whether DTH was an accurate measure of CMI to mycobacterial diseases (Fine *et al.*, 1994). With the advent of gene knock-out technology in mice it has now been possible to resolve this debate. Mice in which the IFN-γ gene was disrupted were unable to control a normally sublethal dose of *M. tuberculosis*, but despite this lack of protective immunity some DTH-like reactivity was elicited (Cooper *et al.*, 1993, Flynn *et al.*, 1993). The antigens that induce these two immune parameters may also differ. Mice immunized with killed BCG developed a strong DTH response but purified T cells did not produce a large amount of IFN-γ when stimulated with various antigen fractions *in vitro* (Kawamura *et al.*, 1994).

In trials where the IFN-γ assay was compared with serological assays, there appeared to be an inverse relationship between IFN-γ responses and antibody production in cattle (Ritacco *et al.*, 1991; Wood *et al.*, 1992). Additionally, Neill *et al.* (1994) found that a significant number of cattle in the early stages of *M. bovis* infection, which failed to respond to skin testing, were positive in the IFN-γ assay. This is further evidence of both the variability within the cellular response and the dichotomy of the cellular versus humoral immune response of tuberculous cattle. These differences in the type of immune response seen in individual animals suggests that the use of a combination of the SIDT, IFN-γ assay and perhaps a highly specific serological test may be necessary to attain maximal diagnostic sensitivity.

FUTURE PROSPECTS OF THE INTERFERON-γ ASSAY

While the relative sensitivity and specificity of the IFN-γ assay compared to the SIDT will continue to be a major issue for debate, the IFN-γ assay offers several additional advantages whether used alone or in combination with the SIDT. The ability to retest animals rapidly without compromising the result, the speed of the assay, 24 h compared with 72 h, the fact that animals do not need to be handled twice, the ability to vary the sensitivity and specificity of the assay and the fact that with a laboratory-based test the potential for subjective analysis or deliberate interference with the test results is significantly reduced, are important features that may be useful in different situations. The requirement to handle animals only once to conduct the IFN-γ assay has been particularly useful in the testing of African buffalo because these animals need to be sedated before being bled or skin tested. It has now been established that the bovine IFN-γ kit can be used for the diagnosis of tuberculosis in both Asian and African buffalo (Wood *et al.*, 1991; Raath *et al.*, 1995) and goats (Vidal *et al.*, 1995).

The introduction of the IFN-γ assay represents a major shift in the diagnosis of bovine tuberculosis and may also be of value for the diagnosis of other ruminant diseases such as John's disease (Billman-Jacobe *et al.*, 1992) and brucellosis (Weynants *et al.*, 1995) as well as human tuberculosis (Davidson *et al.*, 1993) and leprosy (Weir *et al.*, 1994).

EPILOGUE

The host of research that has been conducted worldwide on the cloning and expression of cytokine genes from domestic livestock species has resulted in a large panel of bovine, ovine and porcine recombinant cytokines being available (Wood and Seow, 1996). Although the data on the function of the various cytokines have predominantly come from murine studies, the production of reagents such as cytokine-specific mAbs and EIAs for different veterinary species has provided us with the immunological tools to investigate the functions of these cytokines in large animals' models. The development of the IFN-γ assay highlights the possible practical benefits that can be gained from this type of basic research into cytokines and their function. The cloning and expression of bovine IFN-γ along with studies on the cellular source and functions of this cytokine were pivotal to the development of this assay. The IFN-γ assay is now being used for a wide range of studies as a convenient means of measuring T cell responses to purified antigens (Fifis *et al.*, 1994) and new vaccines (Hodgson *et al.*, 1992).

ACKNOWLEDGEMENTS

This work was supported by grants from the Australian Meat Research Corporation and the National Brucellosis and Tuberculosis Eradication Campaign.

Interferons in Pregnancy of Domestic Animals

4

A.P.F. Flint

Department of Physiology and Environmental Science, Faculty of Agricultural and Food Sciences, University of Nottingham, Sutton Bonington Campus, Loughborough, Leicestershire LE12 5RA, UK

INTRODUCTION

One of the more remarkable findings in IFN biology over the last decade has been the demonstration that in ruminants the trophoblast antiluteolysin is an IFN (for review see Roberts *et al.*, 1992). This compound plays a crucial role in ensuring that the process of luteolysis is blocked in pregnancy, so that the corpus luteum is maintained and circulating progesterone concentrations remain high. Also surprising is the observation that trophoblast IFN is produced in very large quantities, and over a relatively short period in early gestation. Those working on IFNs produced in response to infection are used to dealing with picogram quantities of these compounds, and microgram quantities are only available *in vitro*. It is impressive then that pre-implantation ruminant blastocysts commonly produce hundreds of micrograms of a type I IFN per day and that this material is secreted into a minimal space between the unimplanted blastocyst and the endometrium. The time course of production of this material is surprising in that the synthesis and secretion of IFN by the trophoblast is limited to a precise window (in sheep, for example, synthesis starts between days 8 and 10 after conception, and ceases between days 18 and 20).

The effect of the production of these large quantities of IFN is dramatic, in terms of the endocrinology of the uterus. A key molecule in the endocrinology of the uterus during the late oestrous cycle and early pregnancy is the oxytocin receptor. It is the interaction of circulating oxytocin with this receptor which generates the episodes of prostaglandin F2-α leading to

 Cytokines in Veterinary Medicine (eds V.E.C.J. Schijns and M.C. Horzinek)

luteolysis. In non-pregnant animals at the end of the cycle, concentrations of the oxytocin receptor in the endometrium are high (approximately 700 fmol mg^{-1} protein) whereas in pregnancy the level is seldom higher than 5 fmol mg^{-1} protein. This dramatic difference, which is illustrated in Fig. 4.1, reflects the production by the developing blastocyst of IFN. It can be mimicked by infusion of IFN into the uterus (Parkinson *et al.*, 1992). Furthermore it can be shown in unilaterally pregnant animals that the non-pregnant uterine horn has a high concentration of receptor, but a low concentration of IFN, whereas the pregnant horn has the opposite (Flint *et al.*, 1992).

Why is IFN produced by the blastocyst in such large quantities? Why is it switched on and switched off with such temporal accuracy? And how does

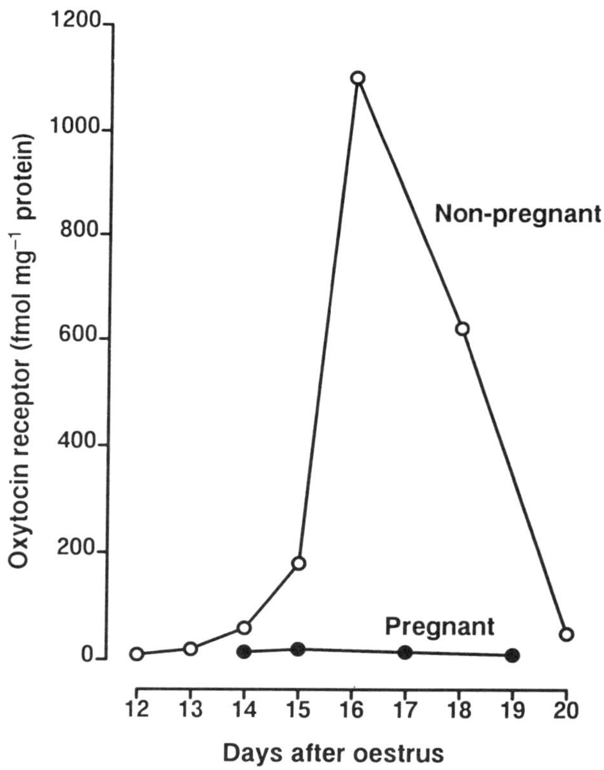

Fig. 4.1. Concentrations of oxytocin receptor in sheep endometrium measured as [^{3}H]oxytocin-binding activity in microsomal extracts obtained by differential centrifugation at various stages after fertile and infertile mating.

it act to inhibit oxytocin receptor expression? These are the questions we shall attempt to answer in this chapter.

CONTROL OF TROPHOBLAST INTERFERON GENE EXPRESSION

The ruminant trophoblast expresses a family of type I IFN genes, which are related to IFN-α_{II} genes, and are named IFN-τ. In asking what controls the expression of these genes, one must examine the upstream genomic DNA sequences controlling gene expression. IFN-τ genes exhibit considerable homologies in their upstream regions (up to approximately 450 bp upstream of the transcription initiation site), and in this region are a number of small sequences related to regulatory sequences thought to control IFN gene expression in adult tissues, in response to virus infection. It was therefore interesting to examine whether processes similar to virus infection may regulate IFN-τ gene expression in the blastocyst. One way in which this can be done is by cloning regulatory (enhancer) sequences in expression vectors which use reporter gene constructs to measure the function of the regulatory sequence. A commonly used reporter gene is chloramphenicol acetyl transferase (CAT), and we used CAT reporter constructs transfected into human choriocarcinoma or mouse fibroblast cells to investigate the response of these sequences to viral infection and also to polyinosinic:polycytidylic acid [poly (I:C)], which mimics viral induction (Guesdon *et al.*, 1996). These experiments showed no activation of CAT expression by either virus (NDV) or poly (I:C), despite pre-exposure to IFN ('priming'; Table 4.1), although endogenous IFN genes in the transfected cells were activated by these agents (Table 4.2). It therefore appears unlikely that viral activation or processes

Table 4.1. Effect of priming with murine IFN-α plus -β on NDV inducibility in L929 cells transfected with –450,+26pCAT-E.

	Per cent CAT expression relative to pCAT-C (= 100%)	
Priming dose (IU ml^{-1})	–NDV	+NDV
0	38	34
10	34	27
100	23	11
1000	3.6	5.9

Source: Guesdon *et al.*, 1996.
pCAT-C = control reporter plasmid with SV40 promoter and enhancer;
pCAT-E = identical plasmid with promoter but no enhancer;
–450,+26pCAT-E = pCAT-E with region –450 bp upstream to 26 bp downstream of a bovine interferon-τ gene transcription start site.

Table 4.2. Interferon production by cells induced by exposure to poly(I : C) or Newcastle disease virus.

Inducer	Cells	Constructs	Interferon concentrations in culture media after 24 h (IU ml^{-1})	
			Control	Induced
Poly(I)(C)	Murine L929[a]	pCAT-C	< 2	8
		–450,+26pCAT-E	< 2	8
	Human JAR[b]	pCAT-C	< 2	41
		–450,+26pCAT	< 2	380
NDV	Murine L929[a]	pCAT-C	< 10	> 10,000
		–450,+26pCAT-E	< 10	> 10,000

Source: Guesdon *et al*., 1996.
[a]IFN titres determined in cytopathic effect inhibition assays using L929 cells and vesicular stomatitis virus.
[b]IFN titres determined using Madin–Darby bovine kidney cells and Semliki Forest virus.

similar to viral activation are involved in stimulating gene expression in the trophoblast. However when reporter gene constructs were prepared with progressively deleted upstream sequences, the activity of the enhancer region was dramatically reduced by removing the sequence between 450 and 338 bp upstream of the transcription start site, suggesting that this sequence masked the activity of an inhibitory region upstream of this area. This was confirmed by further deletion of the region between 338 and 247 bp upstream, when activity returned. When these experiments were carried out in JAR cells (a cell line of human trophoblast carcinoma origin) inhibition returned when the region between 247 and 150 bp was removed, suggesting a further inhibitory sequence between these points.

Examination of the DNA sequences in this part of the gene showed that the regions –357 to –248 and –150 to –71 bp contain sequences identified in adult IFN genes as negative regulatory elements (Guesdon *et al*., 1996). Furthermore the region –351 to –352 bp was found to contain an additional negative regulatory element similar to a sequence found in renin and other genes, and thought to act in those genes as a silencer element.

As noted above, these regulatory elements are highly conserved in regions upstream of IFN-τ genes, in cattle and sheep. They are not in the same places as in adult IFN genes, being considerably further away from the transcription start site. However this would not be expected to affect their regulatory action, as enhancer elements are position independent.

These data suggest that rather than a mechanism involving a viral (type) activation of gene expression, the process by which IFN is controlled may more commonly be via inhibition of gene expression, and that this is reversed or overcome during the crucial 10 days of early pregnancy when the genes are expressed.

The demonstration that specific complexes are formed between the upstream region of the IFN-τ promoter and nuclear proteins from day 21 sheep blastocysts (Leaman *et al.*, 1994) is consistent with these findings. In these conceptuses, in which IFN-τ expression has ceased, the nuclear proteins binding to this region of the gene differ from those found in day 13 to 15 conceptus extracts. They may therefore reflect the synthesis of inhibitory transcription regulating proteins at the time when IFN-τ gene expression is terminated.

INTERFERON CONTROL OF OXYTOCIN RECEPTOR GENE EXPRESSION

The oxytocin receptor gene is an oestrogen responsive gene. That is, it is activated by oestrogen, presumably through oestrogen response elements located in the upstream genomic DNA (Kimura, 1995). Consistent with this is the observation that oxytocin receptor gene expression can be induced prematurely in the uterus by administration of oestrogen to sheep during the mid-luteal phase of the cycle or delayed by treatments that reduce oestrogen secretion (Hixon and Flint, 1987). Furthermore a response to oxytocin is induced in cattle by oestrogen replacement therapy in ovariectomized animals, in proportion to the dose of the oestrogen administered (Mann and Lamming, 1995). IFN may therefore inhibit oxytocin receptor gene expression by reducing oestrogen receptor function in the endometrium. Paradoxically however, IFNs appear to both increase and decrease oestrogen receptor function: IFNs activate oestrogen receptor function in a variety of carcinoma cells in tissue culture (for example see Sica *et al.*, 1993) but in the uterus IFN inhibits oestrogen receptor function (Lamming *et al.*, 1995; Spencer *et al.*, 1995). This biphasic response provides a key to the mechanism by which IFN affects oestrogen receptor function.

It has been known for some time that phorbol esters also activate and inhibit oestrogen receptor binding, the nature of the response being dependent on the time for which cells are exposed to these agents: in the short term they are activatory, but in the long term they are inhibitory. A well-established mechanism of action for phorbol esters on a range of cells is through protein kinase C; again phorbol esters initially activate protein kinase C by causing its translocation to the cell membrane, then after long term exposure to the cells they cause downregulation of protein kinase C activity and gene expression.

Interferons also have biphasic effects on protein kinase C, activating in the short term and downregulating in the long term. Therefore it appears possible that IFNs and phorbol esters may both affect oestrogen receptor function in a biphasic manner through effects on protein kinase C. This may explain the contradictory nature of the responses of the oestrogen receptor to IFN.

This hypothesis is consistent with the observation that the oestrogen receptor is activated by phosphorylation. The human oestrogen receptor is phosphorylated on *N*-terminal serine residues, at positions 104, 106 and 118, and these phosphorylation events are associated with receptor activation (Le Goff *et al.*, 1994). The oestrogen receptor is a substrate for a mitogen activated protein kinase (MAP kinase; Kato *et al.*, 1995), and in some tissues at least, MAP kinase is a component of a kinase cascade activated by protein kinase C. Therefore the possibility exists that protein kinase C downregulation results in decreased oestrogen receptor phosphorylation and hence decreased oestrogen receptor function.

To test this hypothesis we examined oestrogen receptor binding activity and oestrogen receptor phosphorylation in Madin–Darby bovine kidney cells, an epithelioid cell line derived from the distal tubule (Flint *et al.*, 1996). These cells, which express the oestrogen receptor, are commonly used for the assay of IFNs by cytoxicity inhibition assay. In these cells, both IFN and phorbol ester inhibit oestrogen receptor binding activity and oestrogen receptor phosphorylation, in parallel, when the cells are exposed to the agent for 48 h (Fig. 4.2). Conversely, when cells are exposed to phorbol ester for short periods (1 h) both binding and phosphorylation are increased. These responses parallel activation and subsequent downregulation of protein kinase C by either IFN or phorbol ester. The phosphatase inhibitor okadaic acid, which prevents the dephosphorylation of the oestrogen receptor, raises oestrogen receptor binding activity threefold.

These responses to IFN are consistent with the hypothesis outlined above. Furthermore the fact that they are mimicked by phorbol ester, which acts through protein kinase C, strengthens the argument that the response to IFN is exerted through the oestrogen receptor in the manner proposed above. Both type I IFNs and phorbol esters activate protein kinase C in ovine endometrial explants in organ culture, and reduce oxytocin receptor expression (Abayasekara *et al.*, 1995). Type I IFNs therefore appear to be members of a family of cytokines and growth factors that regulate steroid receptor function in a ligand-independent manner (O'Malley *et al.*, 1995; Katzenellenbogen, 1996).

CONCLUSION

As described above, one of the surprising aspects of the action of trophoblast IFN is the fact that it is produced in large quantities over a long time period. Neurotransmitters and other endocrine agents produced in short pulses or episodes are usually thought to use that secretion pattern in order to prevent the development of refractoriness in their target tissue. Conversely an agent

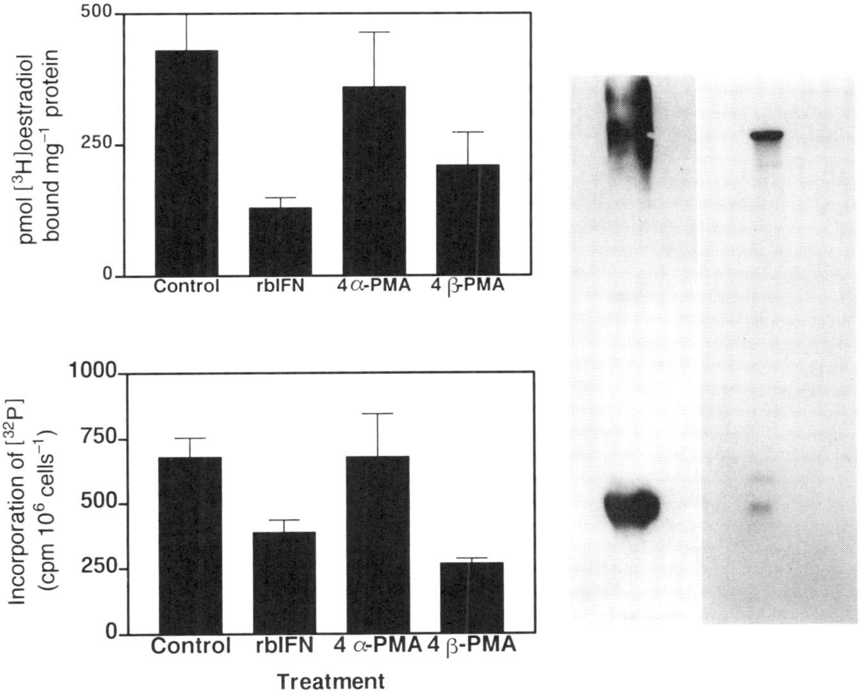

Fig. 4.2. Effects of type I interferon (rBoIFN-α_I, 10^6 IU ml^{-1}) and phorbol myristate acetate (10^{-7} M) on oestrogen receptor binding activity and phosphorylation in Madin–Darby bovine kidney cells. Binding activity was measured as specific [^{3}H]oestradiol binding by whole cells. Phosphorylation was determined by ^{32}P uptake into immunoprecipitated receptor [inset shows Western blot (left) and autoradiograph (right) of receptor after immunoprecipitation].

produced continuously might be expected to induce refractoriness in its target tissue. It is tempting to suggest that the reason for production of such high concentrations of IFNs over a relatively long period is that this release pattern has evolved in order to cause refractoriness (i.e. protein kinase C downregulation) in its target tissue, the uterine endometrium. The data described here therefore provide an explanation for the manner of the onset and cessation of trophoblast IFN production and for the production of high concentrations of this material.

Why trophoblast IFN production should be limited to the period between days 10 and 20 of early gestation in the sheep (probably days 12 to 23 in the cow) may be related to the ability of the uterus to respond to oxytocin. Before day 12, levels of oxytocin receptor in the pregnant or non-pregnant uterus are low (Fig. 4.1); therefore before this time it is not necessary to prevent the expression of the receptor, as the uterus is unresponsive to circulating oxytocin. After day 20, the corpus luteum (the source of circulating oxytocin during early pregnancy in ruminants) ceases to synthesize or to secrete oxytocin, and therefore the animal is unable to mount a luteolytic episode of prostaglandin F2-α secretion. It is therefore unnecessary to produce trophoblast IFN either before day 10 or after day 20 in the sheep (slightly later in cattle). *In vitro* translation experiments show that the IFN messenger RNA represents a large proportion of all RNA molecules in the blastocyst at the time of peak IFN production (Stewart *et al.*, 1989) and therefore it might be expected that production of trophoblast IFN represents a significant drain on the metabolic resources of the developing blastocyst. For this reason selection pressures have presumably resulted in the production of trophoblast IFN during the shortest effective period.

Immunotherapy of Bovine Ocular Squamous Cell Carcinoma and Bovine Vulval Papilloma and Carcinoma with IL-2

5

V.P.M.G. Rutten[1], W.R. Klein[1], W. den Otter[1], F.W.G. Hill[2] and E.J. Ruitenberg[3]

[1]*Veterinary Faculty, Utrecht University, Yalelaan 1, de Uithof, 3584 CL Utrecht, The Netherlands;* [2]*Faculty of Veterinary Science, University of Zimbabwe, Mount Pleasant, Harare, Zimbabwe;* [3]*Central Laboratory of The Netherlands Blood Transfusion Service, Amsterdam, The Netherlands*

INTRODUCTION

Immunotherapy has a longstanding history as a promising treatment modality of tumours. In the past it was often subordinate to surgery or chemotherapy. Upon definition of mediators involved, the use of the original therapeutic agents like Bacillus Calmette–Guérin (BCG) has been succeeded by *in vivo* treatment with purified cytokines like IL-2 or autologous lymphocytes cultured *in vitro* in the presence of these cytokines. In laboratory animal tumour models results were promising. In humans, therapy through high systemic doses of IL-2 appeared to show many side-effects and in general poor results. As an alternative, the authors of this chapter have studied the efficacy of locoregional therapy with IL-2 in a number of laboratory animal tumour models and in larger animals bearing spontaneously occurring tumours. Our studies on IL-2 therapy of bovine ocular squamous cell carcinoma (BOSCC) and bovine vulval papilloma carcinoma complex (BVPC) as well as investigations on the mechanism of induced tumour regression are reviewed in the context of historical data on immunotherapy of BOSCC and aspects of IL-2 therapy of tumours in laboratory animals and humans.

 Cytokines in Veterinary Medicine (eds V.E.C.J. Schijns and M.C. Horzinek)

HISTORICAL REVIEW

Bovine ocular squamous cell carcinoma (BOSCC) or 'cancer eye' is a well known, spontaneously occurring tumour of epithelial origin, which may be located on one of the eyelids and occasionally on the eye limbus of various breeds of cattle. The tumour shows local destructive growth and once it metastasizes (Russell *et al.*, 1956) it leads to death of the affected animal within a few months.

Due to its economical importance in several countries in Europe and Africa (e.g. Zimbabwe) and its potential value as a spontaneously occurring, non-inbred, large animal tumour model, experimental immunotherapy of BOSCC has a long history. Initially extracts of tumour tissue prepared in various ways were used for immunization (Manilla *et al.*, 1972; Van Kampen *et al.*, 1973; Spradbrow *et al.*, 1977; Hoffman *et al.*, 1981). In their studies Spradbow *et al.* (1977) and Hoffman *et al.* (1981) showed positive responses in 39 out of 46 animals and 33 out of 44 animals, respectively, after treatment by intramuscular (i.m.) injection of tumour extract. In the latter study, using a different antigen preparation, 15 tumours regressed totally.

As these observations and spontaneous regressions of large tumours suggest induction of specific antitumour immunity, Kleinschuster *et al.* (1981) decided to treat cows bearing BOSCC by a single intralesional injection of BCG, which is both a potent adjuvant and an immunogen, to potentiate ongoing antitumour responsiveness locally. In this study 17 out of 24 animals treated showed tumour regression of which seven were complete remissions. These findings were further supported by studies of Klein *et al.* (1982, 1986). Rutten *et al.* (1991a) showed that repeated intralesional injection of live BCG improved results. In 64% of the animals treated, lasting total tumour regression was observed. This appeared to be associated with positive skin reactivity to PPD of mycobacteria, indicating the importance of a clear immune response to BCG, implying local cytokine production upon injection.

Bovine vulval papilloma and carcinoma in Friesian cattle in Zimbabwe was also treated by intralesional injection of BCG (Hill *et al.*, 1994a). This tumour complex occurring in the vulval skin of mature cows may be composed of one to numerous papillomas of several stages and, in addition, in more severe cases a solitary carcinoma. Responses to multiple BCG injections included total regression of six out of six carcinomas, which was lasting in three out of four animals which completed the trial. A significant decrease in number of especially the more severe stages of papillomas in 15 out of 30 animals was observed. However, regression appeared to be transient as in ten animals severity and numbers of papillomas increased after 3 months.

In 1984 Rosenberg and co-workers showed that lymphokine activated killer (LAK) cell therapy combined with intravenous (i.v.) injections of high doses of IL-2 was successful in some murine tumour models (Mazumder and Rosenberg, 1984; Lafreniere and Rosenberg, 1985). In various human tumours total or partial regression could be induced using this treatment modality (Rosenberg *et al.*, 1987). However, *in vitro* activation and culture of very high numbers of cells required for this therapy was time consuming and expensive and the systemic injection of high doses of IL-2 caused serious toxic side-effects in patients (Rosenberg *et al.*, 1987).

In addition, Rosenberg *et al.* (1985) reported successful treatment of some established murine tumours by i.v. injection of large amounts (approx. 100,000 IU) of recombinant IL-2 without LAK cells. Talmadge *et al.* (1987) studied the kinetics of the effect of systemic IL-2 therapy on tumours in mice and showed that it has a biphasic dose optimum.

Bubenik *et al.* (1985) and Bubenik and Indrova (1987) were the first to report the beneficial effects of repeated injections of rather low doses of T cell growth factor/IL-2 around the site of a subcutaneous inoculum of cells of a methylcholanthrene-induced fibrosarcoma in mice. Vaage (1987) showed that after subcutaneous implantation of a murine mammary carcinoma in mice, complete cure was achieved by 12 daily peritumoral injections of 1500–10,000 IU of recombinant IL-2. After treatment of human bladder carcinoma by repeated intralesional injections of only 2000–4000 IU of IL-2, Pizza *et al.* (1984) observed complete and partial tumour regression in three out of six, and two out of six patients, respectively. Forni and Giovarelli (1987) observed regression of five out of five inoperable human oral squamous cell carcinomas after daily intralesional injections of 200 IU of IL-2 accompanied by injections of 200 IU of IL-2 around the regional lymph node (RLN) during one or more cycles of 10 days. Den Otter *et al.* (1991, 1996) reported positive results of intra- or peri-lesional immunotherapy in a number of mouse tumour models.

INTERLEUKIN-2 THERAPY OF BOSCC AND BVPC

Intralesional BCG immunotherapy elicits BCG-specific immune responses, which as a bystander effect potentiate tumour-specific reactivity of tumour infiltrating lymphocytes (TIL), abundantly present before, and especially after the initial BCG injection. This is supported by the observation that regression is associated with lasting BCG-specific skin reactivity (Klein *et al.*, 1986; Rutten *et al.*, 1991a). The suggestion that this may be mediated by cytokines especially IL-2 is substantiated by the observation that after treatment of human bladder carcinoma by instillation of BCG high amounts of IL-2, as well as other cytokines, are produced (De Jong *et al.*, 1990) and supported by the studies in laboratory animals and humans mentioned above. Treatment of experimentally induced hepatocellular carcinomas in guinea pigs with BCG or IL-2 (De Jong *et al.*, 1985; Balemans *et al.*, 1993) and IL-2 therapy of transplanted tumours in mice clearly resulted in tumour-specific T cell reactivity (Maas *et al.*, 1991). In addition it was shown that *in vitro* culture of human TIL in the presence of autologous tumour antigen and IL-2 led to tumour-specific cytotoxicity (Rosenberg, 1988, 1990; Bukowski *et al.*, 1991), indicating the presence of these cells in the tumour.

Cows bearing BOSCC were treated with low doses of rHuIL-2 (rHuIL-2). Doses of 2500 IU rHuIL-2 were injected intralesionally at the base of the tumour and in addition into the subparotid RLN once a day during a period of 5 consecutive days, which was repeated after an interval of 2 days. Total regression of tumours, sizes 2–4 cm, was observed in three out of five animals. One cow showed tumour regression (> 80%) accompanied by metastases to the RLN, from the fifth week after initiation of treatment. Compared to untreated tumours, growth of the tumour of the fifth animal was retarded after treatment (Rutten *et al.*, 1989).

Treatment of another ten animals according to this protocol showed similar response percentages. Combination of the BCG and rHuIL-2 treatments in ten animals showed no synergistic effects (unpublished results). In four animals carrying large BOSCC with diameters more than 4 cm, rHuIL-2 therapy alone induced lasting complete remission in one animal and significant (> 50% and > 90% respectively) remission at 3 months after treatment, subsequently progressive growth was observed (Den Otter *et al.*, 1993).

In a subsequent larger study conducted in cattle in Zimbabwe (Den Otter *et al.*, 1995) tumours of 1–3 cm in diameter were treated by intralesional injection of various doses of rHuIL-2, i.e. 500 IU (n = 19), 2000 IU (n = 20), 20,000 IU (n = 18) and saline (n = 14). Side-effects were not noted in any of the animals regardless of the doses used. Usually complete regression took time and first complete regressions were seen after 3 months. Nine months after treatment, high percentages, 89%, 80% and 67% respectively, of tumours in each group showed complete regression. Whereas in the group receiving the highest dosage regression lasted for at least 20 months, tumour recurrences were observed in the two other groups (Table 4.1).

Based on our success in treating BOSCC with IL-2 we expanded our studies to BVPC. Our studies using IL-2 therapy of BVPC, like those concerning BOSCC, focused on the effects of low doses of locally injected rHuIL-2 (Hill *et al.*, 1994b). Again the protocol of therapy during two cycles of 5 days was used. Doses of 2500–5000 IU were injected into two to four papillomas of the bovine vulva. In three animals, carcinomas present in conjunction with many papillomas received additional intralesional injections. In 19 out of 23 cows (83%) treated with IL-2 a clear effect on the tumour load was observed 3 to 5 months after therapy after which tumour progression was observed. One year after therapy, three of these animals showed complete regression of tumours. In all but three animals regression was not restricted to the tumours injected with IL-2.

MECHANISM OF IMMUNOTHERAPY INDUCED REGRESSION OF BOSCC

Lymphocytes present in BOSCC are potential effector cells in BCG- and/or rHuIL-2-induced regression. Moreover due to treatment, influx of inflammatory cells was observed (Rutten, 1992). TIL were isolated from BOSCC before, during and after treatment and their phenotypes were analysed. TIL from five responder and three non-responder animals were cultured in the presence of rHuIL-2 (Rutten, 1992). The phenotype compositions of TIL populations present at the start of treatment varied between tumours and were not predictive for the clinical results. Furthermore, changes of phenotypes of mononuclear cells in the tumour during therapy were not conclusive concerning the cells possibly responsible for regression. The TIL populations from three non-regressing tumours showed $CD4^+$ to $CD8^+$ ratios of > 1. The $CD4^+$ to $CD8^+$ ratios of three out of five regressing tumours were < 1, the other two regressing tumours, however, had $CD4^+$ to $CD8^+$ ratios of > 1, and percentages of $CD8^+$ cells were lower than in progressively growing tumours. CD2 (co)expression was observed on rather high percentages of cells, throughout the experiment. In both responder and non-responder tumours approximately 20% of the TIL were CD2– cells, being B cells, plasma cells, NK cells and γδ T cells.

Culture of TIL, isolated from tumours before treatment, in medium with 10–1000 IU of rHuIL-2 ml^{-1} resulted in cytotoxic activity *in vitro* towards autologous and allogeneic tumour cells as well as control tumour cell lines, i.e. P815 and Yac-1. TIL of both responder and non-responder animals became cytotoxic *in vitro*, suggesting that potential killer cells were present in all tumours (Rutten *et al.*, 1989, 1991b; Rutten, 1992).

FACScan analysis of the majority of cultured TIL showed that $CD4^+$ cells were present in very low percentages, which made it unlikely that they played a role in cytotoxicity *in vitro*. Since $CD4^+$ cells were predominantly present in three out of three non-responder tumours, it is unlikely that these cells are active as effector cells in tumour regression. The predominant presence of $CD2^+$ and/or $CD8^+$ cells in cultures of TIL suggests that these cells are responsible for the cytotoxicity measured *in vitro*. This observation and the presence of $CD8^+$ and $CD2^+$ cells in relatively high quantities in three out of five responding tumours, as observed in freshly isolated TIL populations, suggests an effector role for these cells in tumour destruction. Based on their continuous presence in all tumours, participation of CD2– cells in regression cannot be excluded. As was noted for $CD4^+$ T cells, a role of $\gamma\delta^+$ T cells was never indicated *in vitro*.

Cytotoxicity and phenotype data obtained in cultures of TIL isolated at different times during treatment were comparable to those described above. Again no differences were observed in *in vitro* cytotoxic activity between TIL from animals that did and did not show final regression (Rutten, 1992).

Since autologous as well as allogeneic BOSCC-derived cells were killed *in vitro*, our results were not conclusive with respect to the mechanisms of tumour regression induced by BCG or IL-2 immunotherapy. Both activation of non-specific and tumour-specific immune responsiveness may play a role.

EPILOGUE

Interleukin-2 therapy of BOSCC and BVPC in cattle by way of local injection of low doses of the cytokine may result in transient regression of the tumour as well as lasting total cure.

Transient regression after IL-2 therapy of BVPC in the majority of animals indicates that the low-dose local regimen of IL-2 treatment is effective, though not yet optimal. The high cure rate of BOSCC (> 50%) as well as a variety of mouse tumours (Den Otter *et al.*, 1996) after 'optimal' treatment are promising perspectives for further investigation. Repeated cycles of IL-2 therapy as well as higher local dosage regimens or combined therapy with other immune modulators, like IFNγ, cytostatic agents, etc., may further improve efficacy.

At present we are conducting IL-2 therapy trials of various inoperable canine tumours and equine sarcoid. The potential of IL-2 therapy as a treatment modality for other tumours of veterinary relevance needs to be investigated.

Cytokines and Bovine Herpesvirus 1

6

E. Hanon, C. Hamers and P.-P. Pastoret

Department of Immunology/Vaccinology, Faculty of Veterinary Medicine, University of Liège, B43 bis, B-4000, Liège, Belgium

INTRODUCTION

Like herpes simplex virus (HSV), BHV-1 is a member of the *Alphaherpesvirinae* subfamily, closely related to varicella zoster virus, pseudorabies virus, and equine herpesvirus 1 (Roizman *et al.*, 1992).

Bovine herpesvirus l causes infectious bovine rhinotracheitis (IBR), a worldwide disease of cattle characterized by fever, abortion and respiratory symptoms (Yates, 1982). The lesions consist of microfocal epithelial necrosis located in the upper respiratory tract. These lesions normally resolve in a few days. However, BHV-1 can predispose animals, presumably through immunosuppression (Filion *et al.*, 1983; Bielefeldt Ohmann and Babiuk, 1985), to secondary *Pasteurella haemolytica* bacterial infections which may lead to severe pneumonia, pleuritis and death (Yates, 1982).

The cellular immune response to BHV-1 is principally mediated by activated macrophages, T lymphocytes and PMN (reviewed in Denis *et al.*, 1994). These cells limit the propagation of the virus, directly, by the lysis of BHV-1-infected cells and/or, indirectly, by the production of cytokines which promote the immune response against BHV-1. The humoral immune response is mediated by mucosal and serum antibodies which are involved at neutralizing infectious virus. In addition, BHV-1-specific antibodies and cytotoxic cells expressing receptor for Ig (Fc receptor) participate to specific cellular immune response by antibody-dependent cytotoxicity (ADCC; reviewed in Denis *et al.*, 1994).

 Cytokines in Veterinary Medicine (eds V.E.C.J. Schijns and M.C. Horzinek)

A model of cytokine release (Campos *et al.*, 1993) during infection with a pathogen, like BHV-1, has been proposed in order to explain the sequence of events that result in pathogenesis and immunity to that pathogen. The initial response to pathogens is generally characterized by early inflammatory indications, followed by infiltration and aggregation of inflammatory cells and then by the initiation of specific immune responses mediated by T and B lymphocytes. These responses appear to be orchestrated by the production of cytokines which can be classified, according to their release following infection, into early and late cytokines (Fig. 6.1). Generally, early cytokines are produced by cells at the site of infection and are responsible for the early inflammatory response, regulation of adhesion molecule expression and infiltration of lymphoid cells. Early cytokines include IFN-α and -β, the proinflammatory cytokines such as IL-1β and TNF-α, as well as the chemotactic cytokines. In contrast, late cytokines, such as IL-2, IL-4 and IFN-γ are produced by T lymphocytes following the recognition of antigen in association with MHC molecules present on the surface of antigen-presenting cells. These cytokines are responsible for the

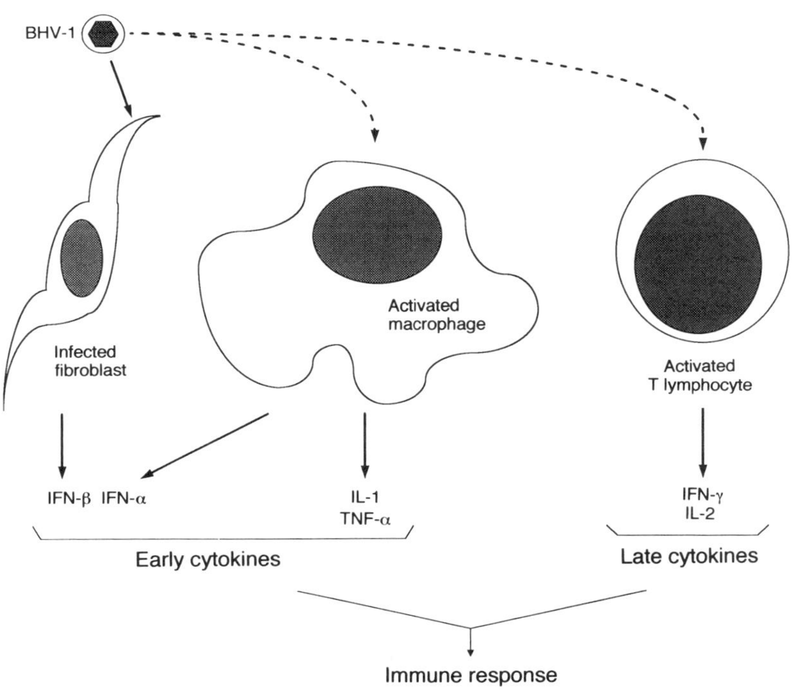

Fig. 6.1. The identified cytokines released during BHV-1 infection.

differentiation and clonal expansion of reactive cells, as well as for the fine tuning of the immune response.

The last few years have seen welcome progress in the characterization of bovine cytokines such as IFN-α, -β and -γ, IL-1β, TNF-α, and IL-2. Several studies investigated, *in vitro* and *in vivo*, their interactions with BHV-1 (reviewed in Myers and Murtaugh, 1995). Therefore, it seems this is a particularly opportune time to summarize the role of these cytokines in the induction of the immune response directed against BHV-1 as well as their possible involvement in the clinico-pathological features of the BHV-1/*Pasteurella haemolytica* respiratory disease.

BOVINE HERPESVIRUS 1 AND THE EARLY CYTOKINES: INTERFERON-α AND -β, INTERLEUKIN-1 AND TUMOUR NECROSIS FACTOR-α

Interferon-α and -β

Early after a primary BHV-1 infection, IFN-α and -β are rapidly produced respectively by leukocytes and fibroblasts present at the site of infection (Bielefeldt Ohmann *et al.*, 1984, 1987) (Fig. 6.1). The local production of IFN-α and -β could induce non-specific protection against BHV-1 through antiviral activity (Straub and Ahl, 1976; Wérenne *et al.*, 1985; Czarniecki *et al.*, 1986). Indeed, rBoIFN-α and -β have been shown to possess antiviral activity *in vitro* by the activation of 2′,5′-oligo(A) synthetase-RNase (2′5′OAS) pathway and the inhibition of cellular and viral protein synthesis (Roberts *et al.*, 1976; Lengyel, 1982; Bielefeldt Ohmann *et al.*, 1987). However, a relative insensitivity of BHV-l to BoIFN, regardless of the IFN type, has been observed (Savan *et al.*, 1979; Vanden Broecke *et al.*, 1985; Czarniecki *et al.*, 1986). This corroborates the *in vivo* findings, where maximal IFN titres in nasal secretion of BHV-l-challenged calves occur simultaneously with maximal virus titres (Todd *et al.*, 1972; LeJan and Asso, 1981; Bielefeldt Ohmann and Babiuk, 1985), and exogenous BoIFN applied intranasally (i.n.) before nasal challenge with BHV-l does not reduce replication to any significant degree, indicating the relative insensitivity of BHV-l *in vivo* to the direct antiviral effects of BoIFN (Babiuk *et al.*, 1985). An IFN-like substance, polyferon, is capable of antiviral activity against BHV-1 (Rouse, 1981; Bielefeldt Ohmann *et al.*, 1989a). *In vitro*, bovine cells infected with BHV-1 or HSV-1 induce the production of this protein by PMN. It is called polyferon since it could not be neutralized by antibodies to bovine IFN-α, -β and -γ. Nevertheless, it probably belongs to the IFN family since it could induce 2′5′OAS in cells normally responsive to IFNs. In addition, polyferon has been shown to be chemotactic for monocytes and PMN.

Interferon-α could also limit the propagation of BHV-1 *in vivo* by activating the phagocytes which exert the first line of cellular defence against invading pathogens. The efficiency whereby these cells can fight the intruders may depend on functions such as migration, generation of reactive oxygen molecules, expression of cellular receptor (Fc receptor and complement receptor), phagocytosis and cellular cytotoxicity. Most of these activities can be modulated by BoIFN *in vitro* (reviewed in Bielefeldt Ohmann *et al.*, 1987). The treatment of alveolar macrophages (AM) with BoIFN-α induced enhanced Fc receptor expression, increased their capacity to mediate ADCC of BHV-1-infected target cells, and augmented their extrinsic antiviral activity as measured by a reduction in the development of plaques in a BHV-1-susceptible cell line (Bielefeldt Ohmann *et al.*, 1984). The *in vivo* administration of rBoIFN-α causes the recovery of neutrophil function which is suppressed following BHV-1 infection as assessed by migration, chemotaxis and generation of reactive oxygen metabolites. This reversal of neutrophil suppression appears to correlate with a reduction in overall clinical disease (Lawman *et al.*, 1987).

Finally, IFN-α could also play a role in the recruitment of leukocytes at the site of infection. Indeed, the production of IFN-α during BHV-1 infection correlates temporally with a virus-induced leukopenia, a selective depletion of blood CD8 T lymphocytes, and an influx of the latter into the lung parenchyma. Moreover, the i.m. injection of rIFN-α to healthy calves led to a similar conclusion (Griebel *et al.*, 1987; Griebel *et al.*, 1989).

Since IFN-α promotes the early protection against BHV-1, it has been tested as a therapeutic agent for BHV-1 infection (Babiuk *et al.*, 1985, 1987). Calves were treated with rBoIFN-α 2 days before challenge with BHV-1. Four days after this virus challenge, calves were inoculated with *Pasteurella haemolytica*. The rBoIFN-α significantly reduced the severity of the disease as compared with placebo-treated controls. In addition, the treated calves were sick for a shorter period and had smaller lung lesions. Administration of IFN-α before viral challenge was more effective than the simultaneous administration with the virus. However, IFN treatment appeared to have no effect on BHV-1 shedding from nasal secretions. This may imply that the IFN exerted its effect through immunomodulation rather than through its antiviral properties.

Interleukin-1β and Tumour Necrosis Factor-α

These cytokines are most probably produced following BHV-1 infection by resident macrophages, endothelial and epithelial cells in the respiratory tract (Bielefeldt Ohmann *et al.*, 1991) (Fig. 6.1). However, BHV-1 can directly affect the production of IL-1β and TNF-α by AM. Indeed, following aerosol infection with BHV-1, the *in vitro* zymosan-stimulated release of IL-lβ by AM decreased progressively from day 1 to day 6 post infection (p.i.)

when compared with controls (Bielefeldt Ohmann and Babiuk, 1986a). Furthermore, after *in vitro* infection of AMs, endotoxin-stimulated release of TNF-α was impaired whereas spontaneous production was increased (Bienhoff *et al.*, 1992). Therefore, it is possible that the BHV-1-induced alteration of the IL-1β and TNF-α production by AM could contribute, in association with the increased prostaglandin E2 release (Bielefeldt Ohmann and Babiuk, 1986a, b), to the apparent immunosuppression observed during BHV-l infection.

Most likely, the overall production of IL-1β and TNF-α in the host, although impaired in AMs, increases and could contribute to the inflammatory symptoms (Bielefeldt Ohmann *et al.*, 1991). These symptoms correspond to the initial fever, within 24–48 h p.i., the hyperaemia, the interstitial oedema, and the massive alveolar infiltration of PMN. Furthermore, the overproduction of these cytokines during secondary bacterial infection could be responsible for the exacerbation of these inflammatory symptoms (Fig. 6.2). Indeed, the i.m. or i.n. injections with rBoTNF-α alone or in combination with rBoIFN-γ 24 h before challenge with BHV-l, enhanced

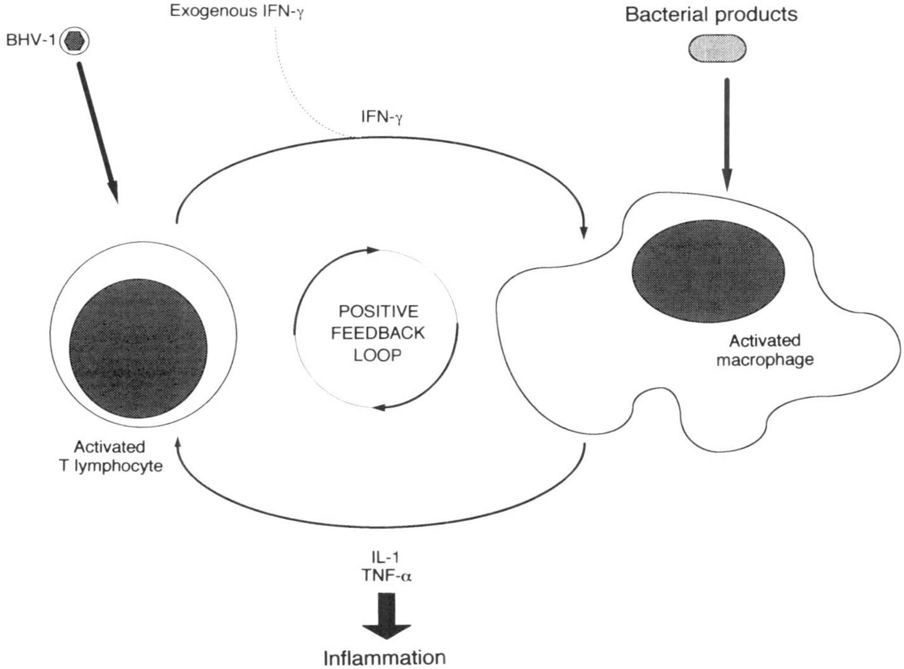

Fig. 6.2. A model of BHV-1 and *Pasteurella haemolytica* synergism leading to exacerbation of inflammation.

virus-induced inflammation, the severity of the infection and subsequent bacterial disease (Bielefeldt Ohmann *et al.*, 1991).

The adjuvanticity of rBoIL-1β was investigated (Reddy *et al.*, 1990). Calves were treated with this cytokine in conjunction with a BHV-1 vaccine and, thereafter, challenged with BHV-1. In rBoIL-l-β-treated calves, serum neutralizing antibody titres and cytotoxic responses against BHV-l-infected bovine kidney fibroblasts were increased while virus excretion was decreased. Moreover, after reactivation with dexamethasone, virus excretion decreased and clinical signs of BHV-l infection were lower in rBoIL-l-β-treated calves. These data suggested that rBoIL-l-β may be an effective adjuvant to BHV-l immunization.

Finally, rBoTNF-α has been shown to possess antiviral activity, which is mediated by IFN-independent mechanisms and by direct lysis of virus-infected cells (Wong and Goeddel, 1986; Campos *et al.*, 1988; Bielefeldt Ohmann *et al.*, 1989c). However, since BHV-1-infected target cells have a low susceptibility to rBoTNF-α, it may be only a minor contribution to direct lysis of BHV-1-infected cells (Campos *et al.*, 1988).

BOVINE HERPESVIRUS 1 AND THE LATE CYTOKINES: BoIFN-γ AND BoIL-2-IFN-γ

In vitro, IFN-γ is produced by T lymphocytes upon BHV-1-antigen presentation (Babiuk and Rouse, 1976; Godfroid *et al.*, 1995; Fig. 6.1). Most likely, respiratory tract T lymphocytes also produce IFN-γ *in vivo* (Bielefeldt Ohmann *et al.*, 1991). This cytokine is strongly implicated in the immune response against BHV-1 (Campos and Rossi, 1985; Campos *et al.*, 1989). IFN-γ regulates the expression of MHC class II molecules, Fc receptor, and complement receptor of monocytes/macrophages (Bielefeldt Ohmann *et al.*, 1986). This cytokine contributes to the generation of non-MHC-restricted cytotoxic responses of cattle to BHV-1 which is the most apparent cellular mediated immune response during BHV-1 infection (Campos *et al.*, 1989). It has been shown to be antibody independent and mediated by cells of the monocyte/macrophage lineage. Moreover, IFN-γ could also play a role in the immune response directed against BHV-1 by promoting the recruitment of CD4 T lymphocytes in the lungs (Bielefeldt Ohmann *et al.*, 1989b).

The i.n. or i.m. injections of rBoIFN-γ 24 h before challenge with BHV-1 was found to have some protective effect (Bielefeldt Ohmann *et al.*, 1991) indicating that IFN-γ could also potentiate the immune response against BHV-1. However, when calves were given i.m. injections of rBolFN-γ after BHV-1 infection but before *Pasteurella haemolytica* infection, the clinical severity increased and survival rates decreased (Bielelfeldt Ohmann *et al.*, 1991). Most probably, exogenous IFN-γ as well as bacterial products increase the production of IL-1 β and TNF-α which enhance further the synthesis of IFN-γ (Newton, 1985). This positive feedback loop leads to the accumulation of proinflammatory cytokines which in turn increase the inflammatory symptoms and the severity of the disease (Fig. 6.2).

INTERLEUKIN-2

Interleukin-2 is produced by activated T lymphocytes following antigen presentation (Fig. 6.1). This well-known cytokine has a central role in orchestrating the immune response. It supports the proliferation of activated T and B lymphocytes, contributes to the generation of lymphokine activated killer cells and influences the oxidative burst by monocytes/macrophages (Ortaldo *et al.*, 1984; Malkovsky *et al.*, 1987). In the context of BHV-1 immunity, rBoIL-2 increased both *in vitro* (Campos *et al.*, 1992) and *in vivo* (Reddy *et al.*, 1989, 1993; Hughes *et al.*, 1991) MHC-restricted cytotoxic responses against BHV-1-infected target cells. In addition, administration of rBoIL-2 to feeder calves at the time of vaccination against BHV-1 enhanced immunity. This was evident by a sixfold increase in serum neutralizing antibody titres to BHV-1 in calves vaccinated and treated with rBoIL-2 compared with calves that were only vaccinated (Reddy *et al.*, 1989). After challenge with BHV-1, calves vaccinated and treated with rBoIL-2 had improved clinical signs compared with calves that were vaccinated but did not receive rBoIL-2. Additional studies using co-administration of low doses of IL-2 with a commercial modified-live BHV-1 vaccine or experimental subunit BHV-1 vaccine also resulted in augmentation of the antiviral response (Reddy *et al.*, 1989, 1993; Hughes *et al.*, 1991, 1992). Mild side-effects (diarrhoea and fever) were observed depending on the administered dose but ceased after IL-2 treatment was stopped (Reddy *et al.*, 1989).

IMMUNOSUPPRESSION INDUCED BY BOVINE HERPESVIRUS 1

Immunosuppression is rapidly observed following BHV-1 infection (Filion *et al.*, 1983; Bielefeldt Ohmann and Babiuk, 1985; Bielefeldt Ohmann *et al.*, 1991). The immunosuppressive mechanisms are considered relevant to the induction of secondary bacterial infections, although the viral-induced inflammatory processes *per se* may also create a favourable environment for bacterial growth (Bielefeldt Ohmann *et al.*, 1991). Virus-induced immunosuppression has been reported in a large number of viral diseases and various mechanisms, both direct and indirect, have been described (McChesney and Oldstone, 1987). *In vitro* studies have demonstrated that

BHV-1 can suppress the PMN functions and alter the AM production of IL-1 and TNF-α (see above). However, the *in vitro* demonstration of virus-induced immunosuppression has focused on the interactions between the virus and the proliferating immune cells (McChesney and Oldstone, 1987). Following BHV-1 infection, the proliferation of mitogen-stimulated PBMC and IL-2-dependent T lymphocytes is inhibited (Carter *et al.*, 1989; Griebel *et al.*, 1990; Hutchings *et al.*, 1990). The mechanisms used by BHV-1 to inhibit the proliferative response to mitogens are poorly understood. However, it has been recently demonstrated that BHV-1 induces apoptotic cell death of mitogen-stimulated PBMC (Hanon *et al.*, 1996). Moreover, inactivated BHV-1 viral particles also had the capacity to induce apoptosis (Hanon *et al.*, 1996). Therefore, it could be postulated that the mechanism of apoptosis induction involves: (i) viral–cellular molecule interactions during the attachment or penetration processes; or (ii) the effect of viral structural protein(s) released into the cell. This could directly induce apoptosis or affect the production of cytokines and indirectly lead to the activation of the apoptotic process.

Alteration of monocytes/macrophages and PMN functions and the destruction of mononuclear cells by BHV-1 most probably affect several *in vivo* immune functions including the local production of cytokines. This mechanism could play a role in the increased susceptibility of the BHV-1-infected host to secondary bacterial infection.

EPILOGUE

In this chapter, it is evident that complex interactions occur between BHV-1, the various host cells that it infects, and the host immune cells. These interactions promote the induction of efficient cellular and humoral immune responses but can also lead to immunosuppression and/or to the exacerbation of the inflammatory symptoms. These processes are, at least in part, regulated through the production of cytokines.

The characterization of the cytokines involved in the immune response is critical if we hope to judiciously apply them to control disease (Blecha, 1991). IFN-α, IFN-γ, IL-1β, and IL-2 have been successfully used to potentiate the cellular and humoral immune responses against BHV-1 (see above). The specific induction of IFN-α production by the BHV-1 antigen *in vitro* can be seen as a diagnostic tool for IBR (Godfroid *et al.*, 1995). Recently, TH1 and TH2 CD4 T lymphocytes have been described in cattle (Brown *et al.*, 1993). These CD4 T lymphocytes produce a different pattern of cytokines which in turn modulate the efficiency of the immune response against a given pathogen. Further studies must now identify their implications in the immune response against BHV-1 and investigate their possible application in vaccinology.

Often BHV-1 infection is complicated by secondary bacterial infection with *Pasteurella haemolytica*. A growing body of evidence supports the contention that the increased susceptibility of the virus-infected host is, at least in part, due to virus-induced alteration of the functional integrity of the host immune system. This may occur either directly by viral interaction with leukocytes or indirectly through virus-induced alteration of cytokine production. The induction of apoptosis and the impairment of monocyte/macrophage and PMN functions by BHV-1 have been described (see above). Understanding how the virus induces these alterations and the dysfunction of the host immunological protection will also be important in designing new vaccines.

Finally, cytokines may contribute to the development of pathological lesions and clinical symptoms either alone or in concert with microbial cytopathic effects. Indeed, the overproduction of proinflammatory cytokines such as IL-1β, TNF-α and IFN-γ (see above) could contribute to the exacerbation of the inflammatory symptoms leading to pneumonia, pleuritis and death (Yates, 1982). Since notable similarities exist between the bovine and human species for virus-induced bacterial pneumonia (Bielefeldt Ohmann *et al.*, 1991), IBR could be seen as a experimental model to study the possible role of cytokines in clinico-pathological features of virus–bacterial respiratory disease in humans.

References for Part I

Abayasekara, D.R.E., Sheldrick, E.L., Flick-Smith, H.C. and Flint, A.P.F. (1995) Role of protein kinase C in the inhibitory action of trophoblast interferon on expression of the oxytocin receptor in sheep endometrium. *Endocrine* 3, 151–158.

Abraham, E. and Shah, S. (1992) Intranasal immunization with liposomes containing IL-2 enhances bacterial polysaccharide antigen-specific secretory antibody response. *Journal of Immunology* 149, 3719–3726.

Adams, L.G., Barthel, R., Feng, J., Qureshi, T., Piedrahita, J. and Templeton, J.W. (1995) Genes associated with innate killing of *Brucella abortus* and *Mycobacterium bovis* by macrophages from genetically resistant cattle. In: Gershwin, L., Lunney, J. and Schore, C. (eds) *Fourth International Veterinary Immunology Symposium.* Davis, California, p. 11.

Anderson, K.P., Fennie, E.H. and Yilma, T. (1988) Enhancement of a secondary antibody response to vesicular stomatitis virus 'G' protein by IFN-γ treatment at primary immunization. *Journal of Immunology* 140, 3599–3604.

Arai, N., Nomura, D., Villaret, D., de Waal Malefijt, R., Seiki, M., Yoshida, M., Minoshima, S., Fukuyama, R., Maekawa, M., Kudoh, J., Shimuzu, N., Yokota, K., Abe, E., Yokota, T., Takebe, Y. and Arai, K. (1989) Complete nucleotide sequence of the chromosomal gene for human IL-4 and its expression. *Journal of Immunology* 142, 274–282.

Aruffo, A. and Seed, B. (1987) Molecular cloning of two CD7 (T-cell leukemia antigen) cDNAs by a COS cell expression system. *European Molecular Biology Organization Journal* 6, 3313–3316.

Atamas, S.P., Choi, J., Yurovsky V. and White, B. (1996) An alternative splice variant of human IL-4, IL-4 delta 2, inhibits IL-4-stimulated T cell proliferation. *Journal of Immunology* 156, 435–441.

(eds V.E.C.J. Schijns and M.C. Horzinek)

Babiuk, L.A. and Rouse, B.T. (1976) Immune interferon production by lymphoid cells: role in the inhibition of herpesviruses. *Infection and Immunity* 13, 1567–1578.

Babiuk, L.A., Bielefeldt Ohmann, H., Gifford, G., Czarniecki, C.W., Scialli, V.T. and Hamilton, E.G. (1985) Effect of bovine alpha interferon on herpesvirus type 1 induced respiratory disease. *Journal of General Virology* 66, 2383–2394.

Babiuk, L.A., Lawman, M.J.P. and Gifford, G.A. (1987) Use of recombinant bovine alpha interferon in reducing respiratory disease induced by bovine herpesvirus type 1. *Antimicrobial Agents and Chemotherapy* 31, 752–757.

Babiuk, L.A., Sordillo, L.M., Campos, M., Hughes, H.P.A., Rossi Campos, A. and Harland, R. (1991) Application of interferons in the control of infectious diseases of cattle. *Journal of Dairy Science* 74, 4385–4398.

Baca-Estrada, M.E., Godson, D.L., Hughes, H.P.A., Van Donkersgoed, J., Van Kessel, A., Harland, R., Shuster, D.E., Daley, M. and Babiuk, L.A. (1995) Effect of recombinant bovine interleukin 1 β on viral bacterial pneumonia in cattle. *Journal of Interferon and Cytokine Research* 15, 431–439.

Balemans, L.T.M., Mattijssen, V., Steerenberg, P.A., Van Driel, B.E.M., De Mulder, P.H.M. and Den Otter, W. (1993) Locoregional therapy with PEG-Interleukin-2 of an intradermally growing hepatocellular carcinoma in the guinea pig induces T-cell mediated antitumour activity. *Cancer Immunology and Immunotherapy* 37, 7–14.

Baroja, M.L. and Ceuppens, J.L. (1987) More exact quantification of interleukin-2 production by addition of anti-Tac monoclonal antibody to cultures of stimulated lymphocytes. *Journal of Immunological Methods* 98, 267–270.

Bielefeldt Ohmann, H. and Babiuk, L.A. (1985) Viral bacterial pneumonia in calves: Effect of bovine herpesvirus-1 on immunological functions. *Journal of Infectious Disease* 151, 937–947.

Bielefeldt Ohmann, H. and Babiuk, L.A. (1986a) Alteration of alveolar macrophage functions after aerosol infection with bovine herpesvirus type 1. *Infection and Immunity* 51, 344–347.

Bielefeldt Ohmann, H. and Babiuk, L.A. (1986b) Alteration of some leukocyte functions following *in vivo* and *in vitro* exposure to recombinant bovine α- and γ-interferon. *Journal of Interferon Research* 6, 123–136.

Bielefeldt Ohmann, H., Gilchrist, J.E. and Babiuk, L.A. (1984) Effect of recombinant DNA-produced bovine interferon alpha (BoIFN-α-1) on the interaction between bovine alveolar macrophages and bovine herpesvirus type 1. *Journal of General Virology* 65, 1487–1495.

Bielefeldt Ohmann, H., Davis, W.C. and Babiuk, L.A. (1986) Surface antigen expression by bovine alveolar macrophages: functional correlation and influence of interferons *in vivo* and *in vitro*. *Immunobiology* 171, 125–142.

Bielefeldt Ohmann, H., Lawman, M.J.P. and Babiuk, L.A. (1987) Bovine interferon: its biology and application in veterinary medicine. *Antiviral Research* 7, 187–210.

Bielefeldt Ohmann, H., Campos, M., Fitzpatrick, D.R., Rapin, N. and Babiuk, L.A. (1989a) A neutrophil-derived antiviral protein: induction requirements and biological properties. *Journal of Virology* 63, 1916–1923.

Bielefeldt Ohmann, H., Campos, M., Harland, R., Griebel, P.J. and Babiuk, L.A. (1989b) 2′,5′-Oligoadenylate synthetase activity in bovine peripheral blood mononuclear cell following bovine herpes type-1 induced respiratory disease: A prognostic indicator? *Journal of Interferon Research* 9, 159–166.

Bielefeldt Ohmann, H., Campos, M., Griebel, P.J. and Babiuk, L.A. (1989c) 2′,5′-Oligo-A-synthetase activity in bovine peripheral blood leukocytes and alveolar macrophages exposed to recombinant interferons and tumor necrosis factor-α. *Canadian Journal Veterinary Research* 53, 161–166.

Bielefeldt Ohmann, H., Campos, M., Griebel, P., Snider, M., Beskorwayne, T., Rapin, N. and Babiuk, L.A. (1989d) Leukocyte trafficking and localization in bovine respiratory tract tissue during primary BHV-1 infection: Implications for regulation by interferons. EOS *Journal of Immunology and Immunopharmacology* 9, 137.

Bielefeldt Ohmann, H., Babiuk, L.A. and Harland, R. (1991) Cytokine synergy with viral cytopathic effects and bacterial products during the pathogenesis of respiratory tract infection. *Clinical Immunology and Immunopathology* 60, 153–170.

Bienhoff, S.E., Allen, G.K. and Berg, J.N. (1992) Release of tumor necrosis factor-α from bovine alveolar macrophages stimulated with bovine respiratory viruses and bacterial endotoxins. *Veterinary Immunology and Immunopathology* 30, 341–357.

Billman-Jacobe, H., Carrigan, M., Cockram, F., Corner, L.A., Gill, I.J., Hill, J.F., Jessep, T., Milner, A.R. and Wood, P.R. (1992) A comparison of the interferon γ assay with the absorbed ELISA for the diagnosis of Johne's disease in cattle. *Australian Veterinary Journal* 69, 25–28.

Blecha, F. (1991) Cytokines: applications in domestic food animals. *Journal of Dairy Science* 74, 328–339.

Bloom, B.R., Modlin, R.L. and Salgame, P. (1992a) Stigma variations: observations on suppressor T cells and leprosy. *Annual Review of Immunology* 10, 453–488.

Bloom, B.R., Salgame, P. and Diamond, B. (1992b) Revisiting and revising suppressor T cells. *Immunology Today* 13, 131–136.

Brown, W.C. and Grab, D.J. (1985) Biological and biochemical characterization of bovine interleukin-2. Studies with cloned bovine T cells. *Journal of Immunology* 133, 3184–3190.

Brown, W.C., Logan, K.S., Wagner, G.G. and Tetzlaff, C.L. (1991) Cell-mediated immune responses to *Babesia bovis* antigens in cattle following infection with tick-derived or cultured parasites. *Infection and Immunity* 59, 2418–2426.

Brown, W.C., Zhao, S., Rice-Ficht, A.C., Logan, K.S. and Woods, V.M. (1992) Bovine helper T cell clones recognize five distinct epitopes on *Babesia bovis* merozoite antigens. *Infection and Immunity* 60, 4364–4372.

Brown, W.C., Palmer, G.H., McElwain, T.F., Hines, S.A. and Dobbelaere, D.A.E. (1993a) *Babesia bovis*: Characterization of the T helper cell response against the 42 kilodalton merozoite surface antigen (MSA-1) in cattle. *Experimental Parasitology* 77, 97–110.

Brown, W.C., Woods, V.M., Dobbelaere, D.A.E. and Logan, K.S. (1993b) Heterogeneity in cytokine profiles of *Babesia bovis*-specific bovine $CD4^+$ T cell clones activated *in vitro*. *Infection and Immunity* 61, 3273–3281.

Brown, W.C., Zhao, S., Woods, V., Tripp, C.A., Tetzlaff, C.L., Heussler, V.T., Dobbelaere, D.A.E. and Rice-Ficht, A.C. (1993c) Identification of two Th1 cell epitopes on the 77 kDa *Babesia bovis* merozoite protein (Bb-1) by use of truncated recombinant fusion proteins. *Infection and Immunity* 61, 236–244.

Brown, W.C., Davis, W.C., Choi, S.H., Dobbelaere, D.A.E. and Splitter, G.A. (1994a) Functional and phenotypic characterization of WC1+ g/d T cells isolated from *Babesia bovis*-stimulated T cell lines. *Cellular Immunology* 153, 9–27.

Brown, W.C., Davis, W.C., Dobbelaere, D.A.E. and Rice-Ficht, A.C. (1994b) $CD4^+$ T cell clones obtained from cattle chronically infected with *Fasciola hepatica* and specific for adult worm antigen express both unrestricted and Th2 cytokine profiles. *Infection and Immunity* 62, 818–827.

Brown, W.C., Woods, V.M., Chitko-McKown, C.G., Hash, S.M. and Rice-Ficht, A.C. (1994c) IL-10 is expressed by bovine type 1 helper (Th1), type 2 helper (Th2) and unrestricted (Th0) parasite-specific T cell clones, and inhibits proliferation of all three subsets in an accessory cell-dependent manner. *Infection and Immunity* 62, 4697–4708.

Brown, W.C., Davis, W.C. and Tuo, W. (1996a) Human IL-12 upregulates proliferation and IFN-γ production by parasite antigen-stimulated Th cell clones and γδ T cells of cattle. *Annals of the New York Academy of Sciences* 795, 321–324.

Brown, W.C., McElwain, T.F., Ruef, B.J., Suarez, C.E., Shkap, V., Chitko-McKown, C.G., Tuo, W., Rice-Ficht, A.C. and Palmer, G.H. (1996b) *Babesia bovis* rhoptry-associated protein 1 is immunodominant for T helper cells of immune cattle and contains T cell epitopes conserved among geographically distant *B. bovis* strains. *Infection and Immunity* 64, 3341–3350.

Bubenik, J. and Indrova, M. (1987) Cancer immunotherapy using local interleukin-2 administration. *Immunology Letters* 16, 305–309.

Bubenik, J., Indrova, M., Perlmann, P., Berzins, K., Mach, O., Kraml, J. and Toulcova, A. (1985) Tumour inhibitory effects of TCGF/IL-2 containing preparations. *Cancer Immunology and Immunotherapy* 19, 57–61.

Bukowski, R.M., Sharfman, W., Murthy, S., Rayman, P., Tubbs, R., Alexander, J., Budd, G.T., Sergi, J.S., Bauer, L., Gibson, V., Stanley, J., Boyett, J., Pontes, E. and Finke, J. (1991) Clinical results and characterization of tumour-infiltrating lymphocytes with or without recombinant interleukin 2 in human metastatic renal cell carcinoma. *Cancer Research* 51, 4199–4205.

Butler, J.E., Maxwell, C.F., Pierce, C.S., Hylton, M.B., Asoksky, R. and Kiddy, C.A. (1972) Studies on the relative synthesis and distribution of IgA and IgG1 in various tissues and body fluids of the cow. *Journal of Immunology* 109, 38–48.

Callard, R.E. (1991) Immunoregulation by interleukin-4 in man. *British Journal of Haemotology* 78, 293–299.

Campos, M. and Rossi, C.R. (1985) Cell-mediated cytotoxicity of bovine mononuclear cells to IBRV-infected cells: dependence on sephadex G-10 adherent cells. *Veterinary Immunology and Immunopathology* 8, 363–375.

Campos, M., Bielefeldt Ohmann, H., Hughes, G., Babiuk, L.A. and Lawman, M.P.J. (1988) Studies on the *in vitro* biological activities of recombinant tumor necrosis factor (rBoTNF) alpha. I. Synergistic antiviral efficacy of rBoTNF alpha, recombinant bovine interferon γ (rBoIFN γ) and their combination. *Antiviral Research* 10, 71–81.

Campos, M., Rossi-Campos, A., Potter, A., Harland, R., Bielefeldt Ohmann, H. and Babiuk, L.A. (1989b) Parenteral administration of interferons protects pigs from acute death caused by the gram-negative bacterium, *Actinobacillus (Haemophilus) pleuropneumoniae*. *Immunobiology* Suppl. 4, 24–28.

Campos, M., Bielefeldt Ohmann, H., Hutchings, D., Rapin, N., Babiuk, L.A. and Lawman, M.J.P. (1989a) Role of interferon-α in inducing cytotoxicity of peripheral blood mononuclear leukocytes to bovine herpesvirus type 1 (BHV-1)-infected cells. *Cellular Immunology* 120, 259–269.

Campos, M., Rossi, C.R., Bielefeldt Ohmann, H., Beskorwayne, T., Rapin, N. and Babiuk, L.A. (1992) Characterization and activation requirements of bovine lymphocytes acquiring cytotoxic activity after interleukin-2 treatment. *Veterinary Immunology and Immunopathology* 32, 205–223.

Campos, M., Godson, D.L., Hughes, H.P.A. and Babiuk, L.A. (1993) Cytokine applications in infectious disease. In: Goddeeris, B.M.L. (ed.) *Ruminant Immunology*. CRC Press, Boca Raton, Florida, 229 pp.

Campos, M., Godson, D.L., Hughes, H.P.A. and Babiuk, L.A. (1994) Cytokine applications in infectious disease. In: Goddeeris, B.M. and Morrison, W.I. (eds) *Cell-Mediated Immunity in Ruminants*. CRC Press, Boca Raton, Florida.

Carter, J.J., Weinberg, A.D., Pollard, A., Reeves, R., Magnuson, J.A. and Magnuson, N.S. (1989) Inhibition of T-lymphocyte mitogenic responses and effects on cell functions by bovine herpesvirus 1. *Journal of Virology* 63, 1525–1530.

Cerretti, D.P., McKereghan, K., Larsen, A., Cosman, D., Gillis, S. and Baker, P.E. (1986) Cloning, sequence, and expression of bovine interferon-γ. *Journal of Immunology* 136, 4561–4564.

Chiang, Y., Roth, J.A. and Andiens, J.J. (1990) Influence of recombinant bovine interferon γ and dexamethasone on pneumonia attributable to *Haemophilus somnus* in calves. *American Journal of Veterinary Research* 51, 759–762.

Chitko-McKown, C.G., Ruef, B.J., Rice-Ficht, A.C. and Brown, W.C. (1995) Interleukin-10 downregulates proliferation and expression of interleukin-2 receptor p55 chain and interferon-γ but not interleukin-2 or interleukin-4 by parasite-specific helper T cell clones obtained from cattle chronically infected with *Babesia bovis* or *Fasciola hepatica*. *Journal of Interferon and Cytokine Research* 15, 915–922.

Clerici, M. and Shearer, G.M. (1994) The Th1–Th2 hypothesis of HIV infection: new insights. *Immunology Today* 15, 575–581.

Cooper, A.M., Dalton, D.K., Stewart, T.A., Griffin, J.P., Russell, D.G. and Orme, I.M. (1993) Disseminated tuberculosis in interferon γ gene-disrupted mice. *Journal of Experimental Medicine* 178, 2243–2247.

Cox, G.J.M., Zamb, T.J. and Babiuk, L.A. (1993) Bovine herpesvirus 1: immune responses in mice and cattle injected with plasmid DNA. *Journal of Virology* 67, 5664–5667.

Cullor, J.S., Smith, W., Zinkl, G., Dellinger, D. and Boone, T. (1992) Hematologic and bone marrow changes after short- and long-term administration of two recombinant bovine granulocyte colony-stimulating factors. *Veterinary Pathology* 29, 521–527.

Czarniecki, C.W., Hamilton, E.B., Fennie, C.W. and Wolf, R.L. (1986) *In vitro* biological activities of *Escherichia coli*-derived bovine interferons-α, -β, and -γ. *Journal of Interferon Research* 6, 29–37.

Daley, M.J., Coyle, P.A., Williams, T.J., Furda, G., Dougherty, R. and Hayes, P.W. (1991) *Staphylococcus aureus* mastitis: pathogenesis and treatment with bovine interleukin-1 β and interleukin-2. *Journal of Dairy Science* 74, 4413–4424.

Daley, M.J., Furda, G., Dougherty, R., Coyle, P.A., Williams, T.J. and Johnston, P. (1992) Potentiation of antibiotic therapy for bovine mastitis by recombinant bovine interleukin-2. *Journal of Dairy Science* 75, 3330–3338.

Daley, M., Williams, T., Coyle, P., Furda, G., Dougherty, R. and Hayes, P. (1993) Prevention and treatment of *Staphylococcus aureus* infections with recombinant cytokines. *Cytokine* 5, 276–284.

D'Andrea A., Rengaraju, M., Valiante, N.M., Chehimi, J., Kubin, M., Aste, M., Chan, S.H., Kobayashi, M., Young, D., Nickbarg, E., Chizzonite, R., Wolf, S.F. and Trinchieri, G. (1992) Production of natural killer cell stimulatory factor (Interleukin 12) by peripheral blood mononuclear cells. *Journal of Experimental Medicine* 176, 1387–1398.

D'Andrea, A., Aste-Amezaga, M., Valiante, N.M., Ma, X., Kubin, M. and Trinchieri, G. (1993) Interleukin 10 (IL-10) inhibits human lymphocyte interferon γ-production by suppressing natural killer cell stimulatory factor/IL-12 synthesis in accessory cells. *Journal of Experimental Medicine* 178, 1041–1048.

Davidson, P.M., Creati, L., Wood, P.R., Roberston, D.M. and Hosking, C.S. (1993) Lymphocyte production of γ-interferon. A test for non-tuberculous mycobacterial lymphadenitis in childhood. *European Journal of Paediatrics* 152, 31–35.

De Jong, W.H., Steerenberg, P.A., Van de Plas, M.M.T., Kruizinga, W. and Ruitenberg, E.J. (1985) T cell involvement in the adoptive transfer of line 10 immunity in the strain 2 guinea pigs. *Journal of the National Cancer Institute* 75, 483–489.

De Jong, W.H., De Boer, E.C., Van de Meijden, A.P.M., Vegt, P., Steerenberg, P.A., Debruyne, F.M.J. and Ruitenberg, E.J. (1990) Presence of interleukin-2 in urine of superficial bladder cancer patients after intravesical treatment with BCG. *Cancer Immunology and Immunotherapy* 31, 182–186.

Del Prete, G., Maggi, E., Parronchi, P., Chretien, I., Tiri, A., Macchia, D., Ricci, M., Banchereau, J., de Vries, J. and Romagnani, S. (1988) IL-4 is an essential factor for the IgE synthesis induced *in vitro* by human T cell clones and their supernatants. *Journal of Immunology* 140, 4193–4198.

Del Prete, G.F., De Carli, M., Mastromauro, M., Biagiotti, R., Macchia, D., Falagiani, P., Ricci, M. and Romagnani, S. (1991) Purified protein derivative of *Mycobacterium tuberculosis* and excretory–secretory antigen(s) of *Toxocara canis* expand *in vitro* human T cells with stable and opposite (type 1 T helper or type 2 T helper) profile of cytokine expression. *Journal of Clinical Investigation* 88, 346–350.

Del Prete, G., De Carli, M., Almerigogna, F., Giudizi, M.G., Biagiotti, R. and Romagnani, S. (1993) Human IL-10 is produced by both type 1 helper (Th1) and type 2 helper (Th2) T cell clones and inhibits their antigen-specific proliferation and cytokine production. *Journal of Immunology* 150, 353–360.

Denis, M., Splitter, G., Thiry, E., Pastoret, P.-P. and Babiuk, L.A. (1994) Infectious bovine rhinotracheitis (bovine herpesvirus 1): helper T cells, cytotoxic T cells, and NK cells. In: Goddeeris, B.M.L. (ed.) *Cell-mediated Immunity in Ruminants*. CRC Press, London and Tokyo, pp. 157–172.

Den Otter, W., Maas, R.A., Koten, J.W., Dullens H.F.J., Bernsen, M., Klein, W.R., Rutten, V.P.M.G., Steerenberg, P.A., Balemans, L., Ruitenberg, E.J., Hill, F.W.G. and Heintz, A.P.M. (1991) Effective immunotherapy with local low-doses of interleukin-2. *In Vivo* 5, 561–565.

Den Otter, W., Hill, F.W.G., Klein, W.R., Koten, J.W., Steerenberg, P.A., De Mulder, P.H.M., Rutten, V.P.M.G. and Ruitenberg, E.J. (1993) Low doses of interleukin-2 can cure large bovine ocular squamous cell carcinoma. *Anticancer Research* 13, 2453–2456.

Den Otter, W., De Groot, J.W., Bernsen, M.R., Heintz, A.P.M., Maas, R., Hordijk, G.J., Hill, F.W.G., Klein, W.R., Ruitenberg, E.J. and Rutten, V.P.M.G. (1996) Optimal regimen for local IL-2 therapy. *International Journal of Cancer* 66, 1–4.

De Waal Malefyt, R., Haanen, J., Spits, H., Roncarlo, M.-G., te Velde, A., Figdor, C., Johnson, K., Kastelein, R., Yssel, H. and de Vries, J.E. (1991) Interleukin 10 (IL-10) and viral IL-10 strongly reduce antigen-specific human T cell proliferation by diminishing the antigen-presenting capacity of monocytes via downregulation of class II major histocompatibility complex expression. *Journal of Experimental Medicine* 174, 915– 924.

Ding, L. and Shevach, E.M. (1992) IL-10 inhibits mitogen-induced T cell proliferation by selectively inhibiting macrophage costimulatory function. *Journal of Immunology* 148, 3133–3139.

Ding, L., Linsley, P.S., Huang, L.-Y., Germain, R.N. and Shevach, E.M. (1993) IL-10 inhibits macrophage costimulatory activity by selectively inhibiting the upregulation of B7 expression. *Journal of Immunology* 151, 1224–1234.

Doherty, M.L., Monaghan, M.L., Bassett, H.F. and Quinn, P.J. (1995) Effects of a recent injection of purified protein derivative on diagnostic tests for tuberculosis in cattle infected with *Mycobacterium bovis*. *Research in Veterinary Science* 58, 217–221.

Domingo, M., Liebana, E., Vilafranca, M., Aranaz, A., Vidal, D., Prats, N., Mateos, A., Casal, J. and Dominguez, L. (1995) A field evaluation of the interferon-γ assay and the intradermal tuberculin test in dairy cattle in Spain. In: Griffin, F. and de Lisle, G. (eds) *Tuberculosis in Wildlife and Domestic Animals*. Otago Conference Series No.3. University of Otago Press, Dunedin, pp. 304–306.

Duits, A.J., van Puijenbroek, A., Vermeulen, H., Hofhuis, F.M., van de Winkel, J.G. and Capel, P.J. (1993) Immunoadjuvant activity of a liposomal IL-6 formulation. *Vaccine* 11, 777–781.

Durum, S.K., Schmidt, J.A. and Oppenheim, J.J. (1985) Interleukin 1: an immunological perspective. *Annual Reviews of Immunology* 3, 263–287.

Estes, D.M., Closser, N.M. and Allen, G.K. (1994) IFN-γ stimulates IgG2 production from bovine B cells costimulated with anti-μ and mitogen. *Cellular Immunology* 154, 287–295.

Estes, D.M., Hirano, A., Heussler, V.T., Dobbelaere, D.A.E. and Brown, W.C. (1995) Expression and biological activities of bovine interleukin 4: Effects of recombinant bovine interleukin 4 on T cell proliferation and B cell differentiation and proliferation *in vitro*. *Cellular Immunology* 163, 268–279.

Fifis, T., Plackett, P., Corner, L.A. and Wood, P.R. (1989) Purification of a major *Mycobacterium bovis* antigen for the diagnosis of bovine tuberculosis. *Scandinavian Journal of Immunology* 29, 91–101.

Fifis, T., Corner, L.A., Rothel, J.S. and Wood, P.R. (1994) Cellular and humoral immune responses of cattle to purified *Mycobacterium bovis* antigens. *Scandinavian Journal of Immunology* 39, 267–274.

Filion, L.G., McGuire, R.L. and Babiuk, L.A. (1983) Nonspecific suppressive effect of bovine herpesvirus type 1 on bovine leukocyte functions. *Infection and Immunity* 42, 106–112.

Fine, P.E.M., Stern, J.A.C., Ponnighaus, J.M. and Rees, R.J.W. (1994) Delayed-type hypersensitivity, mycobacterial vaccines and protective immunity. *The Lancet* 344, 1245–1249.

Finkelman, F.D., Katona, I.M., Mosmann,T.R. and Coffman, R.L. (1988) IFN-γ regulates the isotypes of Ig secreted during *in vivo* humoral immune responses. *Journal of Immunology* 140, 1022–1027.

Fiorentino, D.F., Bond, M.W. and Mosmann, T.R. (1989a) Two types of mouse T helper cell. IV: TH2 clones secrete a factor that inhibits cytokine production by TH1 clones. *Journal of Experimental Medicine* 170, 2081–2095.

Fiorentino, D.F., Zlotnick, A., Vieira, P., Mosmann, T.R., Howard, M., Moore, K.W. and O'Garra, A. (1989b) IL-10 acts on the antigen-presenting cell to inhibit cytokine production by Th1 cells. *Journal of Immunology* 146, 3444–3451.

Firestein, G.S., Roeder, W.D., Laxer, J.A., Townsend, K.S., Weaver, C.T., Hom, J.T., Linton, J., Torbett, B.E. and Glasebrook, A.L. (1989) A new murine $CD4^+$ T cell subset with an unrestricted cytokine profile. *Journal of Immunology* 143, 518–525.

Flint, A.P.F., Stewart, H.J., Lamming, G.E. and Payne, J.H. (1992) Role of the oxytocin receptor in the choice between cyclicity and gestation in ruminants. *Journal of Reproduction and Fertility*, Suppl. 45, 53–58.

Flint, A.P.F., Abayasekara, D.R.E., Riley, P.R., Sheldrick, E.L. and Clarkson, A.M. (1996) Mechanism of action of interferon on oestrogen receptor function. *Biochemical Society Transactions* (in press).

Flynn, J.L., Chan, J., Triebold, K.J., Dalton, D.K., Stewart, T.A. and Bloom, B.R. (1993) An essential role for interferon in resistance to *Mycobacterium tuberculosis* infection. *Journal of Experimental Medicine* 178, 2249–2254.

Forni, G. and Giovarelli, M. (1987) Tumour immunotherapy with interleukin-2 and leucocytes. *Research Monographs in Immunology* 11, 265–282.

Francis, J., Seiler, R.J., Wilkie, I.W., O'Boyle, D., Lumsden, M.J. and Frost, A.J. (1978) The sensitivity and specificity of various tuberculin tests using bovine PPD and other tuberculinis. *Veterinary Record* 103, 420–425.

Gajewski, T.F. and Fitch, F.W. (1988) Anti-proliferative effect of IFN-γ in immune regulation. I. IFN-γ inhibits the proliferation of Th2 but not Th1 murine T lymphocyte clones. *Journal of Immunology* 140, 4245–4252.

Gao, Y., Daley, M.J. and Splitter, G.A. (1995) BHV-1 glycoprotein 1 and recombinant interleukin 1 efficiently elicit mucosal IgA response. *Vaccine* 13, 871–877.

Germann, T., Gately, M.K., Schoenhaut, D.S., Lohoff, M., Mattner, F., Fischer, S., Jin, S.-C., Schmitt, E. and Rude, E. (1993) Interleukin-12/T cell stimulating factor, a cytokine with multiple effects on T helper type 1 (Th1) but not on Th2 cells. *European Journal of Immunology* 23, 1762–1770.

Godfroid, J., Wellemans, G., Vanopdenbosch, E. and Letesson, J.-J. (1995) Infectious bovine rhinotracheitis (IBR) diagnosis based on an *in vitro* bovine herpesvirus-1 (BHV-1) antigen-specific IFN-γ production. *Immunobiology of Viral Infections*. Proc. 3rd Congress Europ. Soc. Vet. Virol., pp. 159–163.

Godson, D.L., Baca-Estrada, M.E., Van Kessel, A., Hughes, H.P.A., Morsey, M.A., Van Donkersgoed, J., Harland, R.J., Shuster, D.E., Daley, M.J. and Babiuk, L.A. (1995) Regulation of bovine acute phase responses by recombinant interleukin-1β. *Canadian Journal of Veterinary Research* 59, 249–255.

Godson, D.L., Campos, M., Attah-Poku, S.K., Redmond, M.J., Cordeiro, D.M., Sethi, M.S., Harland, R.J. and Babiuk, L.A. (1996) Serum haptoglobin as an indicator of the acute phase response in bovine respiratory disease. *Veterinary Immunology and Immunopathology* (in press).

Griebel, P.J. and Hein, W.R. (1996) Expanding the role of Peyer's patches in B-cell ontogeny. *Immunology Today* 17, 30–39.

Griebel, P.J., Qualtiere, L., Davis, W.C., Gee, A., Bielefeldt Ohmann, H., Lawman, M.J.P. and Babiuk, L.A. (1987) T lymphocyte population dynamics and function following a primary bovine herpesvirus type-1 infection. *Viral Immunology* 1, 287–304.

Griebel, P.J., Bielefeldt Ohmann, H., Campos, M., Qualtiere, L., Davis, W.C., Lawman, M.J.P. and Babiuk, L.A. (1989) Bovine peripheral blood leukocyte population dynamics following treatment with recombinant bovine interferon-α1. *Journal of Interferon Research* 9, 245–257.

Griebel, P.J., Bielefeldt Ohmann, H., Lawman, M.J.P. and Babiuk, L.A. (1990) The interaction between bovine herpesvirus type 1 and activated bovine T lymphocytes. *Journal of General Virology* 71, 369–377.

Guesdon, F.M.J., Stewart, H.J. and Flint, A.P.F. (1996) Negative regulatory domains in a trophoblast interferon promoter. *Journal of Molecular Endocrinology* 16 (in press).

Guler, M.L., Gorham, J.D., Hsieh, C.-S., Mackey, A.J., Steen, R.G., Dietrich, W.F. and Murphy, K.M. (1996) Genetic susceptibility to *Leishmania*: IL-12 responsiveness in TH1 cell development. *Science* 271, 984–987.

Hanon, E., Vanderplasschen, A., Lyaku, J., Keil, G., Denis, M. and Pastoret, P.-P. (1996) Inactivated bovine herpesvirus-1 induces apoptotic cell death of mitogen-stimulated bovine peripheral blood mononuclear cells. *Journal of Virology* (in press).

Harboe, M., Wiker, H.G., Duncan, J.R., Garcia, M.M., Dukes, T.W., Brooks, B.W., Turcotte, C. and Nagai, S. (1990) Protein G-based enzyme-linked immunosorbent assay for anti-MPB70 antibodies in bovine tuberculosis. *Journal of Clinical Microbiology* 28, 913–921.

Hash, S.M., Brown, W.C. and Rice-Ficht, A.C. (1994) Characterization of a cDNA encoding bovine interleukin-10: kinetics of expression in bovine lymphocytes. *Gene* 139, 257–261.

Heath, A.W. and Playfair, J.H.L. (1990) Conjugation of interferon-γ to antigen enhances its adjuvanticity. *Immunology* 71, 454–456.

Heath, A.W. and Playfair, J.H.L. (1992) Cytokines as immunological adjuvants. *Vaccine* 10, 427–434.

Hein, W.R. and Mackay, C.R. (1991) The prominence of γδ T cells in the ruminant immune system. *Immunology Today* 12, 30–34.

Heussler, V.T., Eichorn, M. and Dobbelaere, D.A.E. (1992) Cloning of a full-length cDNA encoding bovine IL-4 by the polymerase chain reaction. *Gene* 114, 273–278.

Hill, F.W.G., Rutten, V.P.M.G., Hoyer, M.J., Klein, W.R., Koten, J.W., Steerenberg, P.A., Ruitenberg, E.J. and Den Otter, W. (1994a) Local BCG therapy for bovine vulval papilloma and carcinoma. *Cancer Immunology and Immunotherapy* 39, 49–52.

Hill, F.W.G., Klein, W.R., Hoyer, M.J., Rutten, V.P.M.G., Kock, N., Koten, J.W., Steerenberg, P.A., Ruitenberg, E.J. and Den Otter, W. (1994b) Antitumour effect of locally injected low-doses of recombinant human interleukin-2 in bovine vulval papilloma and carcinoma. *Veterinary Immunology and Immunopathology* 41, 19–29.

Hixon, J.E. and Flint, A.P.F. (1987) Effects of a luteolytic dose of oestradiol benzoate on uterine oxytocin receptor concentrations, phosphoinositide turnover and prostaglandin F2a secretion in sheep. *Journal of Reproduction and Fertility* 79, 457–467.

Hodgson, A.L.M., Krywult, J., Corner, L.A., Rothel, J.S. and Radford, A.J. (1992) Rational attenuation of *Corynebacterium pseudotuberculosis*: Potential cheesy gland vaccine and live delivery vehicle. *Infection and Immunity* 60, 2900–2905.

Hoffmann, D., Jennings, P.A. and Spradbrow, P.B. (1981) Immunotherapy of bovine ocular squamous cell carcinomas with phenol–saline extracts of allogeneic carcinomas. *Australian Veterinary Journal* 57, 159–163.

Hogan, J.S., Smith, K.L., Todhunter, D.A., Schoenberger, P.S. and Shuster, D. (1995) Efficacy of recombinant bovine interleukin-2 as an adjunct to dry cow therapy. *Journal of Dairy Science* 78, 1062–1067.

Howard, M. and O'Garra, A. (1992) Biological properties of interleukin 10. *Immunology Today* 13, 198–200.

Hughes, H.P.A., Campos, M., Godson, D.L., Van Drunen Littel-Van Den Hurk, S., McDougall, L., Rapin, N., Zamb, T. and Babiuk, L.A. (1991) Immunopotentiation of bovine herpes virus subunit vaccination by interleukin-2. *Immunology* 74, 461–466.

Hughes, H.P., Campos, M., Potter, A.A. and Babiuk, L.A. (1992a) Molecular chimerization of *Pasteurella haemolytica* leukotoxin to interleukin-2: effects on cytokine and antigen function. *Infection and Immunity* 60, 565–570.

Hughes, H.P.A., Campos, M., Van Drunen Littel-Van Den Hurk, S., Zamb, T., Sordillo, M., Godson, D.L. and Babiuk, L.A. (1992b) Multiple administration with interleukin 2 potentiates antigen-specific responses to subunit vaccination with bovine herpesvirus-1 glycoprotein IV. *Vaccine* 10, 226–230.

Hutchings, D.L., Campos, M., Qualtiere, L. and Babiuk, L.A. (1990) Inhibition of antigen-induced and interleukin-2-induced proliferation of bovine peripheral blood leukocytes by inactivated bovine herpesvirus 1. *Journal of Virology* 64, 4146–4151.

Jansen, K.E. (1986) Synthetic adjuvants: Avridine and other interferon inducers. In: Nervig, R.M., Gough, P.M., Kaeberle, M.L. and Whetstone, C.A. (eds) *Advances in Carriers and Adjuvants for Veterinary Biologics*. Iowa State University Press, Ames, pp. 79–89.

Jones, S.L., Cox, J.C., Shepard, J.M., Rothel, J.S., Wood, P.R. and Radford, A.J. (1992) Removal of false-positive reactions from plasma in an enzyme immunoasssay for bovine interferon-γ. *Journal of Immunological Methods* 155, 233–240.

Kabelitz, D., Pechhold, K., Bender, A., Wesselborg, S., Wesch, D., Friese, K. and Janssen, O. (1991) Activation and activation-driven death of human γδ T cells. *Immunological Reviews* 120, 71–88.

Kato, S., Endoh, H., Masuhiro, Y., Kitamoto, T., Uchiyama, S., Sasaki, H., Masushige, S., Gotoh, Y., Nishida, E., Kawashima, H., Metzger, D. and Chambon, P. (1995) Activation of the estrogen receptor through phosphorylation by mitogen-activated protein kinase. *Science* 270, 1491–1494.

Katzenellenbogen, B.S. (1996) Estrogen receptors: bioactivities and interactions with cell signaling pathways. *Biology of Reproduction* 54, 286–293.

Kawamura, I., Yang, J., Takaesu, Y., Fujita, M., Nomoto, K. and Mitsuyama, M. (1994) Antigen provoking γ interferon production in response to *Mycobacterium bovis* BCG and functional difference in T-cell responses to this antigen between viable and killed BCG-immunized mice. *Infection and Immunity* 62, 4396–4403.

Kehrli, M.E., Jr, Cullor, J.S. and Nickerson, S.C. (1991) Immunobiology of hematopoietic colony-stimulating factors: potential application to disease prevention in the bovine. *Journal of Dairy Science* 74, 4399–4412.

Kelso, A. (1995) Th1 and Th2 subsets: paradigms lost? *Immunology Today* 16, 374–379.

Kennedy, M.K., Picha, K.S., Shanebeck, K.D., Anderson, D.M. and Grabstein, K.H. (1994) Interleukin-12 regulates the proliferation of Th1, but not Th2 or Th0, clones. *European Journal of Immunology* 24, 2271–2278.

Kimura, T. (1995) Investigation of the oxytocin receptor at the molecular level. In: Ivell, R. and Russel, J.A. (eds), *Oxytocin.* Plenum Press, New York, pp. 259–268.

Klein, W.R., Ruitenberg, E.J., Steerenberg, P.A., De Jong, W.H., Kruisinga, W., Misdorp, W., Bier, J., Tiesjema, R.H., Kreeftenberg, J.J., Teppema, J.S. and Rapp, H.J. (1982) Immunotherapy by intralesional injection of BCG cell walls or live BCG in bovine ocular squamous cell carcinoma. *Journal of the National Cancer Institute* 69, 1095–1103.

Klein, W.R., Steerenberg, P.A., Poelma, F., van der Wiel, E., Rutten, V.P.M.G., Misdorp, W., De Jong, W.H. and Ruitenberg, E.J. (1986) Immune reactivity in cattle with ocular squamous cell carcinoma after intralesional BCG immunotherapy. *Cancer Immunology and Immunotherapy* 22, 87–94.

Kleinschuster, S.J., Rapp, H.J., Green, S.B., Bier, J. and Van Kampen, K.R. (1981) Efficacy of intratumorally administered mycobacterial cell walls in the treatment of cattle with ocular carcinoma. *Journal of the National Cancer Institute* 67, 1165–1171.

Kuhn, R., Rajewsky, K. and Muller, W. (1991) Generation and analysis of interleukin-4 deficient mice. *Science* 254, 707–710.

Lafreniere, R. and Rosenberg, S.A. (1985) Successful immunotherapy of murine experimental hepatic metastases with lymphokine-activated killer cells and recombinant interleukin-2. *Cancer Research* 45, 3735–3741.

Lamming, G.E., Wathes, D.C., Flint, A.P.F., Payne, J.H., Stevenson, K.R. and Vallet, J.L. (1995) Local action of trophoblast interferons in suppression of the development of oxytocin and oestradiol receptors in ovine endometrium. *Journal of Reproduction and Fertility* 105, 165–175.

Lasic, D.D. and Papahadjopoulos, D. (1995) Liposomes revisited. *Science* 267, 1275–1276.

Lassauzet, M.L. and Salamin, P.A. (1993) Lack of effect of recombinant bovine interferon α (I)1 in the treatment of experimentally-induced bovine warts. *Canadian Journal of Veterinary Research* 57, 166–169.

Lawman, M.J., Gifford, G., Gyongyossy-Issa, M., Dragan, R., Heise, J. and Babiuk, L.A. (1987) Activity of polymorphonuclear (PMN) leukocytes during bovine herpes virus-1 induced respiratory disease: effect of recombinant bovine interferon α 1. *Antiviral Research* 8, 225–237.

Leaman, D.W., Cross, J.C. and Roberts, R.M. (1994) Multiple regulatory elements are required to direct trophoblast interferon gene expression in choriocarcinoma cells and trophectoderm. *Molecular Endocrinology* 8, 456–468.

Le Goff, P., Montano, M., Schodin, D.J. and Katzenellenbogen, B.S. (1994) Phosphorylation of the human estrogen receptor. *Journal of Biological Chemistry* 269, 4458–4466.

LeJan, C. and Asso, J. (1981) The local and systemic immune response of calves following experimental infection with IBR virus. *Advances in Experimental Medicine and Biology* 137, 677–692.

Lengyel, P. (1982) Biochemistry of interferons and their actions. *Annual Review of Biochemistry* 51, 251–282.

Lenzini, L., Rotolli, P. and Rotolli, L. (1977) The spectrum of human tuberculosis. *Clinical and Experimental Immunology* 27, 230–237.

Maas, R.A., Van Weering, D.H.J., Dullens, H.F.J. and Den Otter, W. (1991) Intratumoral low-dose interleukin-2 induces rejection of distant solid tumour. *Cancer Immunology and Immunotherapy* 33, 389–394.

Maggi, E., Del Prete, G., Macchia, D., Parronchi, P., Tiri, A., Chretien, I., Ricci, M. and Romagnani, S. (1988) Profiles of lymphokine activities and helper function for IgE in human T cell clones. *European Journal of Immunology* 18, 1045–1050.

Malkovsky, M., Loveland, B., North, M., Asherson, G.L., Gao, L., Ward, P. and Liers, W. (1987) Recombinant interleukin-2 directly augments the cytotoxocity of human monocytes. *Nature* 135, 262–265.

Manetti, R., Gerosa, F., Giudizi, M.G., Biagiotti, R., Parronchi, P., Piccinni, M.-P., Sampognaro, S., Maggi, E., Romagnani, S. and Trinchieri, G. (1994) Interleukin 12 induces stable priming for interferon γ (IFN-γ) production during differentiation of human T helper (Th) cells and transient IFN-γ production in established Th2 cell clones. *Journal of Experimental Medicine* 179,1273–1283.

Manilla, G.T., Kent, R., Van Kampen, K.R. and Marcus, S. (1972) Nucleoprotein in immunotherapy of ocular squamous carcinoma in cattle. *Federation Proceedings* 31, 768.

Mann, G.E. and Lamming, G.E. (1995) Effect of the level of oestradiol on oxytocin-induced prostaglandin F2α release in the ewe. *Journal of Endocrinology* 145, 175–180.

Maruo, S., Toyo-oka, K., Oh-hora, M., Tai, G., Iwata, H., Takenaka, H., Yamada, S., Ono, S., Hamaoka, T., Kobayashi, M., Wysocka, M., Trinchieri, G. and Fujiwara, H. (1996) IL-12 produced by antigen-presenting cells induces

IL-2-independent proliferation of T helper cell clones. *Journal of Immunology* 156, 1748–1755.
Mazumder, A. and Rosenberg, S.A. (1984) Successful immunotherapy of natural killer resistant established pulmonary melanoma metastases by the intravenous adoptive transfer of syngeneic lymphocytes activated *in vitro* by interleukin-2. *Journal of Experimental Medicine* 159, 495–507.
McChesney, M.B. and Oldstone, M.B.A. (1987) Viruses perturb lymphocyte functions: selected principles characterizing virus-induced immunosuppression. *Annual Review of Immunology* 5, 279–304.
Monaghan, M., Collins, D., McGill, K., McMurray, C. and Kelly, A. (1995) Field trials of the γ interferon assay for the diagnosis of bovine tuberculosis in the Republic of Ireland. In: Griffin, F. and de Lisle, G. (eds), *Tuberculosis in Wildlife and Domestic Animals*. Otago Conference Series No.3. University of Otago Press, Dunedin, pp. 319–320.
Morales, T.I. (1994) Transforming growth factor-β and insulin-like growth factor-1 restore proteoglycan metabolism of bovine articular cartilage after depletion by retinoic acid. *Archives of Biochemistry and Biophysics* 315, 190–198.
Morris, E.A., Wilcon, S. and Treadwell, B.V. (1992) Inhibition of interleukin 1-mediated proteoglycan degradation in bovine articular cartilage explants by addition of sodium hyaluronate. *American Journal of Veterinary Research* 53, 1977–1982.
Mosmann, T.R. and Coffman, R.L. (1989a) TH1 and TH2 cells: different patterns of lymphokine secretion lead to different functional properties. *Annual Review of Immunology* 7, 145–173.
Mosmann, T.R. and Coffman, R.L. (1989b) Heterogeneity of cytokine secretion patterns and functions of helper T cells. *Advances in Immunology* 46, 111–147.
Mosmann, T.R. and Sad, S. (1996) The expanding universe of T-cell subsets: Th1, Th2 and more. *Immunology Today* 17, 138–146.
Mosmann, T.R., Cherwinski, H., Bond, M.W., Gieldin, M.A. and Coffman, R.L. (1986) Two types of murine helper T cell clone. I. Definition according to profiles of lymphokine activities and secreted proteins. *Journal of Immunology* 136, 2348–2357.
Muscoplat, C.C., Johnson, D.W., Thoen, C.O., Ayivor, M.D. and Klausner, D.J. (1977) Development of a whole blood stimulation assay for detecting hypersensitivity to PPD in cattle experimentally infected with *Mycobacterium bovis*. *Veterinary Microbiology* 2, 261–265.
Myers, M.J. and Murtaugh, M.P. (1995) *Cytokines in Animal Health and Disease*. Marcel Dekker, New York, Basel, Hong Kong, 465pp.
Neill, S.D., Cassidy, J., Hanna, D., Mackie, D.P., Pollock, J.M., Clements, A., Walton, E. and Bryson, D.G. (1994) Detection of *Mycobacterium bovis* infection in skin test-negative cattle with an assay for bovine interferon-γ. *The Veterinary Record* 135, 134–135.
Neill, S., Hanna, J., Clements, A., Cassidy., Pollock, J. and Bryson, D.G. (1995) Diagnosing tuberculosis in animals. In: Griffin, F. and de Lisle, G. (eds), *Tuberculosis in Wildlife and Domestic Animals*. Otago Conference Series No.3. University of Otago Press, Dunedin, pp. 300–303.
Newton, R.C. (1985) Effect of interferon on the induction of human monocyte secretion of interleukin-1 activity. *Immunology* 56, 441–449.

Nickerson, S.C., Owens, W.E. and Watts, J.L. (1989) Effects of recombinant granulocyte colony-stimulating factor on *Staphylococcus aureus* mastitis in lactating dairy cows. *Journal of Dairy Science* 72, 3286–3294.

Nickerson, S.C., Owens, W.E., Boddie, R.L. and Boddie, N.T. (1992) The effect of chronic immunostimulation of the nonlactating bovine mammary gland with interleukin-2, pokeweed mitogen, and lipopolysaccharide. *Journal of Dairy Science* 1992, 3339–3351.

Nunberg, J.H., Doyle, M.V., York, S.M. and York, C.J. (1989) Interleukin-2 acts as an adjuvant to increase the potency of inactivated rabies virus vaccine. *Proceedings of the National Academy of Sciences, USA* 86, 4240–4243.

O'Malley, B.W., Schrader, W.T., Mani, S., Smith, C., Weigel, N.L., Conneely, O.M. and Clark, J.H. (1995) An alternative ligand-independent pathway for activation of steroid receptors. *Recent Progress in Hormone Research* 50, 333–347.

Ortaldo, J.R., Mason, A.T., Gerard, J.P., Henderson, L.E., Farrar, W., Hopkins, R.F., Herberman, L.B. and Rakiu, H. (1984) Effect of natural and recombinant IL-2 and regulation of IFN-γ production and natural killer activity: lack of involvement of the Tac antigen for these immunoregulatory effects. *Journal of Immunology* 133, 779–783.

Outteridge, P.M. and Lepper, A.W.D. (1973) The detection of tuberculin-sensitive lymphocytes from bovine blood by uptake of radio-labelled nucleosides. *Research in Veterinary Science* 14, 296–305.

Paliard, X., de Waal Malefijt, R., Yssel, H., Blanchard, D., Chretien, I., Abrams, J., de Vries, J. and Spits, H. (1988) Simultaneous production of IL-2, IL-4, and IFN-g by activated human $CD4^+$ and $CD8^+$ T cell clones. *Journal of Immunology* 141, 849–855.

Parkinson, T.J., Lamming, G.E., Flint, A.P.F. and Jenner, L.J. (1992) Administration of recombinant bovine interferon-αI at the time of maternal recognition of pregnancy inhibits prostaglandin F2-α secretion and causes luteal maintenance in cyclic ewes. *Journal of Reproduction and Fertility* 94, 489–500.

Paul, W.E. (1991) Interleukin-4: A prototypic immunoregulatory lymphokine. *Blood* 77, 1859–1870.

Peel, J.E., Kolly, C., Siegenthaler, B. and Martinod, S.R. (1990) Prophylactic effects of recombinant bovine interferon-αI on acute *Salmonella typhimurium* infection in calves. *American Journal of Veterinary Research* 51, 1095–1099.

Pighetti, G.M. and Sordillo, L.M. (1995) Enhanced antigen-specific responses in bovine mammary glands following administration of interleukin-2. *Journal of Dairy Science* 78, 528–537.

Pizza, C., Severini, G., Menniti, D., De Vinci, C. and Corrado, F. (1984) Tumour regression after intralesional injection of interleukin-2 (IL-2) in bladder cancer. *International Journal of Cancer* 34, 359–367.

Plackett, P., Ripper, J., Corner, L.A., Small, K., de Witte, K., Melville, L., Hides, S. and Wood, P.R. (1989) An ELISA for the detection of anergic tuberculous cattle. *Australian Veterinary Journal* 66, 15–19.

Quiroga, G.H., Sordillo, L.M., Adkinson, R.W. and Nickerson, S.C. (1993) Cytologic responses of *Staphylococcus aureus*-infected mammary glands of heifers to interferon γ and interleukin-2 treatment. *American Journal of Veterinary Research* 54, 1894–1900.

Raath, J.P., Bengis, M., Bush, M., Huchzermeyer, H., Keet, D.F., Kernes, D.J., Kriek, N.P.J. and Michel, A. (1995) Diagnosis of tuberculosis due to *Mycobacterium bovis* in the African buffalo *(Syncerus caffer)* in Kruger National Park. In: Griffin, F. and de Lisle, G. (eds), *Tuberculosis in Wildlife and Domestic Animals.* Otago Conference Series No.3. University of Otago Press, Dunedin, pp.313–315.

Rao, K.V.S. and Nayak, A.R. (1990) Enhanced immunogenicity of a sequence derived from hepatitis B virus surface antigen in a composite peptide that includes immunostimulatory region from human interleukin 1. *Proceedings of the National Academy of Sciences, USA* 87, 5519–5522.

Raz, E., Watanabe, A., Baird, M., Eisenberg, A., Parr, T.B., Lotz, M., Kipps, T.J. and Carson, D.A. (1993) Systemic immunological effects of cytokine genes injected into skeletal muscle. *Proceedings of the National Academy of Sciences, USA* 90, 4523–4527.

Reddy, P.G., Blecha, F., Minocha, H.C., Anderson, G.A., Morrill, J.L., Fedorka-Cray, P.J. and Baker, P.E. (1989) Bovine recombinant interleukin 2 augments immunity and resistance to bovine herpesvirus infection. *Veterinary Immunology and Immunopathology* 23, 61–74.

Reddy, D.N., Reddy, P.G., Minocha, H.C., Fenwick, B.W., Baker, P.E., Davis, W.C. and Blecha, F. (1990) Adjuvanticity of recombinant bovine Interleukin-1β: influence on immunity infection, and latency in a bovine herpesvirus-1 infection. *Lymphokine Research* 9, 295–307.

Reddy, P.G., Reddy, D.N., Pruiett, S.E., Daley, M.J., Shirley, J.E., Chengappa, M.M. and Blecha, F. (1992) Interleukin 2 treatment of *Staphylococcus aureus* mastitis. *Cytokine* 4, 227–231.

Reddy, D.N., Reddy, P.G., Xue, W., Minocha, H.C., Daley, M.J. and Blecha, F. (1993) Immunopotentiation of bovine respiratory disease virus accines by interleukin-1β and interleukin-2. *Veterinary Immunology and Immunopathology* 37, 25–38.

Reeves, R., Spies, A.G., Nissen, M.S., Buck, C.D., Weinberg, A.D., Barr, P.J., Magnuson, N.S. and Magnuson, J.A. (1986) Molecular cloning of a functional bovine interleukin 2 cDNA. *Proceedings of the National Academy of Science, USA* 83, 3228–3232.

Ritacco, V., Lopez, B., Barrera, L., Nader, A., Fliess, E. and de Kantor, I.N. (1990) Further evaluation of an indirect enzyme-linked immunosorbent assay for the diagnosis of bovine tuberculosis. *Journal of Veterinary Medicine, Series B* 37, 19–27.

Ritacco, V., Lopez, B., De Kantor, I.N., Barrera, L., Errico, F. and Nader, A. (1991) Reciprocal cellular humoral immune responses in bovine tuberculosis. *Research in Veterinary Science* 50, 365–367.

Roberts, K. and Shevach, E.M. (1991) Immunoregulatory role of gd T cells. *Annals of the New York Academy of Sciences* 636, 1–8.

Roberts, W.K., Hovanessian, A., Brown, R.E., Clemens, M.J. and Kerr, B.M. (1976) Interferon-mediated protein kinase and low-molecular weight inhibitor of protein synthesis. *Nature (London)* 264, 477–480.

Roberts, R.M., Cross, J.C. and Leaman, D.W. (1992) Interferons as hormones of pregnancy. *Endocrine Reviews* 13, 432–452.

Robinson, H.L., Hunt, L.A. and Webster, R.G. (1993) Protection against a lethal influenza virus challenge by immunization with a haemagglutinin-expressing plasmid DNA. *Vaccine* 11, 957–960.

Rodriguez, S.D., Palmer, G.H., McElwain, T.F., McGuire, T.C., Ruef, B.J., Chitko-McKown, C.G. and Brown, W.C. (1996) CD4$^+$ T helper lymphocyte responses against *Babesia bigemina* rhoptry-associated protein-1 (RAP-1). *Infection and Immunity* 64, 2079–2087.

Roizman, B., Desrosiers, R.C., Fleckenstein, B., Lopez, C., Minson, A.C. and Studdert, M.J. (1992) The family of herpesviridae: an update. *Archives of Virology* 123, 425–449.

Romagnani, S. (1994) Lymphokine production by human T cells in disease states. *Annual Review of Immunology* 12, 227–257.

Romano, M.F., Lamberti, A., Petrella, A., Bisogni, R., Tassone, P., Formisano, S., Venuta, S. and Turco, M.C. (1996) IL-10 inhibits nuclear factor-kB/Rel nuclear activity in CD3-stimulated human peripheral T lymphocytes. *Journal of Immunology* 156, 2119–2123.

Rosenberg, S.A. (1988) Immunotherapy of cancer using interleukin 2: current status and future prospects. *Immunology Today* 9, 58–62.

Rosenberg, S.A. (1990) Adoptive immunotherapy of cancer. *Scientific American* 262, 34–41.

Rosenberg, S.A., Mule, J.J., Spiess, P.J., Reichert, C.M. and Schwarz, S.L. (1985) Regression of established pulmonary and subcutaneous tumour mediated by systemic administration of recombinant interleukin-2. *Journal of Experimental Medicine* 161, 1169–1188.

Rosenberg, S.A., Lotze, M.T., Muul, L.M., Leitman, S., Chang, A.E., Ettinghausen, S.E., Matory, Y.L., Skibber, J.M., Shiloni, E., Vetto, J.T., Seipp, C.A., Simpson, C. and Reichert, C.M. (1987) A progress report on the treatment of 157 patients with advanced cancer using lymphokine-activated killer cells and interleukin-2 or high dose interleukin-2 alone. *New England Journal of Medicine* 316, 889–897.

Roth, J.A. and Dagmar, E.F. (1989) Recombinant bovine interferon-γ as an immunomodulator in dexamethasone-treated and nontreated cattle. *Journal of Interferon Research* 9, 143–151.

Rothel, J.S., Jones, S.L., Corner, L.A., Cox, J.C. and Wood, P.R. (1990) A sandwich enzyme immunoassay for bovine interferon-g and its use for the detection of tuberculosis in cattle. *Australian Veterinary Journal* 67, 134–137.

Rothel, J.S., Jones, S.L., Corner, L.A., Cox, J.C. and Wood, P.R. (1992) The γ-interferon assay for diagnosis of bovine tuberculosis in cattle: conditions affecting the production of γ-interferon in whole blood culture. *Australian Veterinary Journal* 69, 1–4.

Rouse, B.T. (1981) Role of neutrophils in antiviral immunity. *Advances in Experimental Medicine and Biology* 137, 263–278.

Ruef, B.J., Tuo, W., Rodriguez, S.D., Roussel, A.J., Chitko-McKown, C.G., Palmer, G.H., McElwain, T.F., Canals, A., Zarlenga, D.S., Gasbarre, L.C. and Brown, W.C. (1997) Immunization with *Babesia bigemina* rhoptry-associated protein 1 induces a type 1 cytokine response. *Journal of Interferon and Cytokine Research* 17, 45–54.

Russell, W.O., Wynne, E.S., Loquvam, G.S. and Mehl, D.A. (1956) Studies on bovine ocular squamous cell carcinoma ('Cancer eye'). *Cancer* 9, 1–53.
Rutten, V.P.M.G. (1992) Immunotherapy of Bovine Ocular Squamous Cell Carcinoma. PhD Thesis, Utrecht University, The Netherlands (ISBN 90-393-0171-9), 184pp.
Rutten, V.P.M.G., Klein, W.R., De Jong, W.A.C., Misdorp, W., Steerenberg, P.A., De Jong, W.H. and Ruitenberg, E.J. (1989) Local IL-2 therapy in bovine ocular squamous cell carcinoma. *Cancer Immunology and Immunotherapy* 30, 165–169.
Rutten, V.P.M.G., Klein, W.R., De Jong, W.A.C., Steerenberg, P.A., De Jong, W.H., Den Otter, W. and Ruitenberg, E.J. (1991a) Immunotherapy of bovine ocular squamous cell carcinoma by repeated intralesional injections of live Bacillus Calmette–Guérin (BCG) or BCG cell walls. *Cancer Immunology and Immunotherapy* 34, 186–190.
Rutten, V.P.M.G., De Jong, W.A.C., Klein, W.R., Den Otter, W., Steerenberg, P.A. and Ruitenberg, E.J. (1991b) Immunotherapy of bovine ocular squamous cell carcinoma: Isolation, culture and characterisation of lymphocytes present in the tumour. *Anticancer Research* 11, 1259–1264.
Ryan, T.J. (1994) Report to the N.Z. Animal Health Board. MAF Quality Management, Ruakura, Hamilton, New Zealand.
Salmon, M., Kitas, G.D. and Bacon, P.A. (1989) Production of lymphokine mRNA by CD45R$^+$ and CD45R$^-$ helper T cells from human peripheral blood and by human CD4$^+$ T cell clones. *Journal of Immunology* 143, 907–912.
Savan, M., Angulo, A.B. and Derbyshire, J.B. (1979) Interferon, antibody responses and protection induced by an intranasal infectious bovine rhinotracheitis vaccine. *Canadian Veterinary Journal* 20, 207–210.
Seder, R.A. and Paul, W.E. (1994) Acquisition of lymphokine-producing phenotype by CD4$^+$ cells. *Annual Review of Immunology* 12, 635–673.
Shuster, D.E. and Kehrli, M.E., Jr (1995) Administration of recombinant human interleukin 1 receptor antagonist during endotoxin-induced mastitis in cows. *American Journal of Veterinary Research* 56, 313–320.
Shuster, D.E., Kehrli, M.E., Jr and Stevens, M.G. (1993) Cytokine production during endotoxin-induced mastitis in lactating dairy cows. *American Journal of Veterinary Research* 54, 80–85.
Sica, G., Iacopino, F., Lama, G., Amadori, D., Baroni, M., Lo Sardo, F., Malarcarne, P., Marchetti, P., Pellegrini, A., Zaniboni, A. and della Cuna, G.R. (1993) Steroid receptor enhancement by natural interferon-β in advanced breast cancer. *European Journal of Cancer* 29A, 329–333.
Sordillo, L.M. and Babiuk, L.A. (1991) Controlling acute *Escherichia coli* mastitis during the periparturient period with recombinant bovine interferon γ. *Veterinary Microbiology* 28, 189–198.
Sordillo, L.M. and Peel, J.E. (1992) Effect of interferon-γ on the production of tumor necrosis factor during acute *Escherichia coli* mastitis. *Journal of Dairy Science* 75, 2119–2125.
Sorg, R.V., Enczmann, J., Sorg, U.R., Schneider E.M. and Ernst, P. (1993) Identification of an alternatively spliced transcript of human interleukin-4 lacking the sequence encoded by exon 2. *Experimental Hematology* 21, 560–563.

Spencer, T.E., Becker, W.C., George, P., Mirando, M.A., Ogle, T.F. and Bazer, F.W. (1995) Ovine interferon-τ inhibits estrogen receptor up-regulation and estrogen-induced luteolysis in cyclic ewes. *Endocrinology* 136, 4932–4944.

Spits, H. and de Waal Malefyt, R. (1992) Functional characterization of human IL-10. *International Archives of Allergy and Immunology* 99, 8–15.

Spradbrow, P.B., Wilson, B.E., Hoffmann, D., Kelly, W.R. and Francis, J. (1977) Immunotherapy of bovine ocular squamous cell carcinomas. *Veterinary Record* 100, 376–378.

Stewart, H.J., McCann, S.H.E., Northrop, A.J., Lamming, G.E. and Flint, A.P.F. (1989) Sheep antiluteolytic interferon: cDNA sequence and analysis of mRNA levels. *Journal of Molecular Endocrinology* 2, 65–70.

Straub, O.C. and Ahl, R. (1976) Lokale Interferonbildung beim Rind nach intranasaler Infektion mit avirulentem IBR-IPV-Virus und deren Wirkung auf eine anschließende Infektion mit Maul- und Klauenseuche-Virus. *Zentralblatt für Veterinärmedizin, Reihe B* 23, 470–482.

Street, N.E., Schumacher, J.H., Fong, A.T., Bass, H., Fiorentino, D.F., Leverah, J.A. and Mosmann, T.R. (1990) Heterogeneity of mouse helper T cells. Evidence from bulk cultures and limiting dilution cloning for precursors of Th1 and Th2 cells. *Journal of Immunology* 144, 1629–1639.

Taga, K., Mostowski, H. and Tosato, G. (1993) Human interleukin-10 can directly inhibit T-cell growth. *Blood* 81, 2964–2971.

Talmadge, J.E., Philips, H., Schindler, J., Tribble, H. and Pennington, R. (1987) Systemic preclinical study on the therapeutic properties of recombinant human interleukin 2 for the treatment of metastatic disease. *Cancer Research* 47, 5725–5732.

Tang, D., DeVit, M. and Johnston, S.A. (1992) Genetic immunization is a simple method for eliciting an immune response. *Nature* 356, 152–154.

Templeton, J.W., Estes, D.M., Price, R.E., Smith III, R. and Adams L.G. (1990) Immunogenetics of natural resistance to bovine brucellosis. *Proceedings of the 4th World Congress on Genetics Applied to Livestock Production*, Edinburgh, pp. 396–399.

Thoen, C.O., Jarnagin, J.L., Muscoplat, C.C., Cram, L.S., Johnson, D.W. and Harrington, R. (1980) Potential use of lymphocyte blastogenic responses in diagnosis of bovine tuberculosis. *Comparative Immunology and Microbiology of Infectious Diseases* 3, 355–361.

Thorns, C.J. and Morris, J.A. (1983) The immune spectrum of *Mycobacterium bovis* infections in some mammalian species: a review. *Veterinary Bulletin* 53, 543–550.

Todd, J.D., Volenec, F.J. and Paton, I.M. (1972) Interferon in nasal secretions and sera of calves after intranasal administration of avirulent infectious bovine rhinotracheitis virus: association of interferon in nasal secretions with early resistance to challenge with virulent virus. *Infection and Immununity* 5, 699–706.

Trinchieri, G. (1995) Interleukin 12: A proinflammatory cytokine with immunoregulatory functions that bridge innate resistance and antigen-specific adaptive immunity. *Annual Review of Immunology* 13, 251–276.

Ulmer, J.B., Donnelly, J.J., Parker, S.E., Rhodes, G.H., Felgener, P.L., Dwarki, V.J., Gromkowaski, S.H., Deck, R.R., DeWitt, C.M., Friedman, A., Hawe, L.A.,

Leander, K.R., Martinez, D., Perry, C., Shiver, J.W., Montgomery, D.L. and Liu, M.A. (1993) Heterologous protection against influenza by injection of DNA encoding a viral protein. *Science* 259, 1745–1749.

Umetsu, D.T., Jabara, H.H., DeKruyff, R.H., Abbas, A.K., Abrams, J.S. and Geha, R.S. (1988) Functional heterogeneity among human inducer T cell clones. *Journal of Immunology* 140, 4211–4216.

Vaage, J. (1987) Local and systemic effects during interleukin-2 therapy of mouse mammary tumours. *Cancer Research* 47, 4295–4298.

Van Kampen, K.R., Crisp, W.E., De Martini, J.C. and Ellsworth, H.S. (1973) The immunologic therapy of squamous cell carcinoma – a preliminary report. *American Journal of Obstetrics and Gynaecology* 116, 569–574.

Van Kessel, A.G., Godson, D.L., Baca-Estrada, M.E., Morsy, M.A., Hughes, H.P.A., Harland, R.J., Shuster, D.E., Daley, M.J. and Babiuk, L.A. (1996) Effect of recombinant bovine interleukin-1β in normal calves and in calves infected with bovine herpesvirus type 1. *Journal of Interferon and Cytokine Research* 16, 881–890.

Vanden Broecke, C., Pirak, M., Thiry, E., Schwers, A., Koch, K., Thein, P., Plasman, P.O., Wérenne, J. and Pastoret, P.-P. (1985) Use of recombinant human interferon (Hu-IFN-γ) in bovine herpesvirus 1 infection in cattle. In: Pastoret, P.-P., Thiry, E. and Saliki, J. (eds), *Immunity to Herpesvirus Infections of Domestic Animals.* Office for Official Publications of the European Communities, Brussels–Luxembourg, EUR 9737 EN, pp. 303–316.

Vidal, D., Domingo, M., Aranaz, A., Liebana, E., Prats, N., Marco, A., Casal, J., Dominguez, L. and Mateos, A.M. (1995) Eradication of tuberculosis from goat herds by means of interferon IFN-γ assay and the single intradermal comparative skin test. In: Griffin, F. and de Lisle, G. (eds), *Tuberculosis in Wildlife and Domestic Animals*. Otago Conference Series No.3. University of Otago Press, Dunedin, pp.328–330.

Weir, R.E., Morgan, A.R., Britton, W.J., Butlin, C.R. and Dockrell, H.M. (1994) Development of a whole blood assay to measure T cell responses to leprosy: a new tool for immuno-epidemiological field studies of leprosy immunity. *Journal of Immunological Methods* 176, 93–101.

Wérenne, J., Vanden Broecke C., Schwers, A., Goossens, A., Bugyaki, L., Maenhoudt, M. and Pastoret, P.-P. (1985) Antiviral effect of bacterially produced human interferon (Hu-IFN-γ) against experimental vaccinia infection in calves. *Journal of Interferon Research* 5, 129–136.

Weynants, V., Godfroid, J., Limbourg, B., Saegerman, C. and Letesson, J-J. (1995) Specific bovine brucellosis diagnosis based on *in vitro* antigen-specific γ interferon production. *Journal of Clinical Microbiology* 33, 706–712.

Whipple, D.L., Bolin, C.A., Davis, A.J., Jarnagin, J.L., Johnson, D.C., Nabors, R.S., Payeur, J.B., Saari, D.A., Wilson, A.J. and Wolf, M.M. (1995) Comparison of sensitivity of the caudal fold skin test and a commercial γ-interferon assay for diagnosis of bovine tuberculosis. *American Journal of Veterinary Research* 56, 415–419.

Wong, G.H.W. and Goeddel, D.V. (1986) Tumor necrosis factors α and β inhibit virus replication and synergize with interferons. *Nature* 323, 819–822.

Wood, P.R. and Rothel, J.S. (1994) *In vitro* immunodiagnostic assays for bovine tuberculosis. *Veterinary Microbiology* 40, 125–135.

Wood, P.R. and Seow, H.-F. (1996) T cell cytokines and disease prevention. *Veterinary Immunology and Immunopathology* 54, 33–44.
Wood, P.R., Corner, L.A. and Plackett, P. (1990a) Development of a simple, rapid *in vitro* cellular assay for bovine tuberculosis based on the production of γ interferon. *Research in Veterinary Science* 1990, 49, 46–49.
Wood, P.R., Rothel, J.S., McWaters, P.G.D. and Jones, S.L. (1990b) Production and characterization of monoclonal antibodies specific for bovine γ-interferon. *Veterinary Immunology and Immunopathology* 25, 37–46.
Wood, P.R., Corner, L.A., Rothel, J.S., Baldock, C., Jones, S.L., Cousins, D.B., McCormick, B.S., Francis, B.R., Creeper, J. and Tweedle N.E. (1991) Field comparison of the interferon-γ assay and the intradermal tuberculin test for the diagnosis of bovine tuberculosis. *Australian Veterinary Journal* 68, 286–290.
Wood, P.R., Corner, L.A., Rothel, J.S., Ripper, J.L., Fifis, T., McComick, B.S., Francis, B., Melville, L., Small, K., De Witte, K., Tolson, J., Ryan, T.J., de Lisle, G.W., Cox, J.C. and Jones, S.L. (1992) A field evaluation of serological and cellular diagnostic tests for bovine tuberculosis. *Veterinary Microbiology* 31, 71–79.
Yanagida, T., Kato, T., Igarashi, O., Inoue, T. and Nariuchi, H. (1994) Second signal activity of IL-12 on the proliferation and IL-2R expression of T helper cell-1 clone. *Journal of Immunology* 152, 4919–4928.
Yates, W.D.G. (1982) A review of infectious bovine rhinotracheitis, shipping fever pneumonia and viral–bacterial synergism in respiratory disease of cattle. *Canadian Journal of Comparative Medicine* 46, 225–263.
Yssel, H., de Waal Malefyt, R., Roncarlo, M.-G., Abrams, J.S., Lahesmaa, R., Spits, H. and de Vries, J.E. (1992) IL-10 is produced by subsets of human $CD4^+$ T cell clones and peripheral blood cells. *Journal of Immunology* 149, 2378–2384.
Zarlenga, D.S., Canals, A. and Gasbarre, L. (1995) Method for constructing internal standards for use in competitive PCR. *Biotechniques* 19, 324–326.
Zurawski, G. and de Vries, J.E. (1994) Interleukin 13, an interleukin 4-like cytokine that acts on monocytes and B cells, but not on T cells. *Immunology Today* 15, 19–26.
Zurawski, S.M., Vega, F., Jr, Huyghe, B. and Zurawski, G. (1993) Receptors for interleukin-13 and interleukin-4 are complex and share a novel component that functions in signal transduction. *European Molecular Biology Organization Journal* 12, 2663–2670.

II

Ovine Cytokines

Edited by C.J. McInnes

A Current Perspective of Ovine Cytokine Research

7

C.J. McInnes, E.A. Innes, J. Brown and G. Entrican

Division of Immunobiology, Moredun Research Institute, 408 Gilmerton Road, Edinburgh EH17 7JH, UK

INTRODUCTION

The first human and murine cytokines were cloned and expressed in the early 1980s. Although veterinary researchers were able to use some of the recombinant human and murine reagents generated they soon realized that many were not biologically cross-reactive in their systems and therefore embarked on cloning programmes of their own. The need to develop homologous reagents has meant that much of the work in the last few years has concentrated on the cloning of ovine cytokine genes and cDNAs and on the expression of the recombinant proteins, with the result that the application of these reagents is still in its infancy. However in the next few years we should expect to see an exponential increase in research applying these reagents to fundamental immunological research as well as disease studies and vaccinology.

This chapter will attempt to present the current state of ovine cytokine research in general while also highlighting the potential use of recombinant cytokines as adjuvants for vaccination.

One area of cytokine research where veterinary researchers have led the way was in the discovery of IFN-τ, an antiviral protein produced by the conceptus and involved in the maternal recognition of pregnancy. Together with the role of other cytokines during early pregnancy and in reproduction generally, this has been reviewed elsewhere (see Bazer *et al.*, 1994; Hansen, 1995; Thatcher *et al.*, 1995).

THE CLONING OF OVINE CYTOKINE cDNAs AND GENES

Although a few of the ovine cytokines have been cloned from cDNA libraries, the advent of the polymerase chain reaction (PCR) allowed researchers to clone ovine cytokines rapidly using primer pairs based on areas of conserved sequence between cDNAs from other species. In general, workers found that the ovine cytokines showed greatest identity to cytokines from other ruminant species (usually cattle), but where no other ruminant cytokine had been cloned then the ovine cytokines showed greatest identity to the human orthologues. When a new human cytokine sequence is reported in the literature the corresponding ovine cDNA is usually cloned soon afterwards using the PCR and primers based on the human sequence. A list of the ovine cytokine cDNAs which have been cloned to date is given in Table 7.1. The predicted amino-acid homologies of the ovine cytokines with the corresponding bovine, human and murine proteins are shown in Table 7.2.

In addition to the cDNAs listed in Table 7.1, the genes for ovine IFN-γ (C. McInnes, unpublished results), IL-5 (C. Bryson, EMBL accession numbers U17052-3), erythropoietin (EPO; Fu *et al.*, 1993), IL-3 and GM-CSF (McInnes *et al.*, 1993) have been cloned. In each case the overall structure of the gene, in terms of the number and the relative positions of the introns, shows little difference to the corresponding human gene. The promoters too show a high degree of conservation with the corresponding regions of the human genes, but since the exact sequences required for controlling the expression of these cytokines have still to be fully elucidated one cannot rule out the possibility that there are important differences between these regions. Although not cloned, the genes for transforming growth factor-β1 (TGF-β; Woodall *et al.*, 1994), the IL-2 receptor (γ-subunit; Ansari *et al.*, 1995), the IFN-γ receptor (Broad *et al.*, 1995a) and IL-7 (Broad *et al.*, 1995b) have been assigned to sheep chromosomes by using either fluorescence *in situ* hybridization or sheep × hamster cell hybrids.

EXPRESSION OF RECOMBINANT OVINE CYTOKINES

Recombinant ovine cytokine proteins have been produced in mammalian, insect, yeast and bacterial expression systems. Details of individual cytokines can be obtained in the references given in Tables 7.1 and 7.3. The yield and purity from each system vary with the individual cytokine being expressed as does the biological activity of the expressed protein. The choice of expression system is usually dictated by the requirement for yield or biological activity. In general, recombinant protein derived from the mammalian expression systems is highly active, but usually the yield is low. On the other hand yield from the bacterial and yeast systems is generally

Table 7.1. Cloned ovine cytokine cDNAs.

Cytokine	Accession no.	Reference
IL-1α	X56754	Andrews *et al.*, 1991; Fiskerstrand *et al.*, 1992
IL-1β	X54796	Fiskerstrand and Sargan, 1990; Seow *et al.*, 1991; Andrews *et al.*, 1991
IL-2	X53934	Goodall *et al.*, 1990; Seow *et al.*, 1990; Bujdoso *et al.*, 1995
IL-3	Z18291	McInnes *et al.*, 1994
IL-4	Z11897	Engwerda and Sandeman 1992; Seow *et al.*, 1993
IL-5	U35038	Seow *et al.*, 1996a
IL-6	X62501	Andrews *et al.*, 1993; Ebrahimi *et al.*, 1995
IL-7	U10089	Barcham *et al.*, 1995
IL-8	X78306	Cordier *et al.*, 1993; Seow *et al.*, 1994b; Legastelois *et al.*, 1994
IL-10	U11421	Martin *et al.*, 1993; Dutia *et al.*, 1994; Martin *et al.*, 1995
IL-12	–	H.-F. Seow, personal communication.
IFN-ω	X59067-8	Whaley *et al.*, 1991
IFN-γ	X52640	McInnes *et al.*, 1990; Radford *et al.*, 1991
GM-CSF	X53561	O'Brien *et al.*, 1991; McInnes and Haig, 1991
G-CSF	L07939	O'Brien *et al.*, 1994
EPO	Z24681	Fu *et al.*, 1993
SCF	Z50743	C.J. McInnes, 1996, unpublished results.
TNF-α	X55152	Young *et al.*, 1990; Green and Sargan, 1991; Nash *et al.*, 1991
TNF-β	–	H.-F. Seow, personal communication.
MCP-1	–	T. Yoshimura, unpublished results.
MIP-1α	–	T. Yoshimura, unpublished results.
RANTES	–	T. Yoshimura, unpublished results.
TGF-β1	X76916	Woodall *et al.*, 1994
VEGF	X89506	D.A. Redmer, 1995, unpublished results.
KGF	Z46236	C.J. McInnes, 1996, unpublished results.

IL, interleukin; IFN, interferon; GM-CSF, granulocyte-macrophage colony stimulating factor; G-CSF, granulocyte-colony stimulating factor; EPO, erythropoietin; SCF, stem cell factor; TNF, tumour necrosis factor; MCP, monocyte chemotactic factor; MIP, migration inflammatory protein; RANTES, regulated on activation, normal T-cell expressed and secreted; TGF, transforming growth factor; VEGF, vascular endothelial growth factor; KGF, keratinocyte growth factor.

high, but problems have been encountered with solubility and subsequent biological activity.

The availability of large amounts of recombinant protein has allowed the production of cytokine-specific mAbs and the subsequent development of both cytokine-specific ELISAs and bioassays. Table 7.3 presents a list of those cytokines for which mAbs have been produced and bioassays developed. Methods for detecting and assaying of ovine cytokines have recently been reviewed (Nash *et al.*, 1996).

Table 7.2. Relative homologies between the predicted amino-acid sequences of ovine cytokines with the corresponding bovine, human and murine proteins.

Cytokine	Bovine	Human	Murine
IL-1α	97	73	62
IL-1β	95	61	58
IL-2	97	65	50
IL-3	85	36	30
IL-4	92	57	37
IL-5	–	64	56
IL-6	88	53	42
IL-7	95	76	74
IL-8	–	78	–
IL-10	93	78	73
IFN-γ	96	63	47
GM-CSF	80	80	57
G-CSF	96	82	68
EPO	96	82	78
SCF	–	87	81
TNF-α	90	79	72

Amino-acid identities calculated from the predicted sequences of the ovine, bovine, human and murine proteins. Data are not available for all the cytokines listed in Table 7.1. For abbreviations see Table 7.1.

IN VITRO BIOLOGY

The evaluation of the *in vitro* biological activities of the majority of ovine cytokines is in its early stages since many have only recently been cloned and expressed. However, in general, those that have been studied show little difference from their human counterparts. The colony stimulating factors (GM-CSF, IL-3 and stem cell factor: SCF), for example, exhibit a range of activities in soft agar bone-marrow cell clonogenic assays that is similar to that of the human proteins. Recombinant ovine GM-CSF stimulates the production of mixed neutrophil/macrophage colonies as well as eosinophil, macrophage and neutrophil colonies (McInnes and Haig, 1991). Stevenson and Jones (1994) showed that rOvGM-CSF, as well as rOvIL-3 and rHuIL-5, enhanced the survival of ovine eosinophils from bone marrow, as measured by sustained eosinophil peroxidase and arylsulphatase activities, over a 7 day period in liquid culture. In another study Haig *et al.* (1995b), demonstrated that rOvGM-CSF together with rOvTNF-α provided both survival and proliferative signals for afferent lymph dendritic cells in culture. These cells represent a subpopulation of migrating dermal dendritic cells involved in antigen presentation and T cell activation. While unable to support the growth and differentiation of these cells on its own, rOvGM-CSF synergized with rOvTNF-α to increase numbers of cells surviving after

Table 7.3. Assays for ovine cytokines

Cytokine	Antibodies	ELISA[b]	Bioassay[c]	References
IL-1α	+	+	Cartilage degradation	[b]Egan *et al.*, 1994a;
IL-1β	+	+	Thymocyte proliferation assays	[c]Fiskerstrand *et al.*, 1992;
			Murine NOB-1/CTLL assays	[c]Seow *et al.*, 1994a
IL-2	+		Con A lymphoblast proliferation	[c]Nash *et al.*, 1993; [c]Seow *et al.*, 1996b
IL-3	+		Bone-marrow cell soft agar	[b]Entrican *et al.*, 1996;
GM-CSF	+	+	clonogenic assay/liquid	[c]Haig *et al.*, 1990
SCF			bone-marrow assay	
Type 1 IFN			Antiviral activity on homologous	[a,b]Rothel *et al.*, 1990;
Type 2 IFN	+	+	adherent cell lines	[b]Zhu and Ott 1994; [c]Entrican *et al.*, 1989; [c]Entrican *et al.*, 1992
IL-4	+		Con A lymphoblast proliferation	[c]Seow *et al.*, 1996c
TNF-α	+	+	PK(15)-1512 fibroblast, WEHI-164, L929 cells cytotoxicity assays Thymocyte proliferation assays	[b]Egan *et al.*, 1994b; [c]Nash *et al.*, 1991; [c]Green *et al.*, 1993; [c]Pauli *et al.*, 1994; [c]Seow *et al.*, 1995
IL-6			Immunoglobulin production	[c]Ebrahimi *et al.*, 1995
IL-7			Thymocyte co-cultivation assays	[c]Barcham *et al.*, 1995
IL-8	+	+	Neutrophil chemotaxis	[b]Haig et al., 1996a; [c]Seow *et al.*, 1994b
IL-10	+	+	Inhibition of cytokine production	[b]Martin, PhD Thesis 1995; [c]Martin *et al.*, 1995

[a]Reference gives details of an ELISA based on bovine IFN-γ, but which is fully cross-reactive in sheep.
[b]Reference gives details of an ELISA.
[c]Reference gives details of a bioassay.
For abbreviations see Table 7.1.

14 days in culture compared with growth in TNF-α alone. Interestingly, rOvIFN-γ was found to inhibit both rOvTNF-α and rOvTNF-α plus rOvGM-CSF-mediated survival and growth of afferent lymph dendritic cells.

In the soft agar bone-marrow cell clonogenic assays, rOvIL-3 stimulated the production of eosinophil, macrophage and erythroid colonies as well as mixed colonies, which are the progeny of multi-potential colony forming cells (multi-CFC; McInnes *et al.*, 1993). It also stimulated the production of mast cell colonies, an activity which it shares with rodent IL-3, but not with human IL-3 which is not a mast cell growth factor. In liquid bone-marrow cultures, rOvSCF poorly promotes the development of mast cells from bone marrow precursors although this activity is greatly enhanced when the cells are cultured in both rOvIL-3 and rOvSCF. In contrast, rOvSCF is a potent stimulus for mast cell proliferation using mast cell lines developed from bone marrow precursors, an activity which is also enhanced by rOvIL-3 (C. McInnes, 1996, unpublished results). rOvIL-3 was also shown to synergize with medium conditioned from long-term bone-marrow stromal cell cultures to promote the development of multi-CFCs from their precursors present in bone marrow (Haig *et al.*, 1994). This provided an assay for the precursors of haemopoietic clonogenic CFC in sheep.

IFN-γ has been shown to have properties similar to its human counterpart. It is produced by activated T cells but not B cells, possesses antiviral activity, is both heat and acid labile (Entrican *et al.*, 1989, 1992), and upregulates MHC class I and class II expression on macrophages (Nash *et al.*, 1992a). In addition to its antiviral activity it has been shown to inhibit the replication *in vitro* of other obligate intracellular pathogens such as *Toxoplasma gondii* (Oura *et al.*, 1993), *Neospora caninum* (Innes *et al.*, 1995a) and *Chlamydia psittaci* (Graham *et al.*, 1995).

IN VIVO BIOLOGY

Research into the *in vivo* biology of ovine cytokines to date falls into two categories. First, the detection of cytokines in response to non-specific stimuli, antigen or pathogen challenge and second, the injection of recombinant cytokines or antibodies into the sheep and monitoring of the consequences. A major advantage of studying cytokine biology (and immune and inflammatory responses in general), in the sheep is that the *in vivo* dynamics of local cytokine production can be measured by cannulating the lymphatic vessels of a local lymph node at the site of immunization or infection. This gives a measure of the activation status of cells within the lesion as well as from the different lymph compartments (afferent and efferent). Similarities and differences will be highlighted between cytokine production and utilization in these compartments in inflammatory responses and both primary and anamnestic immune responses.

Cytokine Responses to Challenge with Pathogens or Other Stimuli

A series of studies have been performed on the host inflammatory and immune responses to reinfection with the skin tropic DNA parapoxvirus Orf-virus (OV), which causes a contagious pustular dermatitis in sheep and goats and can also infect man. Despite an apparently normal immune response a long-lasting immunity is not generated and animals can be repeatedly reinfected with the virus (Haig *et al.*, 1996c, 1997). Haig *et al.* (1992) demonstrated in sheep reinfected with OV that IFN-γ was produced by lymphocytes in the afferent lymph in the early stages p.i. whereas it was produced by lymphocytes in the efferent lymph in the later stages, between

4 and 11 days p.i. They postulated that this was consistent with a rapid activation of memory T cells in the afferent lymph to become cytokine-secreting effector cells whereas in the efferent lymph recruitment from the blood of antigen-specific T lymphocytes and activation within the node to produce IFN-γ was delayed by comparison. This was consistent with a previous report, based on the differential expression of CD45R and a range of adhesion molecules, that afferent lymph consisted of > 90% 'memory' T cells whilst efferent lymph contained > 90% 'naive' T cells (MacKay *et al.*, 1990). More recently Haig *et al.* (1996a) found almost equal numbers of 'naive' ($CD45R^+$) and 'memory' ($CD45R^-$) T cells in the efferent lymph and showed that both phenotypes were activated as a result of OV reinfection, but that the $CD45R^-$, $CD4^+$ T cells were the principal source of IFN-γ, GM-CSF and IL-2 after reinfection. The conclusion was that CD45R expression correlated with effector function in T cells (the ability to produce lymphokines), but it is not clear how this relates to T cell memory.

Haig *et al.* (1992) had also demonstrated in sheep reinfected with OV that total colony stimulating activity (as measured in a bone-marrow soft agar clonogenic assay) in both lymph compartments was biphasic, with the first peak of activity coming in the 30 h after reinfection followed by a second peak several days later and coinciding with a major lymphoblast response. The reason for this was unknown, but they postulated that the second peak of activity could be due to viral replication. To test this hypothesis they monitored the phenotype of cells and the production of IL-1β, IL-8, GM-CSF and IFN-γ in the afferent lymph following reinfection with OV and compared this to i.d. injection of UV-inactivated virus (Haig *et al.*, 1996b). They found that in animals reinfected with live virus there was a biphasic lymphoblast response as well as biphasic production of IL-1β, IL-8 and GM-CSF, whereas the inactivated virus elicited a single peak of cytokine production and lymphoblast cells following injection of the dead virus. This was consistent with an early response in the afferent lymph to OV as a recall antigen and suggested that viral replication was indeed required for the second peak of cytokine and lymphoblast production.

Chlamydia psittaci and *T. gondii* are the two most common causes of diagnosed ovine abortion reported annually in the UK. They are obligate intracellular pathogens which can infect a variety of host species and CMI plays a key role in protection (Buxton, 1990; Buxton *et al.*, 1990). Lymphatic cannulation studies in sheep have shown that one feature of the response to experimental subcutaneous infection with live organisms is the appearance of IFN-γ in the efferent lymph draining the site of infection. In both cases, IFN-γ is produced very rapidly in immune sheep – within the first 24 h after reinfection (Graham *et al.*, 1995; Innes *et al.*, 1995b). The kinetics of this response are slower in naive sheep given an experimental primary infection with *T. gondii* (Innes *et al.*, 1995b). Activated lymphoblasts responding to the infection produced IFN-γ following restimulation *in vitro* with *T. gondii* antigen (Innes *et al.*, 1995b). Phenotypic analysis of these cells showed that they were mainly of the $CD4^+$ T cell subset (Innes *et al.*, 1995c). The situation with *C. psittaci* in naive sheep has not been defined. These *in vivo* experiments were performed using the S26/3 abortion strain of *C. psittaci* and the S48 vaccine strain of *T. gondii*. The observation that IFN-γ was produced indicated the importance of this cytokine to the host response to these intracellular pathogens.

Studies performed *in vitro* using rOvIFN-γ have demonstrated its ability to restrict the growth of both of these organisms in ovine fibroblast ST-6 cells (Oura *et al.*, 1993; Graham *et al.*, 1995). It is of particular interest to identify the mechanisms by which IFN-γ exerts these inhibitory effects. IFN-γ is known to induce the expression of nitric oxide synthase and indoleamine 2,3-dioxygenase (IDO) in murine cells (Moncada and Higgs, 1990) and IDO in human cells (De Maeyer and De Maeyer-Guignard, 1988). Nitric oxide (NO) is thought to exhibit its antimicrobial effects via its ability to inhibit key enzymes involved in respiration and DNA synthesis whereas IDO degrades L-tryptophan (L-trp) and can inhibit the multiplication of certain intracellular pathogens unable to synthesize their own L-trp and which are reliant on the available resource within the host cell.

There are few published reports where the inhibition of growth of *C. psittaci* and *T. gondii* by IFN-γ has been directly compared in the same host cell type (none in sheep cells) and these have yielded conflicting results. Murray *et al.* (1989) found that human IFN-γ induced IDO expression in human monocyte-derived macrophages and human AM, and inhibited the growth of both *T. gondii* and the 6BC (avian) strain of *C. psittaci*. This could be reversed with L-trp for both pathogens in both cell types. However, the reversal was more marked in the monocyte-derived macrophages than in the AM, suggesting differences in L-trp catabolism between the cells or that other IFN-γ-induced antimicrobial pathways may be in operation. Thomas *et al.* (1993) studied the growth of the 6BC strain of *C. psittaci* and the PE strain of *T. gondii* in IDO-deficient mutant cells derived from the ME180 human cervical carcinoma cell line. The IDO-deficient cells could partially restrict the growth of both organisms following treatment with IFN-γ but not as efficiently as the wild-type parent cells. L-trp could reverse the inhibition of *T. gondii* in the mutant and wild-type cell lines but could only reverse the inhibition of *C. psittaci* in the wild-type cells.

In ovine ST-6 cells the IFN-γ-mediated inhibition of the S26/3 abortion strain of *C. psittaci* can be reversed by the addition of exogenous L-trp whereas the growth of the S48 strain of *T. gondii* cannot (Fig. 7.1). This suggests that the IFN-γ is exerting its effects against these two pathogens through distinct pathways within the same cell type. In addition the antimicrobial effects of IFN-γ were not shared by rbIFN-α which did not restrict the growth of either pathogen in ST-6 cells when matched for antiviral activity with IFN-γ (Fig. 7.2). Murine IFN-α and IFN-β have been reported

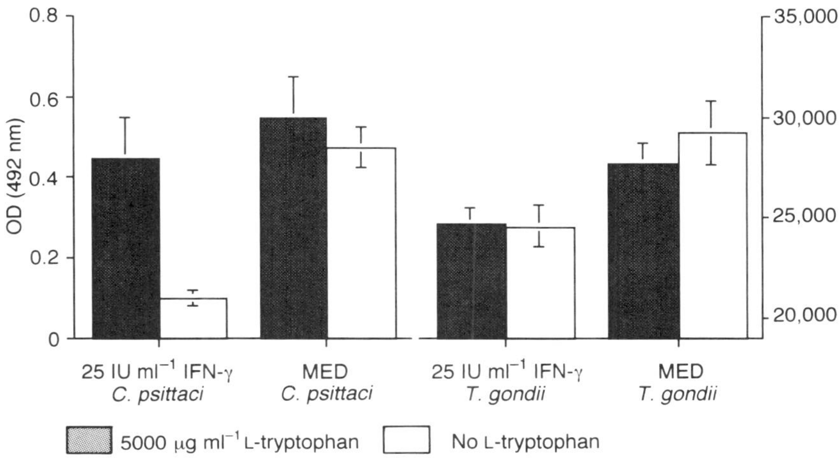

Fig. 7.1. ST-6 cells were pretreated with rOvIFN-γ (25 IU ml^{-1}) for 24 h and infected with *C. psittaci* at a multiplicity of infection of 0.5 or *T. gondii* at a parasite to cell ratio of 4 : 1. L-tryptophan was added at the time of infection. Supernatants were harvested from cells infected with *C. psittaci* 7 days p.i. and analysed for chlamydial LPS by ELISA. This protocol has been fully described elsewhere (Graham *et al.*, 1995). Multiplication of *T. gondii* in culture over a 24-h period was quantified by the differential uptake of tritiated uracil which was added for the final 18 h of the assay (Oura *et al.*, 1993). Control cultures received no rOvIFN-γ, only medium (MED).

to induce IDO in murine lung cells, whereas human IFN-α and IFN-β failed to induce IDO expression in human fibroblasts. This latter observation was not a result of the inability of these cells to express IDO since it could be induced by human IFN-γ (reviewed by De Maeyer and De Maeyer-Guignard, 1988). Expression of IDO does not always serve as a marker of antimicrobial activity. Human IFN-α and IFN-β can stimulate monocyte-derived macrophages to express IDO and degrade L-trp, but do not subsequently restrict the growth of either *C. psittaci* or *T. gondii* (Murray *et al.*, 1989).

A major factor in all of this is the requirement of the pathogen for L-trp. In a study of aa requirements of several strains of *C. psittaci* growing in murine McCoy cells, Allan and Pearce (1983) found that their ovine abortion strain did not have an absolute requirement for L-trp in the culture medium whereas it is clear that the S26/3 ovine abortion strain does (Fig. 7.1).

An alternative IFN-γ-induced antimicrobial pathway is NO, and recent experimental data have shown this pathway itself may influence the IDO pathway. Nitric oxide has been shown to inhibit IDO activity in human PBMC treated with human IFN-γ and in murine peritoneal macrophages

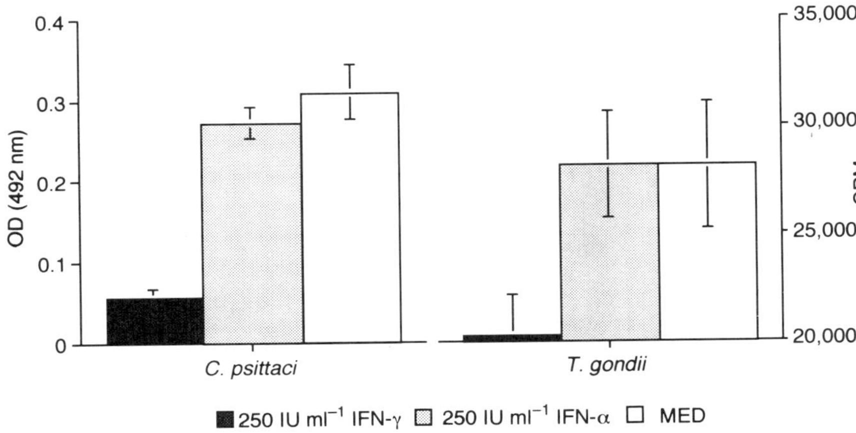

Fig. 7.2. ST-6 cells were pretreated with rOvIFN-γ or rBoIFN-α1 which were matched for their antiviral activity on these cells (250 IU ml^{-1}). Controls received medium (MED) alone. The cells were infected with *C. psittaci* at a multiplicity of infection of 0.5 or with *T. gondii* at a parasite to cell ratio of 4 : 1. Supernatants were harvested from cells infected with *C. psittaci* 7 days p.i. and analysed for chlamydial LPS by ELISA. This protocol has been fully described elsewhere (Graham *et al.*, 1995). Multiplication of *T. gondii* in culture over a 24-h period was quantified by the differential uptake of tritiated uracil which was added for the final 18 h of the assay (Oura *et al.*, 1993).

treated with murine IFN-γ (Thomas *et al.*, 1994). This could help to explain the conflicting results with L-trp experiments. At the moment it is not known if NO induced by IFN-γ plays a role in the restriction of intracellular pathogens in sheep. Preliminary data have shown that IFN-γ-activated ovine fibroblast ST-6 cells or ovine AM did not produce detectable quantities of nitrate in the supernatant. Also, addition of N^{G}MMA, a competitive inhibitor of L-arginine, did not reverse the inhibition of *T. gondii* (S48 strain) multiplication within ovine AM activated with rOvIFN-γ (Innes, Marks, Sternberg and Sutherland, unpublished observations). This suggests that the inhibitory mechanism in this situation was not due to L-arginine-dependent production of NO.

Both *T. gondii* and *C. psittaci* are able to replicate in a wide variety of cell types and IFN-γ is an important component of protective immunity against both pathogens. The data emerging from *in vitro* studies have highlighted several mechanisms which may be responsible for the observed inhibition of intracellular multiplication of both pathogens in IFN-γ-activated cells. The relative importance of any of these mechanisms is likely to depend on the animal species, cell type, strain of pathogen used and the interaction of various host immune factors.

In two separate studies the kinetics of cytokine production in response to infection with the ectoparasite *Lucilia cuprina* has been studied in ovine skin at the site of the lesion and in the efferent lymph draining the proximal node. Elhay *et al.* (1994) demonstrated that the expression of the mRNA for IL-1α, Il-1β, IL-6 and IL-8 increased in the skin following infection with first instar larvae and suggested that this expression may be responsible for the influx of granulocytes and other cells to the lesion. They also showed that IL-2 mRNA expression increased in the 48 h following infection in parallel with an increase in γ/δ-TCR$^+$ T cells, although they did not confirm that these cells were the source of the IL-2. Interestingly they provided evidence that IL-1β, TNF-α and IL-6 are all expressed constitutively in sheep skin. In the efferent lymph following both primary and secondary infection with first instar larvae, Bowles *et al.* (1994) demonstrated an increase in both IFN-γ and soluble IL-2 receptor production. This was concomitant with a relative increase in the percentage of CD8$^+$ T cells present in the efferent lymph. However, the cellular source of the IFN-γ and soluble IL-2 receptor was not determined and it is unclear whether these proteins were produced by cells in the efferent lymph or whether they originated from within the node.

Cytokine gene expression has also been studied at the inoculation site and in the RLNs of sheep infected with three strains of *Corynebacterium pseudotuberculosis* which differed in their virulence (Pepin *et al.*, 1996, unpublished results). *C. pseudotuberculosis* is the causative agent of caseous lymphadenitis, a disease which is characterized by the formation of pyogranulomas in the lymph nodes and lung tissue. The function of these pyogranulomas is to limit the spread of infection and to facilitate the eventual destruction of the bacteria. In this study, however, no significant differences in cytokine gene expression were found despite the fact that the two attenuated strains of bacteria did not induce pyogranuloma development in the draining lymph nodes.

It has been suggested that TNF-α is responsible for many of the toxic effects of LPS (Beutler and Cerami, 1988). It has been shown that infusion of endotoxin rapidly leads to elevated plasma levels of TNF-α (Sloane *et al.*, 1992), and that its source may be the leukocytes (neutrophils and mononuclear phagocytes) which accumulate in the lung and pulmonary circulation during endotoxaemia (Cirelli *et al.*, 1995). Ellis *et al.* (1991) demonstrated that AM infected *in vitro* with *C. pseudotuberculosis* or *Pasteurella haemolytica* produce TNF-α, and have further shown *in vivo* that in both experimental and natural infections with *C. pseudotuberculosis*, TNF-γ is detected in macrophage pyogranulomas in the sheep lung (Ellis *et al.*, 1995). Having previously shown that rBoTNF-α has no effect on the replication of *C. pseudotuberculosis* in cultured ovine AM but significantly enhances ovine lentivirus replication they speculated that expression of TNF-α as a result of an opportunistic bacterial infection may have important implications for the progression of lentiviral-induced pneumonia in sheep (Ellis *et al.*, 1994). Administration of endotoxin via an intravascular bolus has also been shown to increase significantly the production of pituitary hormones (Coleman *et al.*, 1993) – an effect which may be mediated by TNF-α (Nash *et al.*, 1992b).

Administration of Recombinant Cytokines

Intradermal injection of recombinant human and ruminant cytokines has been used to study migration and recruitment of particular cell phenotypes into the skin of sheep. Haig *et al.* (1995a) reported that i.d. injection of rOvGM-CSF resulted in an accumulation of MHC class II^+ dendritic cells the majority of which were $CD1^-$. These cells form a minor population in normal skin, but interestingly are the predominant dendritic cell type which accumulates in the lesion produced by OV infection (Lear *et al.*, 1996). Haig *et al.* (1995a) also showed that intradermal injection of rOvTNF-α only led to a transient accumulation of dendritic cells, but when injected together with the rOvGM-CSF resulted in greatly enhanced accumulation of MHC class II^+, $CD1^+$ dendritic cells compared to either cytokine alone. Both rOv cytokines supported the recruitment of lymphocytes and neutrophils to the site of injection and rOvGM-CSF additionally supported eosinophil recruitment.

Kalaaji *et al.* (1989) found that rbTNF-α recruited lymphocytes from the blood to the site of intradermal injection in a dose-dependent manner, an activity which Colditz and Watson (1992) also found with rHuTNF-α, rHuIL-1α and rbIFN-γ. However, these authors reported that the rHuTNF-α, rHuIL-1α and rHuIL-8 preferentially recruited neutrophils into skin sites while rbIFN-γ was the only cytokine to favour lymphocyte recruitment. In a study using IFN-γ-depleted animals, however, Emery and Davey (1995) have suggested that IFN-γ may not be obligatory for DTH reactions in the skin as reactions to PPD of sheep immunized with BCG were reduced by less than 40% in the IFN-γ-depleted animals. Colditz *et al.* (1994) tried to relate neutrophil recruitment in response to IL-1β, TNF-α and IL-8 to differences in susceptibility to fleece rot and ectoparasite infestation, but found no significant differences between animals bred for relative resistance and susceptibility except in the numbers of IgE+ cells present in the skin.

CONCLUSIONS

The structure and biological function of the majority of ovine cytokines have proved, in the main, to be similar to their human counterparts. However there are notable exceptions, for example ovine IL-3 is a mast cell growth factor whereas human IL-3 is not. Also, although some biological cross-reactivity between certain human and ovine cytokines has been demonstrated, it is clear that there is sufficient variation between some of the human and ovine proteins to require the use of the homologous protein to determine precise biological functions.

The sheep, along with ruminants generally, offers an opportunity to study natural diseases of comparative medical importance in their natural target species. The ability to cannulate lymphatic vessels around single lymph nodes, not easily achieved with other species, gives a unique insight into the distribution, recirculation and activation status of cells involved in immune and inflammatory responses. This should aid our understanding of the function of particular cytokines in the process of lymphocyte traffic and effector cell recruitment and activation.

ACKNOWLEDGEMENTS

The authors are supported by funds from the Scottish Office Agriculture, Environment and Fisheries Department (SOAEFD). They would also like to thank Ciba Geigy for the kind gift of rBoIFN-α.

Recombinant Ovine Interleukin-1β as an Adjuvant for Multivalent Bacterial Vaccines

8

M. Elhay[1], G. Barcham[2], A. Cameron[3], A. Andrews[2] and A. Nash[2]

[1]*TB Research Unit, Statens Seruminstitut, Artillerivej, 5, Copenhagen, 2300S, Denmark;* [2]*Centre for Animal Biotechnology, The University of Melbourne, Parkville, Victoria 3052, Australia;* [3]*CSL Ltd, 45 Poplar Road, Parkville, Victoria 3052, Australia*

INTRODUCTION

The multifaceted nature of the immune response allows for development of protective immunity against a variety of pathogens, each of which may use a different strategy to infect and multiply within the host. To be effective, vaccination must induce an immune response of both sufficient magnitude and of an effector phenotype appropriate for rejection of the pathogen in question. With the exception of attenuated live vaccines, and some whole cell formulations, the magnitude and phenotype of the response to vaccination is determined primarily by the choice of adjuvant. Currently, aluminium based adjuvants such as aluminium hydroxide [$Al(OH)_3$] gel are the only adjuvants licensed for use in man and the most commonly used adjuvants for animal vaccines. While these compounds have been shown to be relatively safe, i.m. administration can be associated with nodule or granuloma formation (Gupta and Relyveld, 1991) and there may be other less well characterized systemic effects related to the use of aluminium-based compounds. In addition, induction of humoral and particularly cell-mediated immunity (CMI) by aluminium adjuvants is poor compared with induction by more reactive oil-based adjuvants. The response induced by $Al(OH)_3$ is typical of the Th2 type (Grun and Maurer, 1989; Lise and Audibert, 1989) and while this may be appropriate for some pathogens this will not always be the case. Despite intense analysis of alternatives including mycobacterial fractions such as muramyl dipeptide, saponins and their derivatives, block-copolymer gels and hydrocarbon derivatives (reviewed in

 Cytokines in Veterinary Medicine (eds V.E.C.J. Schijns and M.C. Horzinek)

Cox and Coulter, 1992) our ability to regulate successfully both the magnitude and phenotype of the immune response using traditional approaches to adjuvant technology, remains limited.

RECOMBINANT CYTOKINES AS NEW GENERATION VACCINE ADJUVANTS

Underpinning recent advances in our understanding of the nature and regulation of protective immunity has been an appreciation of the central role played by cytokines, particularly with respect to Th cell differentiation and the subsequent development of apparently mutually exclusive immune effector mechanisms (Romagnani, 1992). In addition to this key issue, cytokines also mediate a number of other aspects of the response to vaccination such as traffic of APC between the site of vaccination and local secondary lymphoid tissue (Zimmerman *et al.*, 1992). The accumulation of data on the role of specific cytokines during the development of immunity contrasts with our poor understanding of the mode of action of most commercial and experimental adjuvants, and this has prompted speculation that recombinant cytokines may find application in rational vaccine design and formulation.

In rodent models, cytokines such as IL-1, IL-2 and IFN-γ have been demonstrated to serve as effective adjuvants for humoral and cellular responses to a variety of antigens including model proteins (Staruch and Wood, 1983; Heath and Playfair, 1991), viral vaccines based on inactivated preparations (Weinberg and Merigan, 1988; Nunberg *et al.*, 1989) or specific recombinant subunits (Weinberg and Merigan, 1988) and crude experimental parasite vaccines (Playfair and DeSouza, 1987). More recently, liposome-based formulations of IL-6 (Duits *et al.*, 1993) and IL-7 (Bui *et al.*, 1994) have also been demonstrated to exert significant adjuvant activity. While adjuvant activity in the vast majority of these studies was measured as an increase in antigen-specific antibody titre, recent studies using a variety of murine parasite models have demonstrated that the Th polarizing cytokines, IL-4 and IL-12, are able to direct the immune response towards a protective phenotype when used in both a therapeutic (Else *et al.*, 1994; Finkelman *et al.*, 1994) and adjuvant (Wynn *et al.*, 1995) context.

Analysis of the adjuvant activity of cytokines in large animal models has been restricted by the lack of available species-specific recombinant cytokines. In recent years, however, a large number of ruminant cytokines have been cloned and a limited number expressed in quantities sufficient for application in adjuvant trials (Lofthouse *et al.*, 1996). Recombinant bovine (rBo) IL-1β has been used in cattle in conjunction with a commercial modified live BHV-1 vaccine (Reddy *et al.*, 1990), a BHV-1/

parainfluenza(PI)-3 vaccine and a killed bovine diarrhoea virus (BVD) vaccine (Reddy *et al.*, 1993). rBoIL-2 has also been used in conjunction with the modified live BHV-1 vaccine as well as an experimental BHV subunit vaccine (Reddy *et al.*, 1990; Hughes *et al.*, 1991, 1992). In each case, cattle vaccinated with the cytokine-adjuvanted preparations exhibited improved serum neutralizing antibody titres as well as decreased virus shedding and increased protection following challenge. In the ovine model we have expressed a variety of cytokines including rOvIL-1α, rOvIL-1β and rOvIL-2, in quantities sufficient for analysis of adjuvant activity (Andrews *et al.*, 1991, 1994; Nash *et al.*, 1993; Lofthouse *et al.*, 1995a). Using the model protein avidin we have optimized the application of rOvIL-1β with respect to a variety of parameters including dose, route of administration and formulation. These studies demonstrated that when delivered appropriately rOvIL-1β was able to enhance antigen-specific antibody titres by up to tenfold and when used in combination with $Al(OH)_3$ it invoked strong DTH responses (Andrews *et al.*, 1994). In more recent studies rOvIL-1β has been demonstrated to be an important component of an experimental vaccine that protects sheep against larvae of the ectoparasite *Lucilia cuprina* (Bowles *et al.*, 1996), although the mechanism of protection in this case remains ill-defined.

RECOMBINANT OVINE IL-1β AS AN ADJUVANT FOR CLOSTRIDIAL VACCINES

Despite extensive analysis of the adjuvant activity of recombinant IL-1β in both sheep and cattle a number of issues remain to be addressed. Firstly, the range of antigens with which ruminant IL-1β has been tested is limited to model proteins, a small number of viral vaccines and, in one case, an experimental parasite vaccine. Application with bacterial vaccines, and in particular toxoid-based preparations, has not been examined. Secondly, the efficacy of IL-1β as an adjuvant for individual components of multivalent vaccine preparations has not been reported. This issue is of some significance as vaccines for application in livestock species will almost inevitably be multivalent.

A mutivalent toxoid-based vaccine is used to protect sheep and cattle against the anaerobic diseases (e.g. Black disease, tetanus, pulpy kidney and enterotoxaemia) caused by pathogenic clostridial species. The vaccine consists of a balanced mixture of purified toxoids of *Clostridium novyi*, *C. perfringens*, *C. septicum*, *C. tetani* and a formol culture of *C. chauvoei*, formulated in $Al(OH)_3$. The preparation may also be supplemented with a purified *Corynebacterium pseudotuberculosis* (cheesy gland) toxoid depending on location and target species. Protective immunity is mediated by antibody, and while the vaccine is generally considered to be effective, yearly vaccination is required to maintain immunity and even when such a rigorous schedule is adhered to, immunity to some components, such as *C. perfringens*, remains problematic. On this basis, we have investigated the application of rOvIL-1β as an adjuvant for mono- and multivalent clostridial vaccines either alone as part of a soluble formulation or as a supplement to standard $Al(OH)_3$-precipitated preparations.

Recombinant Ovine Interleukin-1β as an Adjuvant for a Monovalent *Clostridium novyi* Vaccine

In trial 1, groups of 20 6- to 10-week-old merino lambs of mixed sex were vaccinated subcutaneously (s.c.) or i.m. with a monovalent *C. novyi* vaccine as one of the following:

1. A soluble preparation;
2. A soluble preparation supplemented with 100 μg per dose of rOvIL-1β;
3. An $Al(OH)_3$-precipitated preparation;
4. An $Al(OH)_3$-precipitated preparation supplemented with 100 μg per dose of rOvIL-1β.

rOvIL-1β was added to vaccine preparations 3–5 h before administration. All lambs received an identical secondary vaccination 4 weeks after the primary. Lambs were bled approximately 2 weeks after each vaccination, and pooled sera from each group assayed by ELISA on *C. novyi* toxin-coated 96-well microtitre plates. For statistical analysis (one-way ANOVA and Fisher's technique of pairwise comparison) sera from 15–20 animals within each group were assayed on an individual basis (not shown).

For all vaccine preparations the post primary response displayed little or no reactivity above a background determined using a pool of prebleed sera sampled from a large number of the trial lambs. Group midpoint titres for the post secondary bleeds are shown in Fig. 8.1. *C. novyi* toxoid without any adjuvant was unable to stimulate a significant response using either route of administration. In contrast, the inclusion of $Al(OH)_3$ into the preparation resulted in a strong secondary response using either the s.c. or the i.m. route.

Completely soluble preparations containing rOvIL-1β induced virtually an identical response using the i.m. route and a significant ($P < 0.05$) twofold greater response when administered via the s.c. route. While the combination of $Al(OH)_3$ and rOvIL-1β did not further enhance the s.c. response above that induced by IL-1β alone, i.m. administration of this formulation resulted in a response significantly greater ($P < 0.05$) than that observed with either adjuvant alone and equivalent to the maximum response observed using the s.c. route.

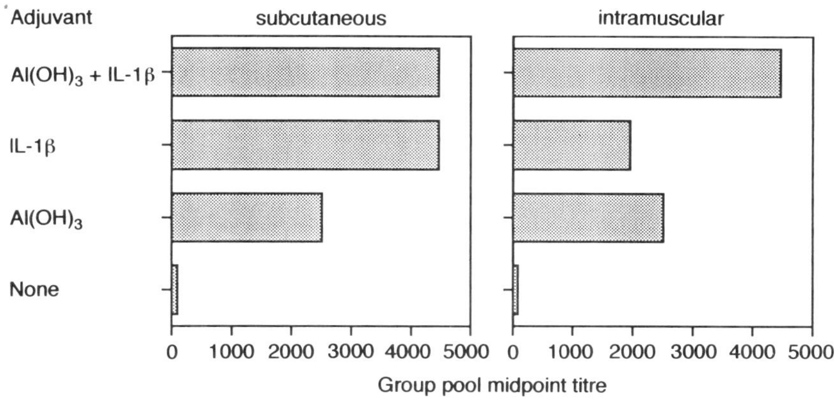

Fig. 8.1. rOvIL-1β as an adjuvant for a *C. novyi* vaccine. Groups of 20 6- to 10-week old merino lambs of mixed sex were vaccinated via a s.c. (neck/shoulder region) or i.m. (hind leg) route with a *C. novyi* toxoid preparation formulated without adjuvant, with $Al(OH)_3$, with 100 μg of rOvIL-1β or with a combination of $Al(OH)_3$ and 100 μg of rOvIL-1β. The vaccines contained toxoid and, where indicated, $Al(OH)_3$ at a concentration equivalent to that found in standard commercial preparations. All lambs received a second vaccination 4 weeks later. Two weeks after each vaccination lambs were bled and sera pooled within each group for assay by indirect ELISA on *C. novyi* toxin-coated 96-well plates. Results presented are the group pool midpoint titres for the 2-week post-secondary response.

Recombinant Ovine Interleukin-1β as an Adjuvant for Limited Multivalent Clostridial Vaccines

While the first trial had established the effectiveness of rOvIL-1β as an adjuvant for a bacterial toxoid-based vaccine, the question of the impact of multiple vaccine components remained. To address this issue a second trial was designed in which a divalent vaccine containing *C. novyi* and *C. perfringens* toxoids and a trivalent vaccine containing *C. novyi*, *C. tetani* and *C. perfringens* toxoids were administered. The trial design was similar to that of trial 1 with the exception that vaccination was restricted to the i.m. route. In addition, lambs were bled again at 57 days after the secondary vaccination to assess the longevity of the response. For technical reasons ELISA-based analysis of serum antibody titres was restricted to the *C. novyi*- and *C. tetani*-specific responses.

As was observed in trial 1 primary response antibody titres were not significantly greater than that determined for the prebleed pool. Post secondary *C. novyi* specific group midpoint titres for lambs vaccinated with the divalent vaccine are shown in Fig. 8.2. Inclusion of the *C. perfringens* component into the *C. novyi* vaccine clearly had a negative impact on the

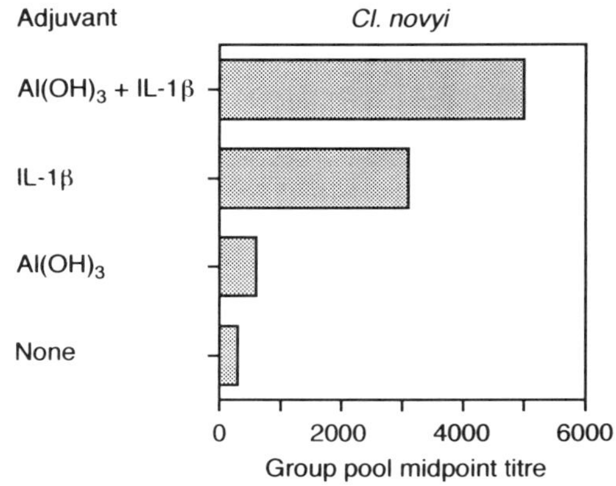

Fig. 8.2. rOvIL-1β as an adjuvant for a divalent vaccine containing *C. novyi* and *C. perfringens* toxoids. Groups of 20 6- to 10-week-old merino lambs of mixed sex were vaccinated i.m. with the divalent preparation formulated without adjuvant, with $Al(OH)_3$, with 100 mg of rOvIL-1β or with a combination of $Al(OH)_3$ and 100 mg of rOvIL-1β. The vaccines contained toxoids and, where indicated, $Al(OH)_3$ at a concentration equivalent to that found in standard commercial preparations. All lambs received a secondary vaccination 4 weeks later. Two weeks after each vaccination lambs were bled and sera pooled within each group for assay by indirect ELISA on *C. novyi* toxin coated 96-well plates. Results presented are the group pool midpoint titres for the 2-week post-secondary response.

ability of $Al(OH)_3$ to act as an adjuvant for the *C. novyi*-specific response, with the titre only marginally above that observed using a soluble preparation containing no adjuvant at all. rOvIL-1β, however, retained its adjuvant activity with the cytokine-supplemented soluble preparation inducing a *C. novyi*-specific antibody titre within the same range observed in trial 1, and some five- to sixfold greater than the response induced by $Al(OH)_3$ alone ($P < 0.05$). As was the case in trial 1 for i.m. administration, the combination of $Al(OH)_3$ and rOvIL-1β induced a further significant ($P < 0.05$) increase in the *C. novyi*-specific titre over that observed with rOvIL-1β alone.

An essentially identical profile of *C. novyi*-specific serum antibody titres was observed when the divalent vaccine was further supplemented with *C. tetani* toxoid (Fig. 8.3). $Al(OH)_3$ alone induced a poor secondary response, the soluble preparation containing rOvIL-1β alone a titre some three- to fourfold greater, and the preparation containing both $Al(OH)_3$ and rOvIL-1β a titre significantly greater again ($P < 0.05$). Analysis of sera sampled 57 days after the secondary vaccination indicated that the high *C. novyi* titres induced by rOvIL-1β at day 15 were reflected at this later time point, with

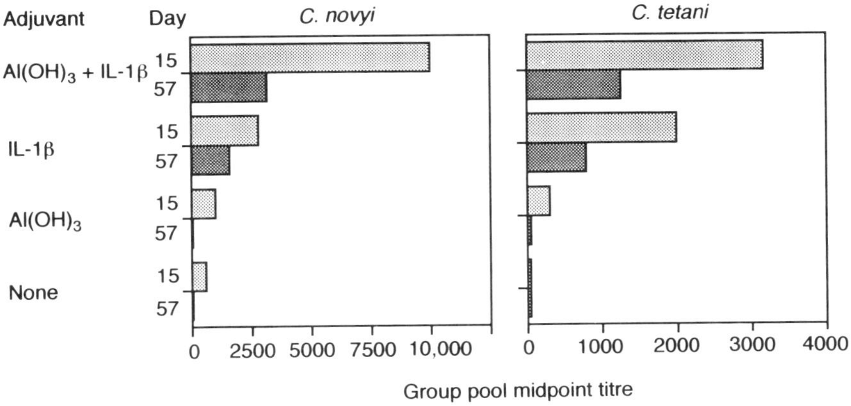

Fig. 8.3. rOvIL-1β as an adjuvant for a trivalent vaccine containing *C. novyi*, *C. perfringens* and *C. tetani* toxoids. Groups of 20 6- to 10-week-old merino lambs of mixed sex were vaccinated i.m. with the trivalent preparation formulated without adjuvant, with $Al(OH)_3$, with 100 mg of rOvIL-1β or with a combination of $Al(OH)_3$ and 100 mg of rOvIL-1β. The vaccines contained toxoids and, where indicated, $Al(OH)_3$ at a concentration equivalent to that found in standard commercial preparations. All lambs received a secondary vaccination 4 weeks later. Two weeks after each vaccination lambs were bled and sera pooled within each group for assay by indirect ELISA on *C. novyi* and *C. tetani* toxin-coated 96-well plates. Results presented are the *C. novyi* and *C. tetani* specific group pool midpoint titres for the 15-day and 57-day post-secondary response.

the titre remaining greater than that induced by $Al(OH)_3$ at day 15. The adjuvant activity of both $Al(OH)_3$ and rOvIL-1β for the response to the *C. tetani* component of the trivalent vaccine was similar to that observed for the *C. novyi*-specific response (Fig. 8.3). The soluble preparation containing rOvIL-1β induced a *C. tetani* response four- to fivefold greater than that observed with the vaccine containing $Al(OH)_3$ and the combination of rOvIL-1 and $Al(OH)_3$ induced a further significant increase ($P < 0.05$). As was the case for the *C. novyi*-specific response, the high *C. tetani*-specific titres detected at day 15 were reflected at day 57.

Recombinant Ovine Interleukin-1β as an Adjuvant for a Commercial Multivalent Clostridial Vaccine

Having established the efficacy of rOvIL-1β as an adjuvant for mono- and limited multivalent clostridial vaccines, a trial was undertaken to assess efficacy with a commercial multivalent vaccine consisting of toxoids of four clostridial species (*C. novyi*, *C. septicum*, *C. tetani* and *C. perfringens*), a formol culture of a fifth (*C. chauvoei*) and, in addition, a toxoid of *C. pseudotuberculosis* (CSL Ltd, Australia, Glanvac 6 in 1 vaccine). Trial design incorporated similar features to trials 1 and 2 with the addition of a dose–response analysis for rOvIL-1β in soluble preparations. In contrast to observations with the di- and trivalent vaccines, $Al(OH)_3$ was able to induce strong *C. novyi*- and *C. tetani*-specific secondary responses in lambs vaccinated with the 6 in 1 vaccine (Fig. 8.4). $Al(OH)_3$-induced titres were in the order of tenfold greater than titres induced by the equivalent soluble preparation containing no adjuvant. This observation serves to highlight the dynamic interactions between multiple components, with varying degrees of endogenous stimulatory or inhibitory activity, and a selected adjuvant such as $Al(OH)_3$. Dose–response analysis indicated that vaccines supplemented with 100 μg per dose of rOvIL-1β, but not 10 or 1 μg per dose (100 μg response significantly greater than 10 or 1 μg ($P < 0.05$)), were able to induce secondary *C. novyi*- and *C. tetani*-specific responses equivalent to those induced by $Al(OH)_3$. Once again the vaccine preparation supplemented with both $Al(OH)_3$ and rOvIL-1β induced both *C. novyi*- and

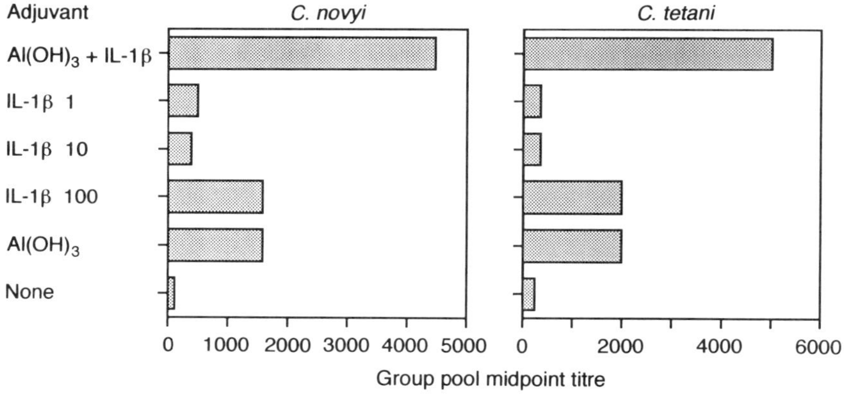

Fig. 8.4. rOvIL-1β as an adjuvant for a commercial multivalent clostridial vaccine containing *C. novyi, C. perfringens, C. tetani, C. septicum* toxoids, a formol culture of *C. chauvoei* and formalin-treated fermentation products of *C. pseudotuberculosis*. Groups of 20 6- to 10-week-old merino lambs of mixed sex were vaccinated i.m. with the multivalent preparation formulated without adjuvant, with $Al(OH)_3$, with 100, 10 or 1 μg of rOvIL-1β or with a combination of $Al(OH)_3$ and 100 μg of rOvIL-1β. All lambs received a secondary vaccination 4 weeks later. Two weeks after each vaccination lambs were bled and sera pooled within each group for assay by indirect ELISA on *C. novyi* and *C. tetani* toxin-coated 96-well plates. Results presented are the *C.novyi* and *C. tetani* specific group pool midpoint titres for the 2-week post-secondary response.

C. tetani-specific responses significantly greater than those observed with either adjuvant alone ($P < 0.05$).

Recombinant Ovine Interleukin-1β Induces Protective Neutralizing Antibodies

Protective immunity conferred by the 6 in 1 vaccine is mediated by neutralizing antibody directed against specific toxins or, in the case of *C. chauvoei* and *C. pseudotuberculosis*, against other less well defined bacterial antigens. To confirm that rOvIL-1β was able to induce antibody with appropriate neutralizing activity, pooled post secondary serum samples from a number of groups in trial 3 were tested for the ability to neutralize *C. tetani*, *C. novyi* and *C. perfringens* toxin following adoptive transfer to mice (Sterne *et al.*, 1962; Webster and Frank, 1985). Results presented in Table 8.1 demonstrated that vaccines adjuvanted with $Al(OH)_3$, rOvIL-1β and a combination of $Al(OH)_3$ and rOvIL-1β were all able to induce neutralizing antibodies.

OTHER ISSUES RELATED TO THE APPLICATION OF RECOMBINANT CYTOKINES AS VACCINE ADJUVANTS

Safety

Numerous adjuvants are restricted in usage due to their ability to induce localized pain, inflammation and tissue damage at the injection site and systemic reactions such as fever (Gupta *et al.*, 1993). The retention in tissue of toxic materials such as oils, leading to the contamination of carcasses

Table 8.1. Effect of different adjuvant formulations on the induction of neutralizing antitoxin antibodies by a commercial multivalent clostridial vaccine.[1]

	Antitoxin titres (IU ml^{-1}) Adjuvant formulation		
Challenge toxin	$Al(OH)_3$	rOvIL-1β	$Al(OH)_3$ + rOvIL-1β
Cl. novyi	24.0	95.0	48.0
Cl. tetani	12.0	21.0	21.0
Cl. perfringens[2]	2.8	2.8	5.7

[1]Groups of 20 6- to 10-week-old merino lambs of mixed sex were vaccinated i.m. with a multivalent preparation, formulated without adjuvant, containing *C. novyi*, *C. tetani*, *C. perfringens* and *C. septicum* toxoids, and a formol culture of *C. chauvoei*, with $Al(OH)_3$, with 100 μg of rOvIL-1β or with a combination of $Al(OH)_3$ and 100 μg of rOvIL-1β. Sera sampled 2 weeks after the second vaccination were pooled within each group and transferred to mice for determination of neutralizing titres using a standard procedure (Sterne *et al.*, 1962; Webster and Frank, 1985)

[2]*C. perfringens* type D.

with residues represents a further limitation. While, as biologicals, cytokines such as rOvIL-1β are likely to be inherently less harmful, their application must be subject to all routine safety considerations. This is especially important in view of the fact that not all known biological properties of any given cytokine may be compatible with administration of exogenous recombinant material. Excessive endogenous IL-1 synthesis, for example, is associated with chronic inflammatory diseases and fatal septic shock.

As part of an ongoing analysis of the safety of rOvIL-1β we carried out a histological analysis of muscle tissue following i.m. injection of multivalent clostridial vaccines formulated with $Al(OH)_3$ or rOvIL-1β. As expected, vaccines formulated with $Al(OH)_3$ gave rise to granulomas (aluminium-based veterinary vaccines are usually administered s.c. to reduce granuloma formation). Routine histological staining with haematoxylin and eosin and Masson's trichrome stains revealed that the granulomas included a capsule consisting of plump epithelioid cells that resembled macrophages. In the central region there was evidence of necrosis throughout which an amorphous eosinophilic substance was present. Occasional lymphocytes were present throughout the granuloma and foci of CD4 T cells, CD8 T cells and B cells were detected adjacent to the fibrous capsule, both within and outside the granuloma. The fibrous capsule stained an intense green after staining with Masson's trichrome indicating collagen. Staining with Aluminon reagent revealed red staining aluminium particles apparently engulfed by macrophages and as free standing bodies, particularly in the centre of necrotic regions. In stark contrast, i.m. injection of soluble vaccine preparations containing rOvIL-1β was not associated with any adverse tissue reactivity (not shown).

These observations were consistent with a previous report describing immunohistological analysis of local tissue reactions following i.d. immunization of sheep with a model protein formulated with $Al(OH)_3$, rOvIL-1β or rOvIL-1β and $Al(OH)_3$ (Lofthouse *et al.*, 1995a). In that study, i.d. administration of $Al(OH)_3$ adjuvanted preparations was associated with a massive infiltration of cells into the skin including B cells, T cells and macrophages. In contrast, rOvIL-1 caused no obvious changes in local skin immunohistology nor did it appear to exacerbate the extensive infiltration caused by the $Al(OH)_3$. In addition to tissue reactivity, the impact of adjuvant active doses of rOvIL-1β on a variety of parameters including body temperature and total and differential white blood cell counts has also been reported (Lofthouse *et al.*, 1995b). rOvIL-1β delivered by the i.m. route induced rises in body temperature of 1°C, with temperatures returning to normal by 24 h p.i. Goff *et al.* (1992) reported similar transient changes in body temperatures of cattle following i.m. injection with rBoIL-1β at doses of approximately 30 μg per animal. Following injection of sheep with rOvIL-1β total white cell counts dropped to one-quarter normal levels within 2 h, followed by an immediate increase in cell count to reach three times normal levels by 20 h post injection. White cell counts declined thereafter to reach pre-injection levels by 3 days post-injection. These fluctuations in white cell count were shown to be entirely due to the change in the number of circulating neutrophils. Such changes are often observed following immunization with many commercial live vaccines and are considered acceptable providing that the fluctuations are transient.

Stability

Cytokines are polypeptides whose functional activity is based on high affinity interaction with specific receptors on target cell populations. This interaction is dependent on both the primary aa sequence and tertiary structure of the cytokine. Protease activity leading to alterations in the primary aa sequence or other molecular interactions that impact on the tertiary conformation can have drastic consequences for the activity of cytokines. Given the relatively crude nature of the 6 in 1 and related vaccines, and the fact that such vaccines are often stored for considerable periods before use, the stability and thus efficacy of rOvIL-1β under such conditions is an important consideration. We have addressed this issue using two different approaches. Firstly, we have mixed rOvIL-1β with multivalent clostridial vaccines for periods of up to 2 weeks at 4°C, then assessed IL-1 structure and function in the laboratory using an IL-1 specific bioassay and Western blot analysis (Egan *et al.*, 1994). Pre-incubation of rOvIL-1β with vaccine did not reduce the activity of IL-1β in bioassays when compared to control, unmixed rOvIL-1β. In addition Western blot analysis with the mAb 2.93 demonstrated that rOvIL-1β was present and had not degraded (data not shown).

In the second approach the 6 in 1 vaccine with and without $Al(OH)_3$ was formulated with rOvIL-1β and allowed to stand at 4°C for periods of up to 2 months before administration. Results shown in Fig. 8.5 demonstrated that formulation of soluble 6 in 1 vaccine antigens with rOvIL-1β 1 day or 2 weeks before administration did not diminish the adjuvant activity of the cytokine. The response induced by these preparations was equivalent to, or greater than, that induced by the same vaccine prepared immediately before use and equivalent to the response induced by $Al(OH)_3$. In contrast, soluble vaccines formulated with rOvIL-1β 4 or more weeks before use appeared to have a reduced capacity for induction of *C. tetani*-specific antibody responses. This decrease in adjuvant activity was not apparent when the formulated vaccines were lyophilized and reconstituted immediately before use. Interestingly, the inclusion of $Al(OH)_3$ into the formulation appeared to stabilize cytokine activity as, in keeping with previous observations, the combined formulation invoked a *C. tetani*-specific response in excess of that induced by either $Al(OH)_3$ or rOvIL-1β alone, even when the vaccine was prepared 8 weeks before use. The profile of the *C. novyi*-specific responses

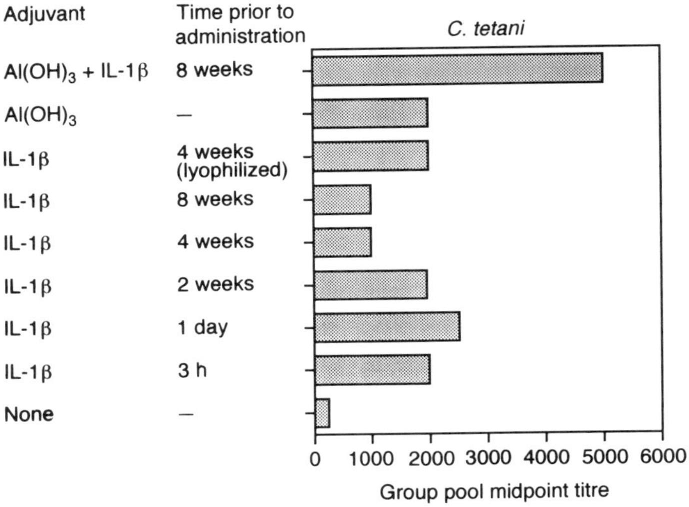

Fig. 8.5. Stability of rOvIL-1β adjuvant activity in a commercial multivalent clostridial vaccine formulated at the indicated time points prior to administration. Groups of 20 6- to 10-week-old merino lambs of mixed sex were vaccinated i.m. with the multivalent preparation formulated without adjuvant, with $Al(OH)_3$, with 100 μg of rOvIL-1β at the indicated time point prior to administration or with a combination of $Al(OH)_3$ and 100 μg of rOvIL-1β also at the indicated time point prior to administration. All lambs received a second vaccination 4 weeks later. Two weeks after each vaccination lambs were bled and sera pooled within each group for assay by indirect ELISA on *C. novyi* and *C. tetani* toxin coated 96-well plates. Results presented are the *C. tetani* specific group pool midpoint titres for the 2-week post-secondary response.

induced by these vaccines was essentially identical to those for *C. tetani* (not shown).

These data serve to demonstrate potential limitations to the incorporation of bioactive proteins into existing commercial products. Nevertheless, the ability of $Al(OH)_3$ to stabilize the activity of rOvIL-1β and retention of activity in the lyophilized preparation suggest that the application of rOvIL-1β will need to be considered on a case-by-case basis. For example, there is currently considerable interest in the application of slow and pulsatile release devices to animal vaccination. It is speculated that such devices may be able to reduce vaccination schedules, perhaps ultimately facilitating 'single shot' regimes. Where such devices require pelletization of vaccine antigens it is difficult to envisage the application of oil-based adjuvants. In these cases a lyophilizable adjuvant such as a recombinant cytokine may represent the ideal alternative.

Cost

In general, the profit margin for commercial vaccines intended for application in livestock species such as sheep is low. As a result, the cost of production of a recombinant cytokine for incorporation into a commercial vaccine intended for the sheep industry may limit its application. Clearly production costs will be dependent on the yield of recombinant material from a given expression system and the processes required to purify the material to an acceptable level. rOvIL-1β used in the current study was expressed in *Escherichia coli* from a heat inducible Celltech vector to a level of approximately 30% of total cell protein. The recombinant cytokine remained predominantly within the soluble fraction and purified as a biologically active protein to greater than 95% using a simple and cheap one-step anion exchange protocol (Andrews *et al.*, 1994). While this would place the cost of rOvIL-1β at an acceptable level the costs associated with other cytokines requiring more extensive purification procedures, including refolding of insoluble material from inclusion bodies, may be prohibitive.

CONCLUSIONS

The application of recombinant cytokines in vaccination strategies for livestock species such as sheep offers a variety of potential benefits. Chemical adjuvants used in existing vaccines, such as the clostridial preparations described above, could be replaced with biological agents that are both more effective and less problematic in terms of carcass damage and residual contamination. Clearly, rOvIL-1β was suited to application with the clostridial vaccines as its impact on antibody titres was compatible with the mechanism of protection against these diseases. Cytokines other than IL-1, however, may also be useful for application with vaccines that aim to induce high antibody titres. Recently we have shown that rOvIL-2, if delivered appropriately, is also able to enhance antibody titres to model protein antigens. While this adjuvant activity was reduced compared to that observed with IL-1β we have observed a synergistic effect when a combination of both rOvIL-1β and rOvIL-2 is used (A. Andrews and A. Nash, 1995, unpublished results).

With respect to the sheep industry, particularly within Australia, current efforts in vaccine development are focused on a variety of endo- and ectoparasite diseases. We have recently described an experimental ectoparasite (*Lucilia cuprina*) vaccine that is based on a novel formulation of four native antigens and rOvIL-1β formulated in an oil-in-water emulsion and which confers protection against wound initiation to a level of 60% (Bowles *et al.*, 1996). While we envisage that improved protection levels might be achieved through rational incorporation of further selected cytokines, we are restricted by a limited understanding of the effector mechanisms mediating protection. In contrast, the mechanisms of protection induced by two distinct classes of sheep helminth parasite vaccines are well characterized. The 'hidden antigen' approach relies on the induction of what is essentially neutralizing humoral immunity against critical components of the parasite not normally seen by the host (Tavernor *et al.*, 1992; Munn *et al.*, 1993). Clearly, in this case, cytokines such as rOvIL-1β may be of some use. Alternative helminth vaccination strategies appear to be based on the induction of immune effector mechanisms (mucosal mast cells, eosinophils and IgE) that, in rodents at least, are typical of a Th2 type response. While the application of rOvIL-1β with such a vaccine exacerbates infection (Nash *et al.*, 1993) it is possible that cytokines which support development of Th2 responses (IL-4) or antagonize development of Th1 responses (IL-4 and possibly IL-10) may be more useful.

Assuming that issues such as safety, stability and cost are not prohibitive there is sufficient evidence in both small and large animal models to suggest that recombinant cytokines may, at some stage, represent valuable tools in rational vaccine design and formulation. It is apparent, however, that for the potential of these molecules to be fully realized, there will need to be concurrent advances in our understanding of the effector mechanisms that impact on parasitic diseases of livestock species such as sheep and cattle.

EPILOGUE

Ovine cytokine research has evolved in the last 10 years from a position of no sheep-specific reagents to one where over 20 cytokines have now been cloned and the majority expressed as recombinant proteins. Several ovine cytokine-specific ELISAs and bioassays are available and more are in the development phase. Although the application of these reagents is still in its infancy, their availability means that researchers are now in a position to answer specific questions about cytokine biology in the sheep under physiological and pathological conditions. The ability to measure cytokine production and utilization both systemically and locally will advance our understanding of protective and exacerbating effector mechanisms in ovine immune and inflammatory responses to infectious agents. The use of cytokines as adjuvants in rational vaccine design is also an exciting area of research.

Once the involvement of specific cytokines has been established in the host response to specific pathogens it may be possible to design cytokine-based pathogen-protective therapies or preventive measures to limit immune-mediated tissue damage. In mice the existence of separate populations of Th cells which produce mutually exclusive subsets of cytokines capable of directing the immune response *in vivo* either towards a predominantly cell-mediated response (Th1; involving IFN-γ and IL-12) or towards a humoral response (Th2; involving IL-4 and IL-10) have been demonstrated (Mosmann and Coffman, 1989). The polarization of the immune response in this fashion may be important for protection against certain pathogens. Although similar T cell populations have not been demonstrated in the sheep it is likely that the same subsets of cytokines that differentiate the two T cell populations in mice will have the same effect on the polarization of the ovine immune response. Thus the use of selected cytokines as adjuvants may result in increased vaccine efficiency. The function of IL-12 in this respect will be of particular interest. The future use of rOv cytokines, both as adjuvants and potentially as therapeutic agents, will depend on a number of factors, the most important being efficacy, safety and cost. Due to the low profit margins associated with livestock farming, and sheep farming in particular, even if these recombinant proteins are found to offer both greater safety and efficiency over currently used reagents, their use will be limited if a cheap source of cytokines cannot be found. However the lessons learned from such research may be important in comparative medical studies where cost restraints are not so important.

ACKNOWLEDGEMENTS

This work was supported by the Australian Research Council through the Collaborative Research Grants Scheme and by the International Wool Secretariat. We are particularly indebted to the wool growers who provided access to the large number of lambs used in the field trials described in this study. In particular we acknowledge the assistance of Graham and Wendy Harvey (Poplars), Ian Taylor (Warranbeen) and Murray Drobinski (Eynesbury Park).

References for Part II

Allan, I. and Pearce, J.H. (1983) Amino acid requirements of strains of *Chlamydia trachomatis* and *C. psittaci* growing in McCoy cells: Relationship with clinical syndrome and host origin. *Journal of General Microbiology* 129, 2001–2007.

Andrews, A.E., Barcham, G.J., Brandon, M.R. and Nash, A.D. (1991) Molecular cloning and characterisation of ovine IL-1α and Il-1β. *Immunology* 74, 453–460.

Andrews, A.E., Barcham, G.J., Ashman, K., Meeusen, E.N., Brandon, M.R. and Nash, A.D. (1993) Molecular cloning and characterization of a ruminant interleukin-6 cDNA. *Immunology and Cell Biology* 71, 341–348.

Andrews, A.E., Lofthouse, S.A., Bowles, V.M., Brandon, M.R. and Nash, A.D. (1994) Production and *in vivo* use of recombinant ovine interleukin-1β as an immunological adjuvant. *Vaccine* 12, 14–24.

Ansari, H.A., Pearce, P.D., Maher, D.W. and Broad, T.E. (1995) Human chromosome-10 loci map to 3 different sheep chromosomes. *Mammalian Genome* 6, 46–48.

Barcham, G.J., Andrews, A.E. and Nash, A.D. (1995) Cloning and expression of a cDNA encoding ovine interleukin-7. *Gene* 154, 265–269.

Bazer, F.W., Ott, T.L. and Spencer, T.E. (1994) Pregnancy recognition in ruminants, pigs and horses – signals from the trophoblast. *Theriogenology* 41, 79–94.

Beutler, B. and Cerami, A. (1988) Tumor necrosis, cachexia, shock, and inflammation: A common mediator. *Annual Review of Biochemistry* 57, 505–519.

Bowles, V.M., Meeusen, E.N., Chandler, K., Verhagen, A., Nash, A.D. and Brandon, M.R. (1994) The immune response of sheep infected with larvae of the sheep blowfly *Lucilia cuprina* monitored via efferent lymph. *Veterinary Immunology and Immunopathology* 40, 341–352.

Bowles, V.M., Meeuson, E., Young, A.R., Andrews, A.E., Nash, A.D. and Brandon, M.R. (1996) Vaccination of sheep against larvae of the sheep blowfly (*Lucilia cuprina*). *Vaccine* (in press).

Broad, T.E., Burkin, D.J., Cambridge, L.M., Lewis, P.E., Maher, D.W., Ansari, H.A., Pearce, P.D. and Jones, C. (1995a) 4 Human chromosome-3q and chromosome-4 human chromosome-21 loci map onto sheep chromosome-1q. *Mammalian Genome* 6, 202–205.

Broad, T.E., Burkin, D.J., Cambridge, L.M., Maher, D.W., Lewis, P.E., Ansari, H.A., Pearce, P.D. and Jones, C. (1995b) Assignment of five loci from human chromosome 8q onto sheep chromosome 9. *Cytogenetics and Cell Genetics* 68, 102–106.

Bui, T., Faltynek, C.R. and Ho, R.J.Y. (1994) Biological response of recombinant interleukin-7 on herpes simplex virus infection in guinea pigs. *Vaccine* 12, 646–652.

Bujdoso, R., Williamson, M., Roy, D., Hunt, P., Blacklaws, B., Sargan, D. and Mcconnell, I. (1995) Molecular cloning and expression of DNA encoding ovine interleukin 2. *Cytokine* 7, 223–231.

Buxton, D. (1990) Ovine toxoplasmosis; a review. *Journal of the Royal Society of Medicine* 83, 509–511.

Buxton, D., Barlow, R.M., Finlayson, J., Anderson, I.E. and MacKellar, A. (1990) Observations on the pathogenesis of *Chlamydia psittaci* infection of pregnant sheep. *Journal of Comparative Pathology* 102, 221–237.

Cirelli, R.A., Carey, L.A., Fisher, J.K., Rosolia, D.L., Elsasser, T.H., Caperna, T.J., Gee, M.H. and Albertine, K.H. (1995) Endotoxin infusion in anesthetized sheep is associated with intrapulmonary sequestration of leukocytes that immunohistochemically express tumor necrosis factor-α. *Journal of Leukocyte Biology* 57, 820–826.

Colditz, I.G. and Watson, D.L. (1992) The effect of cytokines and chemotactic agonists on the migration of T lymphocytes into skin. *Immunology* 76, 272–278.

Colditz, I.G., Lax, J., Mortimer, S.I., Clarke, R.A. and Beh, K.J. (1994) Cellular inflammatory responses in skin of sheep selected for resistance or susceptibility to fleece rot and fly strike. *Parasite Immunology* 16, 289–296.

Coleman, E.S., Elsasser, T.H., Kemppainen, R.J., Coleman, D.A. and Sartin, J.L. (1993) Effect of endotoxin on pituitary hormone secretion in sheep. *Neuroendocrinology* 58, 111–122.

Cordier, G., Greenland, T., Arnaud, P. and Mornex, J.F. (1993) Isolation of complementary DNA from ovine interleukin 8. *Veterinary Research* 24, 365–367.

Cox J.C. and Coulter A.R. (1992) Advances in adjuvant technology and application. In: Yong, W.K. (ed.), *Animal Parasite Control Utilizing Biotechnology*. CRC Press, Boca Raton, pp. 49–112.

De Maeyer, E. and De Maeyer Guignard, J. (1988) *Interferons and other Regulatory Cytokines*. John Wiley & Sons, New York.

Duits, A.J., Puijenbroek, A., Vermeulen, H., Hofhuis, F.M.A., van de Winkel, J.G.J. and Capel, P.J.A. (1993) Immunoadjuvant activity of a liposomal IL-6 formulation. *Vaccine* 11, 777–781.

Dutia, B.M., Hunt, P., Sargan, D.R., Dalziel, R.G. and Hopkins, J. (1994) Sequence of the sheep interleukin-10-encoding cDNA. *Gene* 149, 393–394.

Ebrahimi, B., Roy, D.J., Bird, P. and Sargan, D.R. (1995) Cloning, sequencing and expression of the ovine interleukin 6 gene. *Cytokine* 7, 232–236.

Egan, P.J., Andrews, A.E., Barcham, G.J., Brandon, M.R. and Nash, A.D. (1994a) Production and application of monoclonal antibodies to ovine interleukin-1α and interleukin-1β. *Veterinary Immunology and Immunopathology* 41, 241–257.

Egan, P.J., Rothel, J.S., Andrews, A.E., Seow, H.-F., Wood, P.R. and Nash, A.D. (1994b) Characterization of monoclonal antibodies to ovine tumor necrosis factor-α and development of a sensitive immunoassay. *Veterinary Immunology and Immunopathology* 41, 259–274.

Elhay, M.J., Hanrahan, C.F., Bowles, V.M., Seow, H.-F., Andrews, A.E. and Nash, A.D. (1994) Cytokine mRNA expression in skin in response to ectoparasite infection. *Parasite Immunology* 16, 451–461.

Ellis, J.A., Lairmore, M.D., O'Toole, D.T. and Campos, M. (1991) Differential induction of tumor necrosis factor-α in ovine pulmonary alveolar macrophages following infection with *Corynebacterium pseudotuberculosis*, *Pasteurella haemolytica*, or lentiviruses. *Infection and Immunity* 59, 3254–3260.

Ellis, J.A., Russell, H.I. and Du, C.W. (1994) Effect of selected cytokines on the replication of *Corynebacterium pseudotuberculosis* and ovine lentiviruses in pulmonary macrophages. *Veterinary Immunology and Immunopathology* 40, 31–47.

Ellis, J.A., Campos, M., Snyder, M., Chelak, B. and Haines, D.M. (1995) Local production of tumor necrosis factor-α in corynebacterial pulmonary lesions in sheep. *Veterinary Pathology* 32, 68–71.

Else, K.J., Finkelman, F.D., Maliszewski, C.R. and Grencis, R.K. (1994) Cytokine mediated regulation of chronic intestinal helminth infection. *Journal of Experimental Medicine* 179, 347–351.

Emery, D.L. and Davey, R.J. (1995) An analysis, using monoclonal antibodies, of the role of interferon-γ in ovine immune responses. *Immunology and Cell Biology* 73, 146–152.

Engwerda, C.R. and Sandeman, R.M. (1992) The isolation and sequence of sheep interleukin-4. *DNA Sequence* 3, 111–113.

Entrican, G., Haig, D.M. and Norval, M. (1989) Identification of ovine interferons: differential activities derived from fibroblast and lymphoid cells. *Veterinary Immunology and Immunopathology* 21, 187–195.

Entrican, G., McInnes, C.J., Rothel, J.S. and Haig, D.M. (1992) Kinetics of ovine interferon-γ production: detection of mRNA and characterisation of biological activity. *Veterinary Immunology and Immunopathology* 33, 171–178.

Entrican, G., Deane, D., MacLean, M., Inglis, L., Thompson, J., McInnes, C. and Haig, D.M. (1996) Development of a sandwich ELISA for ovine granulocyte-macrophage colony-stimulating factor. *Veterinary Immunology and Immunopathology* (in press).

Finkelman, F.D., Madden, K.B., Cheever, A.W., Katona, I.M., Morris, S.C., Gately, M.K., Hubbard, B.R., Gause, W.C. and Urban, J.F. (1994) Effects of interleukin 12 on immune responses and host protection in mice infected with intestinal nematode parasites. *Journal of Experimental Medicine* 179, 1563–1572.

Fiskerstrand, C. and Sargan, D. (1990) Nucleotide sequence of ovine interleukin-1β. *Nucleic Acids Research* 18, 7165.
Fiskerstrand, C.E., Roy, D.J., Green, I. and Sargan, D.R. (1992) Cloning, expression and characterization of ovine interleukins 1 α and β. *Cytokine* 4, 418–428.
Fu, P., Evans, B., Lim, G.B., Moritz, K. and Wintour, E.M. (1993) The sheep erythropoietin gene: molecular cloning and effect of hemorrhage on plasma erythropoietin and renal/liver messenger RNA in adult sheep. *Molecular and Cellular Endocrinology* 93, 107–116.
Goff, J.P., Naito, Y., Kehru, M.E., Hayes, P. and Daley, M. (1992) Physiologic effects of administration of interleukin-1β in cows. *American Journal of Veterinary Research* 53, 1983–1987.
Goodall, J.C., Emery, D.C., Perry, A.C., English, L.S. and Hall, L. (1990) cDNA cloning of ovine interleukin 2 by PCR. *Nucleic Acids Research* 18, 5883.
Graham, S.P., Jones, G.E., MacLean, M., Livingstone, M. and Entrican, G. (1995) Recombinant ovine interferon-γ inhibits the multiplication of *Chlamydia psittaci* in ovine cells. *Journal of Comparative Pathology* 112, 185–195.
Green, I.R. and Sargan, D.R. (1991) Sequence of the cDNA encoding ovine tumor necrosis factor-α: problems with cloning by inverse PCR. *Gene* 109, 203–210.
Green, I.R., Fiskerstrand, C., Bertoni, G., Roy, D.J., Peterhans, E. and Sargan, D.R. (1993) Expression and characterization of bioactive recombinant ovine TNF-α: some species specificity in cytotoxic response to TNF. *Cytokine* 5, 213–223.
Grun, J.L. and Maurer, P.H. (1989) Different T helper cell subsets elicited in mice utilising two different adjuvant vehicles: the role of endogenous interleukin-1 in proliferative responses. *Cellular Immunology* 121, 135–145.
Gupta, R.K. and Relyveld, E.H. (1991) Adverse reactions after injection of adsorbed diphtheria, pertussis, tetanus (DPT) vaccine are not due only to pertussis organisms or pertussis components of the vaccine. *Vaccine* 9, 699–702.
Gupta, R.K., Relyveld, E.H., Lindblad, E.B., Bizzini, B., Ben Efraim, S. and Gupta, C.K. (1993) Adjuvants – a balance between toxicity and adjuvanticity. *Vaccine* 11, 293–306.
Haig, D.M., Brown, D. and MacKellar, A. (1990) Ovine haemopoiesis: the development of bone marrow, derived colony, forming cells *in vitro* in the presence of factors derived from lymphoid cells and helper T-cells. *Veterinary Immunology and Immunopathology* 25, 125–137.
Haig, D., Entrican, G., Yirrell, D., Deane, D., Miller, H.R.P., Norval, M. and Reid, H.W. (1992) Differential appearance of interferon-γ and colony stimulating activity in afferent versus efferent lymph following orf virus infection of sheep. *Veterinary Dermatology* 3, 221–229.
Haig, D.M., Thomson, J. and Percival, A. (1994) The *in vitro* detection and quantitation of ovine bone marrow precursors of multipotential colony- forming cells. *Journal of Comparative Pathology* 111, 73–85.
Haig, D.M., Hutchison, G., Green, I., Sargan, D. and Reid, H.W. (1995a) The effect of intradermal injection of GM-CSF and TNF-α on the accumulation of dendritic cells in ovine skin. *Veterinary Dermatology* 6, 211–220.
Haig, D.M., Percival, A., Mitchell, J., Green, I. and Sargan, D. (1995b) The survival and growth of ovine afferent lymph dendritic cells in culture depends on tumour necrosis factor-α and is enhanced by granulocyte-macrophage colony-

stimulating factor but inhibited by interferon-γ. *Veterinary Immunology and Immunopathology* 45, 221–236.

Haig, D., Deane, D.L., Myatt, N., Thomson, J., Entrican, G., Rothel, J. and Reid, H.W. (1996a) The activation status of ovine CD45R$^+$ and CD45R$^-$ efferent lymph T-cells after orf virus reinfection. *Journal of Comparative Pathology* 115, 163–174.

Haig, D., Deane, D., Percival, A., Myatt, N., Thomson, J., Inglis, L., Rothel, J., Seow, H-F., Wood, P., Miller, H.R.P. and Reid, H.W. (1996b) The cytokine response of afferent lymph following orf virus reinfection of sheep. *Veterinary Dermatology* 7, 11–20.

Haig, D.M., McInnes, C., Deane, D., Lear, A., Myatt, N., Reid, H.W., Rothel, J., Seow, H-F., Wood, P.R., Lyttle, D. and Mercer, A. (1996c) Cytokines and their inhibitors in orf virus infection. *Veterinary Immunology and Immunopathology* 54, 261–267.

Haig, D.M., McInnes, C.J., Deane, D., Reid, H.W. and Mercer, A. (1997) The immune and inflammatory response to orf virus. *Comparative Immunology, Microbiology and Infectious Diseases* (in press).

Hansen, P.J. (1995) Interactions between the immune system and the ruminant conceptus. *Journal of Reproduction and Fertility* Suppl. 49, 69–82.

Heath A.W. and Playfair J.H.L. (1991) Effects of added cytokines on immune response and memory. In: Gregoriadis, G. (ed.), *Vaccines*. Plenum Press, New York, pp. 77–84.

Hughes, H.P., Campos, M., Godson, D.L., van Drunen Littel van den Hurk, S., McDougall, L., Rapin, N., Zamb, T. and Babiuk, L.A. (1991) Immunopotentiation of bovine herpes virus subunit vaccination by interleukin-2. *Immunology* 74, 461–466.

Hughes, H.P., Campos, M., van Drunen Littel van den Hurk, S., Zamb, T., Sordillo, L.M., Godson, D. and Babiuk, L.A. (1992) Multiple administration with interleukin-2 potentiates antigen-specific responses to subunit vaccination with bovine herpesvirus-1 glycoprotein IV. *Vaccine* 10, 226–230.

Innes, E.A., Panton, W.R.M., Marks, J., Trees, A.J., Holmdahl, J. and Buxton, D. (1995a) Interferon-γ inhibits the intracellular multiplication of *Neospora caninum*, as shown by incorporation of H-3uracil. *Journal of Comparative Pathology* 113, 95–100.

Innes, E.A., Panton, W.R.M., Thomson, K.M., Maley, S. and Buxton, D. (1995b) Kinetics of interferon-γ production *in vivo* during infection with the s48 vaccine strain of *Toxoplasma gondii*. *Journal of Comparative Pathology* 113, 89–94.

Innes, E.A., Panton, W.R.M., Sanderson, A., Thomson, K.T., Wastling, J.M., Maley, S.M. and Buxton, D. (1995c) Induction of CD4$^+$ and CD8$^+$ T-cell responses in efferent lymph responding to *Toxoplasma gondii* infection: analysis of phenotype and function. *Parasite Immunology* 17, 151–160.

Kalaaji, A.N., Mccullough, K. and Hay, J.B. (1989) The enhancement of lymphocyte localization in skin sites of sheep by tumor necrosis factor-α. *Immunology Letters* 23, 143–147.

Klekamp, J., Jarzecka, K., Redmond, N., Obermeier, S.K., Hoover, R.L., Summar, M.L. and Perkett, E.A. (1995) Vascular endothelial-cell growth-factor (vegf), cloning of ovine cdna – constitutive expression in pulmonary-artery smooth-muscle cells with regulation by hypoxia and dexamethasone (dex). *Pediatric Research* 37, A 338.

Lear, A., Hutchison, G., Reid, H.W., Norval, M. and Haig, D.M. (1996) Phenotypic characterisation of the dendritic cells accumulating in the ovine dermis following primary and secondary orf virus infections. *European Journal of Dermatology* 6, 135–140.

Legastelois, I., Greenland, T., Arnaud, P., Mornex, J.F. and Cordier, G. (1994) Sequencing of the ovine interleukin-8-encoding cDNA using the polymerase chain reaction. *Gene* 150, 369–369.

Lise, L. and Audibert, F. (1989) Immunoadjuvants and analogues of immunomodulatory bacterial structures. *Current Opinions in Immunology* 2, 269–274.

Lofthouse, S.A., Andrews, A.E., Nash, A.D. and Bowles, V.M. (1995a) Humoral and cellular-responses induced by intradermally administered cytokine and conventional adjuvants. *Vaccine* 13, 1131–1137.

Lofthouse, S.A., Andrews, A.E., Barcham, G.J. and Nash, A.D. (1995b) Parameters related to the application of recombinant ovine interleukin-1β as an adjuvant. *Vaccine* 13, 1277–1287.

Lofthouse, S.A., Andrews, A.E., Elhay, M.J., Bowles, V.M., Meeusen, E.N.T. and Nash, A.D. (1996) Cytokines as adjuvants for ruminant vaccines. *International Journal for Parasitology* (in press).

MacKay, C.R., Marston, W. and Dudler, L. (1990) Naive and memory T cells show distinct pathways of lymphocyte recirculation. *Journal of Experimental Medicine* 171, 801–817.

Martin, H., Nash, A., Brandon, M. and Andrews, A. (1993) The cloning and characterization of ovine interleukin-10. *Journal of Leukocyte Biology* 60.

Martin, H.M., Nash, A.D. and Andrews, A.E. (1995) Cloning and characterisation of an ovine interleukin-10-encoding cDNA. *Gene* 159, 187–191.

McInnes, C.J. and Haig, D.M. (1991) Cloning and expression of a cDNA encoding ovine granulocyte-macrophage colony-stimulating factor. *Gene* 105, 275–279.

McInnes, C.J., Logan, M., Redmond, J., Entrican, G. and Baird, G.D. (1990) The molecular cloning of the ovine γ-interferon cDNA using the polymerase chain reaction. *Nucleic Acids Research* 18, 4012.

McInnes, C., Haig, D. and Logan, M. (1993) The cloning and expression of the gene for ovine interleukin-3 (multi, CSF) and a comparison of the *in vitro* hematopoietic activity of ovine IL-3 with ovine GM-CSF and human M-CSF. *Experimental Hematology* 21, 1528–1534.

McInnes, C.J., Logan, M., Haig, D. and Wright, F. (1994) Cloning of a cDNA encoding ovine interleukin-3. *Gene* 139, 289–290.

Moncada, S. and Higgs, E.A. (eds) (1990) *Nitric Oxide from L-Arginine: A Bioregulatory System*. Elsevier Science Publishers, New York.

Mosmann, T. and Coffman, R. (1989) Th1 and Th2 cells: different patterns of lymphokine secretion lead to different functional properties. *Annual Review of Immunology* 7, 145–173.

Munn, E.A., Smith, T.S., Graham, M., Tavernor, A.S. and Greenwood, C.A. (1993) The potential value of integral membrane proteins in the vaccination of lambs against *Haemonchus contortus*. *International Journal of Parasitology* 23, 261–269.

Murray, H.W., Szuro-Sudol, A., Wellner, D., Oca, M.J., Granger, A.M., Libby, D.M., Rothermel, C.D. and Rubin, B.Y. (1989) Role of tryptophan degradation in respiratory burst-independent antimicrobial activity of gamma interferon-stimulated human macrophages. *Infection and Immunity* 57, 845–849.

Nash, A.D., Barcham, G.J., Brandon, M.R. and Andrews, A.E. (1991) Molecular cloning, expression and characterization of ovine TNF-α. *Immunology and Cell Biology* 69, 273–283.

Nash, A.D., Barcham, G.J., Andrews, A.E. and Brandon, M.R. (1992a) Characterisation of ovine alveolar macrophages: regulation of surface antigen expression and cytokine production. *Veterinary Immunology and Immunopathology* 31, 77–94.

Nash, A.D., Brandon, M.R. and Bello, P.A. (1992b) Effects of tumour necrosis factor-α on growth hormone and interleukin 6 mRNA in ovine pituitary cells. *Molecular and Cellular Endocrinology* 84, R31–37.

Nash A.D., Lofthouse S.A., Barcham G.J., Jacobs H.J., Ashman K., Meeusen E.N.T., Brandon M.R. and Andrews A.E. (1993) Recombinant cytokines as immunological adjuvants. *Immunology and Cell Biology* 71, 367–379.

Nash, A., Andrews, A., Martin, H., Colditz, I., Wood, P. and Seow H-F. (1996) Methods for detection and assay of ovine cytokines. In: *Manual of Immunological Methods*. Academic Press (in press).

Nunberg, J.H., Doyle, M.V., York, S.M. and York, C.J. (1989) Interleukin-2 acts as an adjuvant to increase the potency of inactivated rabies virus vaccine. *Proceedings of the National Academy of Science, USA* 86, 4240–4243.

O'Brien, P.M., Rothel, J.S., Seow, H.F. and Wood, P.R. (1991) Cloning and sequencing of the cDNA for ovine granulocyte-macrophage colony-stimulating factor (GM-CSF), *Immunology and Cell Biology* 69, 51–55.

O'Brien, P.M., Seow, H.F., Rothel, J.S. and Wood, P.R, (1994) Cloning and sequencing of an ovine granulocyte colony-stimulating factor cDNA. *DNA Sequence* 4, 339–342.

Oura, C.A., Innes, E.A., Wastling, J.M., Entrican, G. and Panton, W.R. (1993) The inhibitory effect of ovine recombinant interferon-γ on intracellular replication of *Toxoplasma gondii*. *Parasite Immunology* 15, 535–538.

Pauli, U., Bertoni, G., Duerr, M. and Peterhans, E. (1994) A bioassay for the detection of tumor necrosis factor from eight different species: evaluation of neutralization rates of a monoclonal antibody against human TNF-α. *Journal of Immunological Methods* 171, 263–265.

Playfair, J.H.L. and deSouza, J.B. (1987) Recombinant gamma interferon is a potent adjuvant for a malaria vaccine in mice. *Clinical and Experimental Immunology* 67, 5–10.

Radford, A.J., Hodgson, A.L.M., Rothel, J.S. and Wood, P.R. (1991) Cloning and sequencing of the ovine γ-interferon gene. *Australian Veterinary Journal* 68, 82–84.

Reddy, D.N., Reddy, P.G., Minocha, H.C., Fenwick, B.W., Baker, P.E., Davis, W.C. and Blecha, F. (1990) Adjuvanticity of recombinant interleukin-1β: influence on immunity, infection and latency in a bovine herpesvirus-1 infection. *Lymphokine Research* 9, 295–307.

Reddy, D.N., Reddy, P.G., Xue, W., Minocha, H.C., Daley, M.J. and Blecha, F. (1993) Immunopotentiation of bovine respiratory disease virus vaccines by

interleukin-1β and interleukin-2. *Veterinary Immunology and Immunopathology* 37, 25–38.
Romagnani, S. (1992) Induction of TH1 and TH2 responses: a key role for the 'natural' immune response? *Immunology Today* 13, 379–381.
Rothel, J.S., Jones, S.L., Corner, L.A., Cox, J.C. and Wood, P.R. (1990) A sandwich enzyme immunoassay for bovine interferon-γ and its use for the detection of tuberculosis in cattle. *Australian Veterinary Journal* 67, 134–137.
Seow, H.-F., Rothel, J.S., Radford, A.J. and Wood, P.R. (1990) The molecular cloning of ovine interleukin 2 gene by the polymerase chain reaction. *Nucleic Acids Research* 18, 7175.
Seow, H.-F., Rothel, J.S., David, M.J. and Wood, P.R. (1991) Nucleotide sequence of ovine macrophage interleukin-1β cDNA. *DNA Sequence* 1, 423–426.
Seow, H.-F., Rothel, J.S. and Wood, P.R. (1993) Cloning and sequencing an ovine interleukin-4, encoding cDNA. *Gene* 124, 291–293.
Seow, H.-F., Rothel, J.S. and Wood, P.R. (1994a) Expression and purification of recombinant ovine interleukin-1β from *Escherichia coli*. *Veterinary Immunology and Immunopathology* 41, 229–239.
Seow, H.-F., Yoshimura, T., Wood, P.R. and Colditz, I.G. (1994b) Cloning, sequencing, expression and inflammatory activity in skin of ovine interleukin-8. *Immunology and Cell Biology* 72, 398–405.
Seow, H.-F., Rothel, J.S., Pepin, M., David, M.J. and Wood, P.R. (1995) Expression, biological activity and kinetics of production of recombinant ovine TNF-α. *Veterinary Immunology and Immunopathology* 44, 279–291.
Seow, H.-F., David, M-J., McWaters, P., Hurst, L. and Wood, P.R. (1996a) Cloning and sequencing of an ovine interleukin-5 cDNA. *DNA Sequence* 6, 331–335.
Seow, H.-F., Macreadie, I.G., Mucha, M.-J., Hurst, L., Rothel, J.S. and Wood, P.R. (1996b) Comparison of biological activity of recombinant ovine interleukin-4 expressed from *Escherichia coli* and *Saccharomyces cerevisiae*. *Protein Expression and Purification* (in press).
Seow, H.-F., Mucha, M.-J., Hurst, L., Rothel, J.S. and Wood, P.R. (1996c) Expression of ovine interleukin-2 cDNA in *Escherichia coli*. *Veterinary Immunology and Immunopathology* (in press).
Sloane, P.J., Elsasser, T.H., Spath, J.A., Albertine, K.H. and Gee, M.H. (1992) Plasma tumor necrosis factor-α during long-term endotoxemia in awake sheep. *Journal of Applied Physiology* 73, 1831–1837.
Staruch, M.W. and Wood, D.D. (1983) The adjuvanticity of interleukin-1 *in vivo*. *Journal of Immunology* 130, 2191–2194.
Sterne, M., Batty, I., Thomson, A. and Robertson, J.M. (1962) Immunisation of sheep with multicomponent clostridial vaccines. *Veterinary Record* 74, 909–913.
Stevenson, L.M. and Jones, D.G. (1994) Cross-reactivity amongst recombinant haematopoietic cytokines from different species for sheep bone-marrow eosinophils. *Journal of Comparative Pathology* 111, 99–106.
Tavernor, A.S., Smith, T.S., Langford, C.F., Munn, E.A. and Graham, M. (1992) Vaccination of young Dorset lambs against haemonchosis. *Parasite Immunology* 14, 645–655.

Thatcher, W.W., Meyer, M.D. and Danet Desnoyers, G. (1995) Maternal recognition of pregnancy. *Journal of Reproduction and Fertility*, Supplement 49, 15–28.

Thomas, S.M., Garrity, L.F., Brandt, C.R., Schobert, C.S., Feng, G-S., Taylor, M.W., Carlin, J.M. and Byrne, G.I. (1993) IFN-γ-mediated antimicrobial response. Indoleamine 2,3-dioxygenase-deficient mutant host cells no longer inhibit intracellular *Chlamydia* spp. or *Toxoplasma* growth. *Journal of Immunology* 150, 5529–5534.

Thomas, S.R., Mohr, D. and Stocker, R. (1994) Nitric oxide inhibits indoleamine 2,3-dioxygenase activity in interferon-γ primed mononuclear phagocytes. *Journal of Biological Chemistry* 269, 14457–14464.

Webster, A.C. and Frank, C.L. (1985) Comparison of immune response stimulated sheep, rabbits and guinea pigs by the administration of multi-component clostridial vaccines. *Australian Veterinary Journal* 62, 112–114.

Weinberg, A. and Merigan, T.C. (1988) Recombinant interleukin-2 as an adjuvant for vaccine-induced protection. *Journal of Immunology,* 140, 294–299.

Whaley, A.E., Carroll, R.S. and Imakawa, K. (1991) Cloning and analysis of a gene encoding ovine interferon α-II. *Gene* 106, 281–282.

Woodall, C.J., Mclaren, L.J. and Watt, N.J. (1994) Sequence and chromosomal localization of the gene encoding ovine latent transforming growth-factor-β-1. *Gene* 150, 371–373.

Wynn, T.A., Cheever, A.W., Jankovic, D., Poindexter, R.W., Caspar, P., Lewis, F.A. and Sher, A. (1995) An IL-12-based vaccination method for preventing fibrosis induced by schistosome infection. *Nature* 376, 594–596.

Young, A.J., Hay, J.B. and Chan, J.Y. (1990) Primary structure of ovine tumor necrosis factor α cDNA. *Nucleic Acids Research* 18, 6723.

Zhu, D.M. and Ott, T.L. (1994) Development of an enzyme-linked-immunosorbent-assay (ELISA) for ovine interferon-γ (oifnγ). *Biology of Reproduction* 50, 176.

Zimmerman, G.A., Prescott, S.M. and McIntyre, T.M. (1992) Endothelial cell interactions with granulocytes: tethering and signaling molecules. *Immunology Today* 13, 93–100.

III

Porcine Cytokines

Edited by F. Blecha

Cytokine Applications in Pigs 9

F. Blecha

Department of Anatomy and Physiology, College of Veterinary Medicine, 1600 Denison Avenue, Kansas State University, Manhattan, Kansas 66506-5602, USA

INTRODUCTION

Development and regulation of an immune response is an extremely complex and complicated event involving many different organ systems and cell types of the body. Orchestration of this event is conducted chiefly by cytokines, the protein mediators of the immune system (Wood *et al.*, 1994a). Over the last several years there has been a tremendous increase in understanding how cytokines influence and regulate immunity (Myers and Murtaugh, 1995). Although this knowledge may be incomplete, the availability of purified cytokines has fostered studies designed to devise means of intervening in the regulation of the immune system, with the ultimate goal of enhancing immunity.

Several porcine cytokines have been cloned and expressed (Murtaugh, 1994; Wood *et al.*, 1994a; see Vandenbroeck and Billiau, Chapter 10 in this volume). Unfortunately, studies evaluating the *in vivo* use of porcine cytokines in pigs are still very limited. However, some insight into the application of porcine cytokines can be gained from experiments that have used bovine and human cytokines in pigs. Earlier reviews have discussed the potential and practical use of cytokines in domestic food animals, albeit primarily cattle (Lawman *et al.*, 1989; Bielefeldt-Ohmann and Martinod, 1990; Blecha, 1990a, b, 1991; Kehrli *et al.*, 1991; Bielefeldt-Ohmann, 1995; Sordillo and Daley, 1995). This chapter will focus on the *in vivo* use of cytokines in pigs.

WHY USE CYTOKINES IN PIGS?

The rationale for using cytokines in pigs generally encompasses one of two areas: immunoenhancement of non-specific immune responses and augmentation of specific immunity (Blecha and Charley, 1990). Because of their important regulatory function, cytokines potentially can be used to augment immunity during various periods in the production of pigs when immune function or maturation may be low. Studies have shown that stress-induced increases in cortisol are related to lower lymphocyte proliferative responses in young pigs (Brown-Borg *et al.*, 1993) and the stress of weaning decreases the potential for porcine splenocytes to produce IL-2 (Bailey *et al.*, 1992). Furthermore, some immune cell functions, such as NK cell activity (Huh *et al.*, 1981; Cepica and Derbyshire, 1984; Yang and Schultz, 1986), lymphocyte proliferative responses (Becker and Misfeldt, 1993), and IFN-α production (Nowacki *et al.*, 1993) are low, if not absent, in neonatal pigs. Strategies to increase these immune cell responses in young pigs with cytokines or cytokine-inducing therapies may be possible.

Vaccination programmes have increased the profitability of livestock production (Wood *et al.*, 1994b); however, large economic losses still occur from disease even in animals that have been vaccinated. Cytokines have been used successfully as vaccine adjuvants in laboratory animals (Heath and Playfair, 1992) and may prove to be an effective means of increasing vaccine efficacy in pigs (Blecha, 1991). Collectively, these studies produce expectations that cytokine prophylaxis and/or therapy may be effective in pigs.

CYTOKINE AUGMENTATION OF NON-SPECIFIC IMMUNITY IN PIGS

Four cytokines have been used to investigate the capability of increasing non-specific immune responses or disease resistance in pigs: IFN-α, IFN-γ, IL-1 and IL-2. In addition, the physiological influence of administering proinflammatory cytokines, such as IL-1, IL-6 and TNF-α to pigs has been studied.

Interferons

Antiviral activity in pigs infected with transmissible gastroenteritis virus (TGEV) is due to IFN-α (La Bonnardière and Laude, 1981; Charley and Laude, 1988; Laude *et al.*, 1992; Naidoo and Derbyshire, 1992). Consequently, a number of studies have been conducted to investigate the possibility of using IFN or IFN-inducing agents prophylactically or therapeutically in pigs. Loewen and Derbyshire (1988) have shown that a synthetic IFN inducer, poly (I:C) complexed with poly-L-lysine and carboxymethyl cellulose (poly ICLC), delayed the onset of clinical signs of TGEV in newborn pigs when administered before oral exposure to the virus. The authors attributed this protective effect of poly ICLC to the activation of NK cells (Lesnick and Derbyshire, 1988). This research group has also shown that poly ICLC reduced virus yield when administered parenterally 6 h before TGEV inoculation into ligated intestinal segments of neonatal pigs (Jordan and Derbyshire, 1994). However, administering natural porcine IFN-α or rHuIFN-α2a at the same time as TGEV inoculation did not reduce virus yield in the ligated intestinal segments. Similarly, rPoIFN-α and rPoIFN-γ have been shown to decrease mortality and reduce clinical signs of *Actinobacillus pleuropneumoniae* infection in pigs when administered 24 h before challenge (Campos *et al.*, 1989). Cummins *et al.* (1995) evaluated the efficacy of oral administration of low-dose natural human IFN-α to neonatal pigs that were naturally infected with TGEV. Pigs were treated orally for 4 consecutive days with 1, 10 or 20 IU of human IFN-α. When treatment was initiated within a few hours of birth, oral administration of human IFN-α had no influence on pig survival. However, when treatment was initiated between 1 and 12 days of age, oral administration of human IFN-α significantly increased survival rates during a natural TGEV infection when compared to placebo-treated controls. Collectively, these findings suggest that IFN may be useful prophylactically and that low-dose oral IFN administration may be effective therapeutically.

In vivo use of IFN in pigs requires some understanding of the pharmacokinetics of this cytokine. *In vivo* administration of porcine recombinant IFN-α-1 has been evaluated in newborn and adult pigs (La Bonnardière *et al.*, 1990; Lefevre *et al.*, 1990). Five-day-old pigs injected i.m. with 1×10^6 IU kg^{-1} of IFN-α1 did not display any untoward effects of cytokine administration, including a febrile response and alterations in weight gain. Serum concentrations of IFN-α1 reached a maximum of 1700 IU ml^{-1} within 1 h of injection and had an apparent half-life of about 3.5 h. Adult pigs (80 kg) injected with IFN-α1 (5×10^5 and 1×10^6 IU kg^{-1}) differed from neonatal pigs in two aspects. Whereas the maximum concentration of IFN-α1 in serum and the time required to achieve this concentration were similar to the neonatal pig values, the lower dosage of IFN-α1 had an apparent half-life about twice as long in the adult pigs. In addition, adult pigs injected with the higher dosage of IFN-α1 displayed a significant but transient period of hyperthermia.

Because glucocorticoids are associated with stress and increased susceptibility to disease, glucocorticoid immunosuppression models are often used to evaluate cytokine immunomodulation (Roth and Flaming, 1990). Saulnier *et al.* (1991) evaluated the immunomodulatory effects of rPoIFN-γ in glucocorticoid immunosuppressed pigs. Pigs were immunosuppressed with dexamethasone and treated i.m. with rPoIFN-γ at 1 μg kg^{-1} (1.12×10^6 IU mg^{-1}) on 2 consecutive days. Treatment with IFN-γ reversed the dexamethasone-induced suppression in lymphocyte proliferative responses to phytohaemagglutinin and IL-2 production. In addition, IL-1 production was increased in pigs treated with rPoIFN-γ. These findings suggest that IFN-γ therapy may be effective in immunosuppressed pigs.

Interleukin-1

Binns *et al.* (1992) characterized the tissue infiltration of leukocytes that occurs in pigs after injecting IL-1 into skin. Recombinant human IL-1α induced dose-dependent lymphocyte traffic in pig skin that peaked at 4 h. These investigators extended their findings by evaluating cytokine-induced changes in endothelial cell function (Keelan *et al.*, 1994; Woolley *et al.*, 1995). Intravenous administration of rHuIL-1α (5 $\mu g\ kg^{-1}$ bolus followed by 50 $ng\ kg^{-1}\ min^{-1}$ for 2 h) induced the expression of E-selectin, an endothelial-cell adhesion molecule for leukocytes, in all tissues examined (Keelan *et al.*, 1994). In addition, skin sites injected with IL-1α displayed increased E-selectin expression. These studies suggest that the pig will be a useful model for investigating cytokine-induced changes in endothelial cell function *in vivo*.

Enhancement of non-specific antibacterial resistance by IL-1 is well documented (Czuprynski and Brown, 1987; Czuprynski *et al.*, 1988; Kurtz *et al.*, 1989; McIntyre *et al.*, 1989). Because neutrophil functions have been shown to be lower in neonatal pigs (El-Awar and Hahn, 1991; Markowsda-Daniel *et al.*, 1992), Shi *et al.* (1994) conducted two experiments to determine if IL-1 influenced non-specific immunity and disease resistance in young pigs. Although porcine IL-1α and IL-1β have been cloned and expressed (Maliszewski *et al.*, 1990; Huether *et al.*, 1993; Vandenbroeck *et al.*, 1993b) they are not available in sufficient quantities for *in vivo* experiments. However, porcine IL-1β and bovine IL-1β have a high degree of similarity (Huether *et al.*, 1993), and preliminary studies have shown that rBoIL-1β is biologically active in pigs (Blecha *et al.*, 1993).

Intramuscular injections of rBoIL-1β (5 or 10 $\mu g\ kg^{-1}$) caused a transient increase in rectal temperatures in 9- to 11-day-old pigs (Fig. 9.1). No other untoward effects of IL-1β injection were observed. Similar to other studies that have shown that this cytokine increases circulating leukocyte numbers (Ulich *et al.*, 1987), IL-1 injections in pigs caused an increase in both mature and immature neutrophils (Fig. 9.2), which probably relates to the ability of this cytokine to induce the synthesis of CSF (Mochizuki *et al.*, 1987; Bagby, 1989). Neutrophil bactericidal activity and ADCC were greater in IL-1-treated pigs than in saline-injected controls (Shi *et al.*, 1994). However, NK cell activity and neutrophil production of superoxide anion were not affected by treatment with IL-1. When pigs were challenged with *Streptococcus suis* at 19 days of age, the severity of the streptococcal infection was less in pigs that were treated with IL-1 at 9 to 11 days of age.

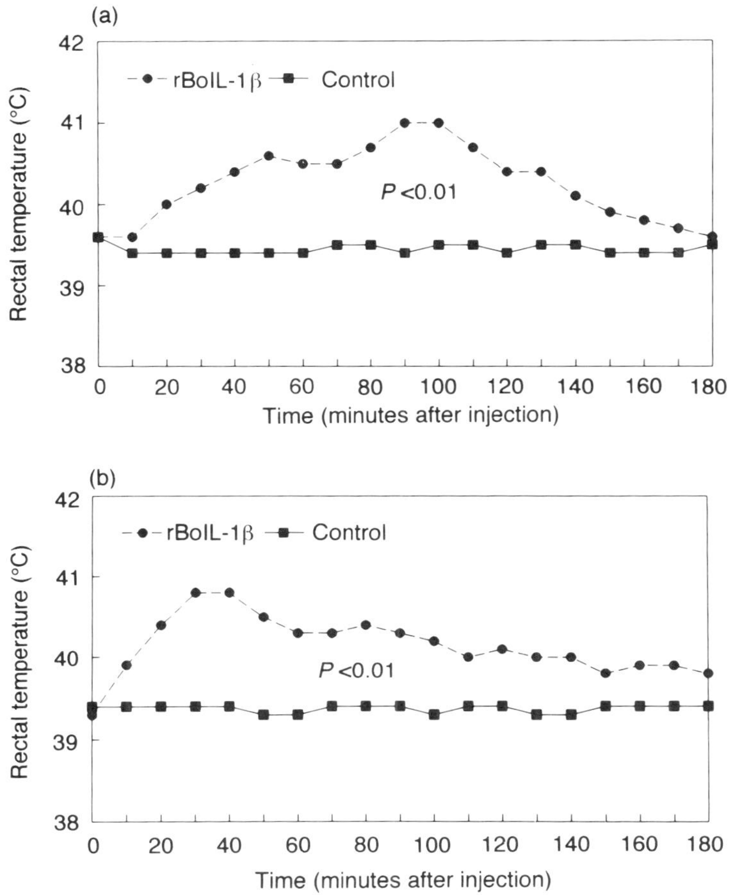

Fig. 9.1. Rectal temperatures of neonatal pigs after an i.m. injection of rBoIL-1β. Panel A: 9-day-old pigs were injected with 5 μg kg^{-1} rBoIL-1β or physiological saline (control); means (n = 5) are different ($P < 0.01$) between 20 and 160 min p.i. Panel B: 11-day-old pigs were injected with 10 μg kg^{-1} rBoIL-1β or physiological saline (control); means (n = 5) are different ($P < 0.01$) between 20 and 140 min p.i. For both 9- and 11-day-old pigs, SEM for control and rBoIL-1β rectal temperature means were ≤ 0.1% and ≤ 0.5% of the mean, respectively. Reprinted from Shi *et al.* (1994) with permission.

These findings suggest that IL-1 may induce granulopoiesis, selectively enhance neutrophil antimicrobial activity and increase resistance to streptococcal infection in neonatal pigs. Thus, defects in non-specific immunity associated with age or early weaning in neonatal pigs may be improved by administration of IL-1.

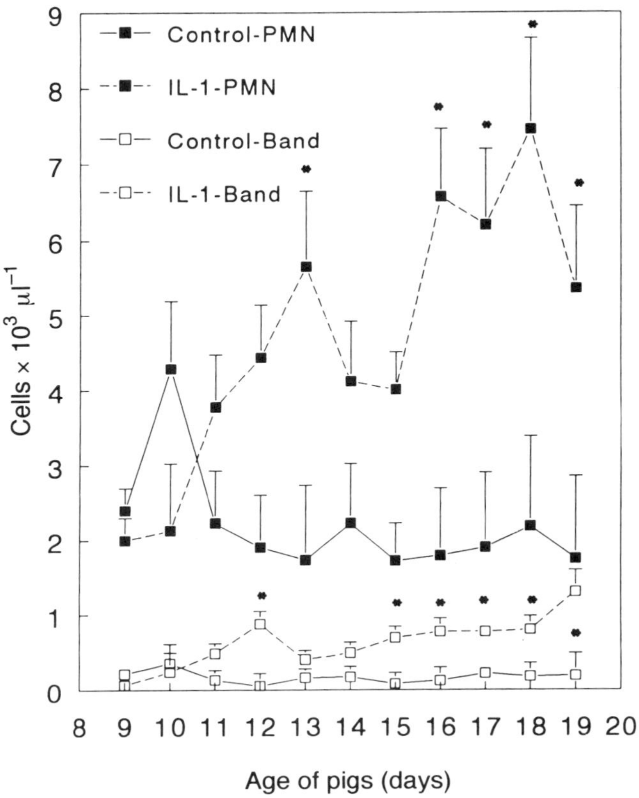

Fig. 9.2. Number of mature (PMN) and immature (Band) neutrophils in peripheral blood of IL-1-treated or control pigs. rBoIL-1β or physiological saline (control) was administered on days 9, 10 and 11. Values are means ±SEM, $n = 5$, *$P < 0.05$. Reprinted from Shi *et al.* (1994) with permission.

Interleukin-2

Several *in vitro* studies have shown that rHuIL-2 augments porcine lymphocyte proliferative responses (Fong and Doyle, 1986; Stott *et al.*, 1986; Bhagyam *et al.*, 1988; Nakajima *et al.*, 1992) and NK cell activity (Charley and Fradelizi, 1987; Bhagyam *et al.*, 1988; Knoblock and Canning, 1992; Nakajima *et al.*, 1992). These findings provided the rationale to determine if *in vivo* administration of rHuIL-2 augments porcine NK cell cytotoxicity. Indeed, two studies have shown that injections of rHuIL-2 increase *in vivo* porcine NK cell activity. Hennessy *et al.* (1990) injected 12 4-week-old pigs with either 10,000 or 100,000 IU kg^{-1} of rHuIL-2 i.m. for 5 consecutive days and Nakajima *et al.* (1992) injected four 12 to 25 kg pigs with either

20,000 or 100,000 IU kg^{-1} of rHuIL-2, i.v. twice daily for 4 days. Both studies showed clear increases in porcine NK cell activity. In addition, Blecha *et al.* (1995) found that pigs injected for 3 consecutive days with rBoIL-2 had increased porcine NK cell activity. Whether IL-2-enhanced NK cell activity increases disease susceptibility in pigs is not known. However, these findings suggest that low NK cell activity may be increased in neonatal pigs by administration of IL-2. Furthermore, recent studies by Gianello *et al.* (1994) suggest that pigs could be used as a preclinical model for studies of the effects of IL-2 on graft-versus-host disease.

Interleukin-6

Researchers at the University of Minnesota have expressed rPoIL-6 in *Escherichia coli* and evaluated the influence of this cytokine on several physiological responses in pigs (Scamurra, 1995). Daily treatment with IL-6 increased serum concentrations of the acute-phase protein, haptoglobin, and decreased concentrations of serum zinc. These responses are characteristic of the proinflammatory properties of this multifunctional cytokine. However, *in vivo* treatment of pigs with rPoIL-6 did not affect body temperature or serum concentrations of adrenocorticotropic hormone.

Tumour Necrosis Factor-α

Proinflammatory cytokines, such as TNF-α, influence adhesion molecule expression and lymphocyte trafficking. Pigs injected with rBoTNF-α displayed an increased expression of the adhesion molecule, E-selectin (Keelan *et al.*, 1994) and lymphocytes that responded to this cytokine were primarily $CD2^+$ (Woolley *et al.*, 1995).

Many physiological and behavioural effects of inflammation are regulated by the central expression of cytokines in the central nervous system (CNS). Megyeri *et al.* (1992) have shown that the permeability of the blood–brain barrier in newborn pigs was increased by intracisternal injection of rHuTNF-α and Warren *et al.* (1995a) have shown that mRNA encoding TNF-α is present in the pig brain. Furthermore, pigs injected intracerebroventricularly with rPoTNF-α display behavioural, metabolic and neuroendocrine responses similar to those of pigs that were injected peripherally with endotoxin (Warren *et al.*, 1995a, b). These findings support the concept that proinflammatory cytokines, including TNF-α, cause many of the behavioural and physiological responses of inflammation in pigs.

USE OF CYTOKINES AS VACCINE ADJUVANTS IN PIGS

Adjuvants enhance immune responses by influencing antigen processing, presentation and elimination, and immunomodulation through the production of cytokines (Dalsgaard *et al.*, 1990; Wood *et al.*, 1994a). A logical extension of this knowledge is the use of recombinant cytokines as adjuvants. Several studies have documented the potential of cytokines as adjuvants in animals (reviewed by Heath and Playfair) including domestic food animals (Blecha, 1991; Wood *et al.*, 1994a).

Studies that have used porcine cytokines as adjuvants are sparse. Scamurra (1995) found that daily injections of rPoIL-6 increased the secondary antibody response to keyhole limpet haemocyanin; no effect was observed on the primary response. However, several earlier studies have shown that heterologous cytokines, specifically, human IL-2, can be used as vaccine adjuvants in pigs (Anderson *et al.*, 1987; Nunberg *et al.*, 1988; Kawashima and Platt, 1989; Hennessy *et al.*, 1990). Because rBoIL-1β and rBoIL-2 have been shown to be effective adjuvants in cattle (Reddy *et al.*, 1989, 1990, 1993) and because the porcine IL-1β nucleotide sequence is 82% similar to the bovine sequence in the protein coding region (Huether *et al.*, 1993), Blecha *et al.* (1995) reasoned that rBoIL-1β and rBoIL-2 might be effective adjuvants in pigs.

Recombinant bovine IL-1β and rBoIL-2 were administered in conjunction with a single *Streptococcus suis* vaccination in pigs (Blecha *et al.*, 1995). Sixty 4-week-old pigs were allotted to eight groups: non-vaccinated controls; vaccinated controls; rBoIL-1β at 0.1, 1.0 and 10 μg kg^{-1}; and rBoIL-2 at 2.5, 25 and 250 μg kg^{-1}. At vaccination, pigs were injected i.m. with their respective cytokine treatments. Pigs received additional cytokine injections on 2 consecutive days. The highest dosage of rBoIL-1β caused profound physiological effects, including vomiting and lethargy, and some pigs displayed CNS disturbances. However, when pigs were challenged with *S. suis*, this dosage of rBoIL-1β was protective. Pigs administered rBoIL-1β at 10 μg kg^{-1} had higher antibody responses to *S. suis*, which included increased responses against *S. suis* antigens that have been associated with protection (Fig. 9.3). In addition, pigs injected with 10 μg kg^{-1} of rBoIL-1β had less severe clinical signs of the disease, better growth performance during the infection, and less severe gross pathological lesions caused by the bacteria. No pigs in this treatment group died from the bacterial challenge. These data show that rBoIL-1β (10 μg kg^{-1}), administered i.m. for 3 consecutive days at vaccination, is more effective than the *S. suis* vaccine alone in protecting pigs against a *S. suis* challenge. However, because of the adverse reaction to the cytokine at the time of administration, 10 μg kg^{-1} of rBoIL-1β cannot be administered to 4-week-old pigs. Clearly, it would be beneficial to find a dosage of rBoIL-1β that produced the same positive results as the highest dose of the cytokine but without the adverse effects at the time of administration. Furthermore, it is possible that porcine IL-1β may be more effective in pigs than rBoIL-1β at a dosage that does not cause adverse effects. Considering the encouraging results of this study, these possibilities should be explored.

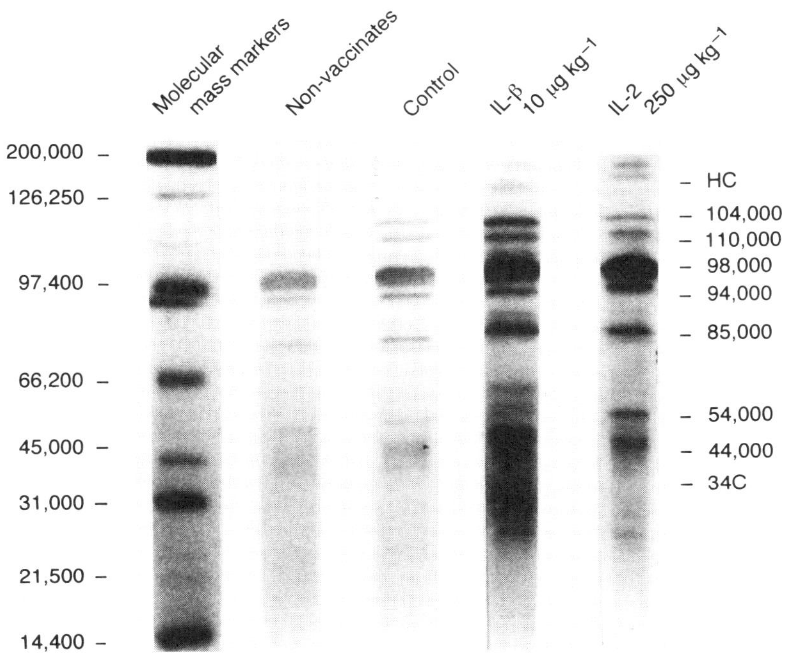

Fig. 9.3. Western blots of *Streptococcus suis* (serotype 2) antigens probed with serum obtained from pigs vaccinated against *S. suis* using IL-1 and IL-2 as adjuvants. Pigs were vaccinated with a commercial *S. suis* (serotypes 1 and 2) vaccine and injected i.m. with rBoIL-1β (10 μg kg^{-1}) or rBoIL-2 (250 μg kg^{-1}) for 3 consecutive days at vaccination. Controls were vaccinated but did not receive IL-1 or IL-2. Molecular mass markers are shown in the left lane. Numbers on the right are molecular mass of the indicated bands; HC = high molecular mass complex, 34C = 34 kDa complex. The 94 kDa protein has been associated with protection from *S. suis* challenge (Holt *et al.*, 1990). Reprinted from Blecha *et al.* (1995) with permission.

CONCLUSIONS

Cytokines, such as IFN-α, IFN-γ, IL-1 and IL-2, have been shown to influence both non-specific and specific porcine immune responses. However, to date, few porcine cytokines have been used *in vivo* in pigs. Given the encouraging findings from studies that have used heterologous cytokines in pigs, it is likely that the use of porcine cytokines may overcome some of the adverse effects that are observed when heterologous cytokines are used. Furthermore, advances in cytokine delivery, such as slow-release devices and delivery via viral vectors, will undoubtedly facilitate the use of porcine cytokines in pigs.

Recent Progress in the Molecular Characterization of Porcine Cytokines

10

K. Vandenbroeck and A. Billiau

Laboratory of Immunobiology, Rega Institute for Medical Research, Minderbroedersstraat 10, B-3000, Leuven, Belgium

INTRODUCTION

In the early 1980s the first human and murine cytokines were succesfully cloned, i.e. the cDNAs and genes coding for IFN-γ, TNF-α and IL-1. The elucidation of the complex nature of these and other cytokines benefited greatly from the milligram amounts of pure recombinant protein becoming available soon thereafter. Since then, it did not take a long time to recognize the tremendous impact that both *in vivo* administration of recombinant cytokines and inhibition by antagonists of endogenously formed cytokines could have on health restoration in various experimental animal models of disease.

It is known that certain murine or human cytokines are not able to induce biological effects in heterologous cells, a finding in contrast to mammalian proteinaceous hormones that generally do act across species barriers. With the nucleotide sequences being available, it became clear that this species specificity was due, at least in part, to high substitution rates in the coding sequences of cytokine genes, that have occurred during mammalian radiation.

From a pragmatic point of view, this 'species specificity' has been one of the major grounds to clone livestock animal cytokines. Application of heterologous recombinant cytokines, of human or murine origin, in domestic food animals might indeed compromise interpretation of results because of their weakened biological potential or their immunogenicity.

At present, many of the nucleotide sequences coding for swine cytokines have been cloned (IFN-γ, IL-1α and -β, TNF-α and -β, IL-2, IL-4, IL-6, IL-10, IL-12, TPO), mostly shortly after the human sequence had become available. Firm progress has been made also in the characterization of porcine chemokines (AMCF-I/IL-8, AMCF-II/GCP-2, NAP-2, MCP-1 and MCP-2), a promising novel class of cytokines, the members of which share the characteristic of being chemoattractant for either monocytes, granulocytes or lymphocytes.

Here we review the attempts undertaken to characterize the structural and molecular features of swine cytokine genes and (recombinant) proteins, their chromosomal location, and the technology used to produce them by means of recombinant DNA approaches. We will pay special attention to those properties of pig cytokines that are currently thought to be unique or unusual as compared to those of other mammals.

LYMPHOKINES AND MONOKINES

Interferon-γ

Interferon-γ differs essentially from the classical type I IFN (reviewed by La Bonnardière *et al.*, 1994) by means of its split gene structure, and an expression pattern being restricted in most mammalian species to activated lymphocytes and NK cells. The pig seems to represent a notable exception, since in this species the gene is transcribed also in trophectodermal cells (reviewed by La Bonnardière, 1993).

Its gene product is a small glycoprotein that exerts a variety of effects among which its antiviral activity, its macrophage activating effect and its MHC class II-inducing properties are the most important ones. Molecular cloning and sequencing of the 5.6 kb genomic porcine IFN-γ gene region (Dijkmans *et al.*, 1990; Table 10.1) revealed a four-exon structure and an overall transcription signal pattern similar to those reported for human IFN-γ (Gray and Goeddel, 1982). The insertion in the third intron of a 230-bp sequence with 82% similarity to the consensus pig repetitive element-1 (PRE-1; Singer *et al.*, 1987), a pig-specific short interspersed repetitive element (SINE), that is missing in the human gene, appeared to be the most remarkable difference between both sequences. Additional minor differences include the deletion in the first intron of a 90-bp stretch containing a 24-bp alternating CA dinucleotide repeat found in the precise location of a DNAase-I-hypersensitivity site, as present in the repressed IFN-γ gene of a human cell line (Hardy *et al.*, 1987), and the substitution in the same intron of a protooncogene c-Rel binding octanucleotide motif [AATTTTCC] for the unresponsive motif [AATCCTCC] (Sica *et al.*, 1992). Given the well-documented engagement of the first intron in transcriptional

Table 10.1. Cloning, chromosomal location of porcine cytokine genes and cDNAs, and recombinant expression systems for production of their proteins.

Cytokine	Chromosomal location	Accession number	DNA type	Comments on cDNA sources and/or recombinant protein production (promoter)	References
IFN-γ	5	X53085	gDNA	Sequence of the genomic locus for swine IFN-γ	Dijkmans *et al.*, 1990
		S63967	cDNA	Semi-artificial cDNA constructed by exon ligation	Vandenbroeck *et al.*, 1991
				Expressed in *E. coli*, refolded from IB *(λPL)*	Vandenbroeck *et al.*, 1993a
				Expressed in baculovirus/Sf9 cell system *(polyhedrin)*	Vandenbroeck *et al.*, 1994a
TNF-α	7	M29709	cDNA	Cloning by PCR from stimulated monocytes	Pauli *et al.*, 1989
		X54001	gDNA	Cloning by PCR using human sequence-based primers	Drews *et al.*, 1990
		X54859	gDNA	Sequence of the genomic locus for TNF-α and TNF-β	Kuhnert *et al.*, 1991
				Omp-A/FLAG fusion protein in *E. coli (trp/lac)*	Su *et al.*, 1992
				Mammalian PK and L929 cells *(metallothionein)*	Von Niederhäusern *et al.*, 1993
IL-1α	3	X52731	cDNA	Cloning of cDNA by cross-hybridization from alveolar macrophage cDNA library	Maliszewski *et al.*, 1990
IL-1β	3	X74568	gDNA	Complete genomic DNA sequence; expressed in *E. coli* as a hexahistidine/factor Xa site fusion protein *(λPL)*	Vandenbroeck *et al.*, 1993b
		M86725	cDNA	Cloning of cDNA from alveolar macrophage library	Huether *et al.*, 1993
IL-2	8	X56750	cDNA	cDNA cloning by PCR from stimulated lymphocytes	Goodall *et al.*, 1991
		X58428	cDNA	cDNA cloning by PCR from lymphocytes	Lefèvre, 1991 *(UP)*
				Expressed in baculovirus/Sf21 cells *(polyhedrin)*	Collins *et al.*, 1994
IL-4	—	X68330	cDNA	Cloning by PCR from lamina propria lymphocyte RNA	Bailey *et al.*, 1993
		L12991	cDNA	Cloning from T lymphocytes	Zhou and Murtaugh, 1993 *(UP)*
IL-6	—	M86722	cDNA	Cloning by cross-hybridization from a synovial fibroblast cDNA library	Richards and Saklatvala, 1991
		M80258	cDNA	Cloning from splenocyte cDNA library	Mathialagan *et al.*, 1992
IL-10	—	L20001	cDNA	Cloning of cDNA from splenocyte cDNA library; transient COS cell expression system (*cytomegalovirus*)	Blancho *et al.*, 1995
IL-12	—	U08317	cDNA	p40 subunit of porcine interleukin-12	Foss and Murtaugh, 1995 *(UP)*
		L35765	cDNA	p35 subunit of porcine interleukin-12	Foss and Murtaugh, 1995 *(UP)*
TPO	—	—	cDNA	Cloning by PCR from kidney RNA	Gurney *et al.*, 1995
IL-8	—	M99367	cDNA	Cloning from alveolar macrophage cDNA library	Goodman *et al.*, 1992
		M86923	cDNA	Cloning from alveolar macrophage cDNA library	Lin *et al.*, 1994
		X61151	cDNA	Cloning from macrophages	Sanjanwala, 1994 *(UP)*
AMCF-II	—	M99368	cDNA	Cloning from alveolar macrophage cDNA library	Goodman *et al.*, 1992
NAP-2	—	X77935	cDNA	Cloning from platelet cDNA library	Power *et al.*, 1994
MCP-1	—	Z48479	cDNA	Cloning from luteal cell-derived cDNA library	Hosang *et al.*, 1994a
		X79416	cDNA	Cloning from endothelial cell cDNA library	Zach *et al.* *(UP)*
MCP-2	—	Z48480	cDNA	Cloning from luteal cell-derived cDNA library	Hosang *et al.*, 1994b

IB: inclusion bodies; *UP*: unpublished.

regulation of the human IFN-γ gene (Chrivia *et al.*, 1990; Ciccarone *et al.*, 1990; Sica *et al.*, 1992), the partial reorganization in the porcine equivalent might be associated with an altered (de-)repression/activation profile with regard to the human gene, and, hence, explain to some extent the species-specific differences in expression pattern of the gene.

A microsatellite marker consisting of a $(GA)_n(AGGA)_n$ compound structure, which is located downstream from the intronic PRE-1 element, has been used to assign the porcine IFN-γ locus to linkage group XI (Ellegren *et al.*, 1993a). The gene was physically localized on the p arm by *in situ* hybridization with a bovine IFN-γ cDNA fragment, close to the centromere of chromosome 5 (Johansson *et al.*, 1993). As such, it belongs to a syntenic group positioned on a large chromosomal fragment that seems to be conserved among humans, cattle and pigs. The insertion of an intronic PRE-1 sequence in the porcine IFN-γ gene might be associated with its near-centromeric position, since Yasue *et al.* (1991) reported that PRE-1 is localized predominantly on centromeric swine chromosome regions. This distribution contrasts to those of both human and murine SINEs in that the latter are known to be correlated with Q-band pattern.

Comparison of the porcine IFN-γ nucleotide and aa sequences with previously reported IFN-γ sequences reveals 75% and 59% similarity with human, and 86% and 77% similarity with both bovine and ovine IFN-γ. The higher degree of homology among porcine, bovine and ovine IFN-γs as well as the conservation of an *N*-glycosylation site at position 83 within these three species, reflect the close phylogenetic relationship between Bovidae and Suidae.

Since recombinant murine and human IFN-γ display apparent failure to act across the species border on porcine cells, production of rPoIFN-γ has become a *conditio sine qua non* for experimentation in swine. A semi-artificial cDNA coding for the mature portion of porcine IFN-γ has been overexpressed in *Escherichia coli* (Vandenbroeck *et al.*, 1991). As expected from previous attempts to overexpress the human IFN-γ cDNA in *E. coli* (Arakawa *et al.*, 1985), nearly all protein formed was found to be deposited in cytoplasmic inclusion bodies. Therefore, an *in vitro* refolding procedure was developed that produced high levels of pure, dimeric and bioactive protein (Vandenbroeck *et al.*, 1993a). Since porcine IFN-γ possesses two *N*-glycosylation sequons, attachment of sugars might modulate biological activity. To obtain glycosylated protein, the precursor cDNA was expressed also in a baculovirus/Sf9 cell system (Vandenbroeck *et al.*, 1994a). Glycoform heterogeneity was observed for the recombinant protein, secreted from its authentic signal peptide into the supernatant and purified by immunoaffinity chromatography. Although the molecular masses and number of protein bands observed on SDS–PAGE (17, 19 and 21 kDa) differed from those reported for natural conceptus-derived IFN-γ (18.5, 20.5, 23 and 24.5 kDa; La Bonnardière, 1993), in both cases all glycoforms appeared to originate from a single 16.5 to 17 kDa un-*N*-glycosylated polypeptide. Surprisingly, lectin labelling studies indicated the presence of sialic acid linked α(2–6) to galactose in the carbohydrate moiety of 19 and 21 kDa insect cell-derived IFN-γ (Vandenbroeck *et al.*, 1994a). This is a rather unexpected finding since the only baculovirus-expressed protein containing sialic acid described is human plasminogen (Davidson and Catellino, 1991). Studies are underway to analyse the sugar chains of baculovirus-derived poIFN-γ in further detail.

Both refolded *E. coli*-derived and baculovirus-expressed porcine IFN-γ conserved the structural integrity of natural IFN-γ as judged by their antiviral activity, acid instability, basic charge and tendency to dimerize (Vandenbroeck et al., 1993a, 1994a). Antiviral activity measured in a VSV on Madin–Darby bovine kidney (MDBK) cells challenge assay corresponded to that showed by an *E. coli*-derived rPoIFN-γ preparation from a different source (unpublished data from Genentech, USA; Charley *et al.*, 1988).

Tumour Necrosis Factors-α and -β

The tumour necrosis factors-α ('cachectin') and -β ('lymphotoxin') are generally considered as crucial mediators of inflammation. While TNF-α is produced by mononuclear phagocytes, lymphocytes are the natural cellular source of TNF-β. Together they regulate a number of antiviral and anti-parasitic defence mechanisms.

Both the cDNA sequence of porcine TNF-α (Pauli *et al.*, 1989), and the 10.2 kb DNA sequence of the complete genomic TNF locus, comprising TNF-β (Kuhnert *et al.*, 1991), have been reported. Drews *et al.* (1990) reported the sequence of a 1.8 kb DNA fragment containing the porcine TNF-α gene, obtained through PCR from genomic DNA. As is the case for mice, rabbits and humans, both TNF genes appeared to be closely physically linked: the more 5′ located TNF-β gene is separated merely by a 1 kb spacer stretch from the TNF-α gene (Kuhnert *et al.*, 1991). Upstream from the swine TNF genes, several stretches of medium and highly repetitive sequences were identified. One such region, containing matches to the ARS consensus sequence of yeast and to matrix-attachment regions, suggests that the TNF locus constitutes one regulation unit. Further studies on the TNF chromatin structure by Kuhnert *et al.* (1992) revealed the absence of DNAase I hypersensitivity (DH) sites in fibroblasts and kidney cells, although they were present in monocytes and macrophages. Thus, transcriptional activity of TNF genes correlates with appearance of a specific and different DH-site pattern for both TNF genes.

The TNF genes have been assigned to swine chromosome 7 by *in situ* hybridization (Solinas *et al.*, 1992). Analysis of a polymorphic $(GCT)_n$ microsatellite marker upstream from the TNF-β gene confirmed assignment of TNF to a site adjacent to or within the SLA complex (Ellegren *et al.*,

1993a). Linkage of TNF-β to the swine SLA by means of an RFLP was formerly reported by Chardon *et al.* (1991). The close vicinity of this highly polymorphic microsatellite marker to the TNF genes might have functional significance, since genetic polymorphisms in both the murine (Freud *et al.*, 1992) and human (McGuire *et al.*, 1994) TNF-α genes have been associated with (parasitic) disease suceptibility.

Similarity of the deduced aa sequence with the corresponding human sequence is 86% for TNF-α and 77% for TNF-β (Kuhnert *et al.*, 1991), implying a biological activity for TNF that is less species-bound than that of IFN-γ. This is best illustrated by studies on antigenic cross-reactivity of anti-cytokine antibodies: porcine and human IFN-γ cannot be neutralized in a mutual cross-neutralization assay (Charley *et al.*, 1988; Kontsek, P., Vandenbroeck, K., Martens, E. and Billiau, A. 1996, unpublished results). On the contrary, anti-human TNF-α mAbs have been described that efficiently neutralize the biological activity of porcine TNF-α (Jesmok *et al.*, 1992).

An *E. coli* expression system has been developed for production of rPoTNF-α as a fusion protein with an OmpA signal peptide (Su *et al.*, 1992). Following purification by affinity chromatography, the recombinant cytokine was released from the fusion protein through enterokinase treatment. Both cleaved and uncleaved forms were found to possess cytotoxic activity on WEHI cells. Another expression system for pig TNF-α, using mammalian cells and a zinc-inducible methallothionein-drive expression vector, has been elaborated by Von Niederhäusern *et al.* (1993). The recombinant protein was secreted in the culture supernatant, had a molecular mass of 17 to 19 kDa and was reactive with a polyclonal anti-murine TNF-α antiserum.

One Interleukin-1α and Two Interleukins-1β

Porcine IL-1 is the only swine cytokine to be identified and characterized before and independently from the human counterpart. The majority of this pioneer work has been done by Saklatvala and co-workers in the early 1980s. Porcine IL-1 was initially named 'catabolin' by virtue of its property to induce degradation of the matrix of live cartilage in organ culture (Saklatvala and Dingle, 1980). It was purified to homogeneity from pig mononuclear leukocytes and shown to be a form of IL-1 because of its pyrogenic, lymphocyte activating and amyloid-A-inducing properties (Saklatvala *et al.*, 1984). Besides the initially reported acidic form (pI 4.5), a basic form (pI 8.3) was isolated from pig leukocytes that appeared to be serologically different from the first, though the molecular masses of both IL-1 species were identical (Saklatvala *et al.*, 1985). The basic form accounted for about half of the total cartilage-resorbing activity produced by leukocytes. An antiserum made to the pI 4.5 form of IL-1 reacted with the original human IL-1α (March *et al.*, 1985; Saklatvala, Cambridge, 1994, personal communication). Hence, an association appeared to exist between the subtype of IL-1 (α or β) and its charge (acidic vs. basic/neutral), that proved to hold true for the IL-1 forms isolated from rabbits and humans as well. The identification of two distinct human IL-1 cDNAs definitively confirmed that the charge differences of the two IL-1 forms were caused by genetically determined aa compositions and not by differences in post-translational modification, such as glycosylation (March *et al.*, 1985).

In 1993 both the genomic DNA covering the 8.8 kb pig prointerleukin-1β gene (Fig. 10.1; called IL-1β1 (see below); Vandenbroeck *et al.*, 1993b) and the cDNA (IL-1β2; Huether *et al.*, 1993) were reported. Surprisingly, comparison of the coding parts of both nucleotide sequences and their inferred aa sequences revealed significant disagreement: 39 out of 267 aa positions were different (86% similarity). The likely existence of two distinct IL-1β genes in the pig genome was confirmed by Southern analysis of the genomic DNA of both domesticated and wild swine showing heterogeneity at the IL-1β locus (Vandenbroeck *et al.*, 1993b), and reinforced by the isolation of a genomic clone that appeared to perfectly match Huether's cDNA sequence (Vandenbroeck *et al.*, 1994b). Thus, the pig seems to represent the only mammalian species which, besides one IL-1α gene (Malizewski *et al.*, 1990), possesses two different IL-1β genes. One of the most

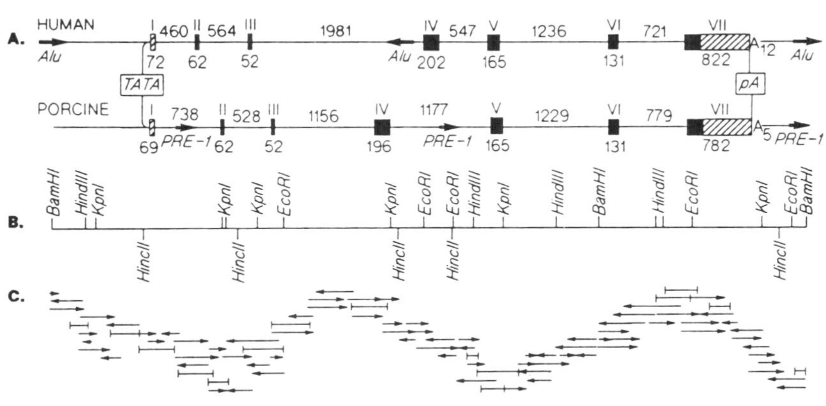

Fig. 10.1. (A) Structural comparison of the porcine prointerleukin-1β1 gene (Vandenbroeck *et al.*, 1993b) and the human prointerleukin-1β gene (Clark *et al.*, 1986). Exons are indicated with roman numerals. Relative intron/exon lengths are to scale. Hatched and closed boxes show the relative location of non-coding and coding exon regions along the gene axis. Arrows locate the position and transcriptional orientation of *Alu* (HuIL-1β) and *PRE-1* (PoIL-1β1) family sequences, found within the respective IL-1 loci. (B) Location of essential endonuclease recognition sites. (C) Sequencing strategy for the porcine IL-1β gene. Horizontal arrows represent the direction and extent of sequencing of subclones. Horizontal lines flanked by two vertical lines represent subclones, sequenced completely on both strands.

remarkable differences between both IL-1β proteins is their charge: IL-1β1 has a theoretical pI of 5.24, whereas that of IL-1β2 amounts to 8.49. This implies that the IL-1 form most probably corresponding to the pI 8.3 basic form isolated by Saklatvala *et al.* (1985) is IL-1β2. It was known already in 1985 that the pI 4.5 form corresponds to IL-1α on the basis of antigenic studies (Saklatvala, Cambridge, 1994, personal communication). However, Saklatvala *et al.* (1983) reported the existence of an additional, though minor, catabolin form with a pI of 4.9, that never has been purified or characterized. Conceivably, this form represents the acidic form of IL-1β (calculated pI 5.24). Taken altogether, the above data suggest that pig synovial/lymphocytic cells can be triggered to produce three, rather than two, dissimilar IL-1 proteins, coded for by separate genes.

The gene coding for prointerleukin-1β1 is interrupted by six intervening sequences (Fig. 10.1; Vandenbroeck *et al.*, 1993b). Three PRE-1-like sequences (Singer *et al.*, 1987) are found in intronic and 3′ extragenic positions. As in man and mouse, the IL-1α and IL-1β genes in swine are tightly linked (Krull *et al.*, 1992), and both map to the near-centromeric region of chromosome 3 (q1.2–q1.3 and q1.1–q1.4, respectively; Mellink *et al.*, 1994). Consensus sequences of many putative *cis*-acting elements are localized in the promoter region (e.g. PU.1, AP2, OCT1/2, CRE, NFIL-1βA), the majority of which are found also in the promoter of the human gene. A 16-bp perfect palindromic sequence (position 444), that is missing in the human gene (Clark *et al.*, 1986), and shares no apparent similarity to any known motifs, is localized in the distal region of the promoter.

Recombinant porcine IL-1β1 was produced in *E. coli* as a fusion protein, consisting of a hexahistidine tag, a faxtor Xa recognition site and the mature form (Vandenbroeck *et al.*, 1993b). The hexahistidine tag allows for rapid purification of the recombinant protein on Ni^{2+}-nitrilotriacetic-acid-agarose by IMAC. Following removal of the histidine tag by factor Xa and additional purification, a preparation of rPoIL-1β was obtained that induced gelatinase B in MDBK cells. The uncleaved form was 20-fold less active than the processed form in a thymocyte proliferation assay (unpublished results).

Interleukin-2

Interleukin-2 is a lymphokine that plays a chief role in stimulation and maintenance of the growth and proliferation of immune cells. Activated T lymphocytes are its major cellular source. The gene was mapped by linkage analysis to swine chromosome 8 (Ellegren *et al.*, 1993b).

The cDNA coding for porcine IL-2 was cloned by PCR from Con A-stimulated lymphocytes, and its aa sequence inferred displays 72% identity with both human and ovine IL-2 (Goodall *et al.*, 1991). However, examination of the contact regions of IL-2 that bind to the p55 and p75 IL-2 receptor chains showed that the porcine sequence comprised within this region exhibits a higher degree of similarity to the corresponding human sequence than to that of sheep or cattle. Since human IL-2 is notably less species-specific than ovine or bovine IL-2, the higher similarity in the p55/p75 contact regions between pig and human IL-2 implies that the action of the pig protein likewise should not be restricted too much by species boundaries. This was indeed illustrated by Collins *et al.* (1994) who produced recombinant porcine and bovine IL-2 by means of the baculovirus expression system in Sf21 cells. Insect cells infected with recombinant virus secreted IL-2 as 15-kDa and 17-kDa proteins into the culture medium. As anticipated, the recombinant porcine protein supported the proliferation of porcine, bovine, human and murine lymphocytes. In contrast, recombinant baculovirus-derived bovine IL-2 only supported growth of homologous cells.

The situation of IL-2 clearly illustrates that the degree of species specificity of cytokines cannot always be predicted solely on the basis of overall homology originating from phylogenetic relationship, but may require more detailed structure/function analysis.

Interleukins-4, -6, -10, -12 and Thrombopoietin

Several swine cytokines exist for which, even though the cDNA has been cloned and sequenced, more elaborate analysis and/or production by recombinant DNA technology still have to be performed or reported. Among these are IL-4, IL-6 and the lately described cytokines IL-10, IL-12 and TPO as well.

Interleukin-4 was cloned by PCR from mitogen-activated lamina propria lymphocytes (Bailey *et al.*, 1993). In the core of the mature protein a stretch of 13 aa is deleted in comparison with human IL-4. An identical 13 aa deletion pattern is found in bovine IL-4. In addition, 7 aa are deleted in precisely the same region of murine IL-4. The (partial) absence of this region in three out of four mammalian IL-4s suggests that it does not contribute essentially to structural or biological integrity of the protein. The preservation of all six cysteines in human, bovine and porcine IL-4s, however, imperatively implies an essential role for disulphide bridges in maintenance of tertiary conformation.

Interleukin-6 has been cloned from an IL-1α-stimulated synovial fibroblast cDNA library (Richards and Saklatvala, 1991), and from a spleen cDNA library as well (Mathialagan *et al.*, 1992). The protein shares 62% aa sequence identity with its human counterpart.

Interleukin-10 is a pleiotropic cytokine that exerts strong suppressive effects on the transcription of a wide range of Th1-specific cytokines in activated T cells. The porcine IL-10 cDNA (Blancho *et al.*, 1995) is the third IL-10 sequence to be characterized, after human and mouse IL-10. Overall similarity between human and pig IL-10 proteins amounts to 74%, and between mouse and pig to 68%. The porcine IL-10 predicted protein sequence shows a 4 aa deletion as compared to human and mouse IL-10, presumably occurring in the signal peptide and, thus, not affecting the structure of the mature protein. Recombinant porcine IL-10 was produced by transiently transfecting COS cells with a relevant expression vector. A polypeptide chain of the expected molecular mass, 18 kDa, was released into the culture medium, together with a 15-kDa protein, supposedly derived from the first by selective proteolytic cleavage. Recombinant pig IL-10 proved to be biologically active in that it blocked transcription of the IFN-γ gene in Con-A-activated porcine T cells (Blancho *et al.*, 1995).

Interleukin-12 is a cysteine-rich disulphide-linked heterodimeric protein composed of 35-kDa and 40-kDa subunits (p35 and p40, each coded for by separate genes). The cytokine was previously termed NK cell stimulatory factor (NKSF) and is a potent inducer of IFN-γ from peripheral blood lymphocytes (PBL) (Kobayashi *et al.*, 1989). The cDNA sequences for both subunits of porcine IL-12 have been cloned recently (Foss and Murtaugh 1994, unpublished results), and display significant similarity to those of man and mice.

Thrombopoietin (TPO) is the lately identified ligand for the c-mpl receptor and a novel cytokine, shown to regulate thrombopoiesis. Its protein sequence comprises an amino terminal domain resembling erythropoietin and an unrelated 179 aa carboxy terminal half that contains multiple *N*-linked glycosylation sequons. Gurney *et al.* (1995) cloned the cDNA coding for the porcine form by PCR from kidney mRNA. The aa sequence of mature porcine TPO is 73% similar to the human one. Moreover, these authors isolated a second splice form, TPO-2, with a 4-aa deletion within the amino terminal half, that is conserved between pigs, humans and mice. Since these 4 aa are encoded at the 5′ junction of exon 6 in the human gene, the TPO-2 mRNA form probably results from alternative splicing of the primary transcript. As to the function of TPO-2, even though the corresponding mRNA appeared to be abundantly expressed in each of the three species, the protein coded for was not biologically active.

THE GROWING FAMILY OF PORCINE CHEMOKINES

Chemokines represent a recently discovered, large family of structurally related chemotactic cytokines that stimulate the migration of leukocytes towards the inflammatory locus. The pathological relevance of chemokines is thought to be enormous, since they may represent the ultimate and direct mediators of inflammation. The structural hallmark for this family of proinflammatory proteins is the conservation of four cysteine residues. They can be divided into two subfamilies depending on the arrangement of the two most amino-terminally located cysteines, i.e. a CC pattern typical for β chemokines, such as the monocyte chemotactic proteins (MCP), and a CXC pattern found predominantly in the α class of granulocyte/neutrophil chemotactic proteins (GCP) to which IL-8 belongs. Chemokines are produced by various cell types following treatment with appropriate physiological inducers, the most important of which are 'regular' cytokines, e.g. IL-1, TNF-α and IFN-γ. In contrast to classical chemotactic agents (C5a, leukotriene B4 and platelet activating factors), chemokines selectively attract and activate distinct leukocyte subpopulations.

Alveolar Macrophage Chemotactic Factor-I/Interleukin-8

The first two porcine chemokines to be isolated and characterized, were both 10-kDa peptides released by LPS-stimulated AM (Goodman *et al.*, 1991). Both, designated alveolar macrophage chemotactic factor-I and -II (AMCF-I and -II), were purified to homogeneity, structurally characterized and found to display potent and efficacious neutrophil – but not monocyte – chemotactic activity. The corresponding full-length cDNAs were isolated and sequenced (Goodman *et al.*, 1992), and appeared to code for distinct proteins containing four cysteines, the first two of which arranged in a CXC pattern. At this stage it became clear that AMCF-I in all probability represents the porcine homologue of human IL-8, because the overall identity between both amounts to 74%. Moreover, functional criteria, such as their similar chemotactic dose–response profiles for homologous PMN and their apparent chemotactic potencies for lymphocytes as well, corroborate their being homologues. With the exception of the amino terminal aa position of the theoretically predicted mature protein [Asp(GAC) versus Ala(GCA)], the sequence of AMCF-I is identical to the porcine IL-8 sequence reported by Lin *et al.* (1994). Furthermore, these authors showed that in pig alveolar macrophages, AMCF-I/IL-8 induction by LPS is dose dependent and direct, i.e. not mediated through the action of TNF or IL-1, but suppressed by IL-4 or dexamethasone.

Alveolar Macrophage Chemotactic Factor-II/ Granulocyte Chemotactic Protein-2

The identity of AMCF-II is less clear. A high degree of structural similarity has been reported (Goodman *et al.*, 1992) to exist with huENA-78 (67%) and members of the GRO subfamily (52–61%). Recently, however, our group purified and characterized two novel granulocyte chemoattractants from bovine and human cells, designated granulocyte chemotactic protein-2 (boGCP-2 and huGCP-2). Both were found to be mutual homologues on the basis of their identical biochemical behaviour during purification, their structural similarity (67% identical residues) and equivalent granulocyte attracting properties (Proost *et al.*, 1993). Alignment of AMCF-II with the aa sequences of both boGCP-2 and huGCP-2 revealed striking similarities of 84% and 63%, respectively (for a total overlapping segment of 75 aa). This observation strongly suggests that GCP-2 and AMCF-II are functional homologues. This is reinforced at the biological level by matching minimum effective doses for homologous neutrophil chemotaxis, i.e. 3 nM for huGCP-2 (Proost *et al.*, 1993) and 1.4 nM for AMCF-II (Goodman *et al.*, 1991). Nevertheless, differences with regard to cellular sources may exist, since boGCP-2 and huGCP-2 were isolated from a kidney cell line and an osteosarcoma cell line, respectively, whereas AMCF-II was purified from macrophages. In addition, boGCP-2 is not species specific since it was at least as active as huGCP-2 on human granulocytes. AMCF-II, on the contrary, attracted porcine PMN but exerted only weak effects on human PMN (Goodman *et al.*, 1991). In conclusion, whatever the actual human molecular homologue of AMCF-II may be, structural and functional similarity data indicate that ENA-78, GCP-2 and AMCF-II constitute a related subgroup within the α subfamily of chemokines.

Neutrophil-activating Peptide-2

In 1983, Rucinski *et al.* reported the isolation from porcine platelets of a low-molecular-mass basic protein that showed significant antigenic cross-reactivity with human β-thromboglobulin (β-TG). In humans, β-TG is the cleavage product of connective tissue-activating peptide III (CTAP-III), that itself is produced extracellularly from platelet basic protein (PBP) by proteolytic cleavage. Walz and Baggiolini (1989) purified a further downstream cleavage product of β-TG, referred to as neutrophil-activating peptide 2 (NAP-2). As opposed to its precursors (PBP, CTAP-III and β-TG), NAP-2 displays pronounced neutrophil activation effects. For that reason, and given its similarity with other chemokines and its typical CXC pattern, NAP-2 appears to belong to the α subfamily of chemokines.

The pig NAP-2 was recently found to contrast with its human homologue in that two, rather than one, active molecular species seem to exist, both generated from an identical precursor (Yan *et al.*, 1993), yet differing in length by the absence (poNAP-2-S) or presence (poNAP-2-L) of a 7-aa stretch at the amino terminus. Thus, most likely, the longer form is a precursor of the shorter one. Both long and short NAP-2 polypeptide forms appeared to possess neutrophil-activating properties, such as induction of mobilization of cytosolic free calcium, and of release of elastase from neutrophils. In contrast, the immediate length precursors of the unique NAP-2 species in humans are β-TG and CTAP-III, both of which are deficient in the promotion of neutrophil activation. Furthermore, as opposed to the situation in humans where inactive NAP-2 precursors are secreted from platelets, porcine platelets directly release the active forms of NAP-2 upon stimulation with thrombin.

Power *et al.* (1994) isolated a heparin-binding protein from porcine platelets that potently attracts neutrophils, and which they named 'CTAP-III/NAP-2' because of its amino terminus being intermediate in length between human CTAP-III and NAP-2. The aa sequences of poNAP-2-L (Yan *et al.*, 1993) and CTAP-III/NAP-2 (Power *et al.*, 1994) are identical. The latter authors demonstrated that this protein can arouse both the bioactivities in human cells that are displayed by huCTAP-III and huNAP-2 separately. Cloning and sequencing of the PBP cDNA revealed that the *N*-terminal region is highly different between humans and pigs (Power *et al.*, 1994), confirming the distinct proteolytic processing pathways for generation of NAP-2 from PBP in both species.

Taken together, the mechanisms of NAP-2 production and platelet–neutrophil interaction appear to diverge between distinct mammalian species, potentially attesting to the existence of discrete species-specific physiological processes for neutrophil recruitment and, hence, inflammation.

Monocyte Chemotactic Proteins-1 and -2

The porcine cDNAs coding for MCP-1 and MCP-2 have been isolated from cDNA libraries derived from the corpus luteum (Hosang *et al.*, 1994a, b). Both chemokines were identified as MCPs on the basis of conserved aa sequence motifs. Their first two cysteines are adjacent. Expression of MCP in luteal cells suggests a physiological participation of chemokines in luteolysis by attracting monocytes and macrophages towards the corpus luteum at the end of a non-fertile cyclus. Although the human cDNA for a third monocyte chemotactic protein (MCP-3) has been cloned and characterized (Opdenakker *et al.*, 1993), identification of a porcine homologue has not been reported.

CONCLUSIONS

Unmitigated understanding of cytokine biology in the pig is determined basically by the availability of probes, homologous recombinant cytokines and their antagonists (mAbs and soluble receptor derivatives). The early 1990s have seen an 'explosion' in available information on swine cytokines. Porcine cytokines and chemokines have been characterized by cloning their cDNAs either through PCR-based methods or through cross-hybridization screening of relevant cDNA libraries or genomic DNA libraries. Detailed approaches for production of rPoIFN-γ, TNF-α, IL-1β, IL-2 and IL-10 in expression systems as diverse as *E. coli*, mammalian cells and insect cells have been described. Yet, few data are accessible on the genomic organization and gene transcription regulation of cytokines in the swine. Analysis of regulatory sequences localized in non-coding promoter or intronic DNA regions should contribute to the elucidation of some peculiar features of cytokine function in the pig, as those described above. In addition, with the genomic DNA cloned, development of transgenic pigs exhibiting controlled and site-specific expression of cytokines or cytokine receptors comes within reach. These might be particularly useful as genetic models of human disease (Petters, 1994).

As a consequence of recent major improvements in molecular biology, the time intervals elapsing between the cloning of novel human cytokines and their porcine counterparts are becoming shorter and shorter. Even though certain important porcine cytokines, such as the homologues of human IL-11, IL-15 and IFN-γ-inducing factor (IGIF), are waiting to be cloned, the assortment of recombinant cytokines available to date should suffice for introductory *in vivo* studies, allowing full evaluation of the advantages, necessity, cost/benefit ratio and feasibility of cytokine use in the pig (see Blecha, Chapter 9 in this volume). Among these should be studies directed to cytokine-based improvement of the swine's condition in terms of economic gains (e.g. vaccination technology, resistance to viral epidemics, general health improvement, fertility and reproduction control) and to the employment of cytokines in porcine models for various human pathological conditions (e.g. wound healing, septic shock, Sinclair miniature swine melanoma model, respiratory inflammation).

With the swine being considered now as exclusive xenograft donor of the (near) future (Sachs, 1994), yet another argument for research on porcine cytokines has emerged. At present, knowledge of immunological and physiological mechanisms ascertaining long-term acceptance and functioning of porcine vascularized donor organs in primate recipients is fragmentary and elusive. As far as immunological parameters are concerned, cytokines might play a key role. Even though several cytokines are capable of acting across the species barrier between humans and pigs (TNF-α, IL-2), some have been shown to be strictly species specific (IFN-γ). Thus, the question arises as to whether in porcine organ recipients the interaction between cytokines secreted by human immune cells and the local porcine endothelium deviates from that seen in normal individuals. Since cytokines can control each other's production – a process in which endothelial cells actively participate – incompetence of even a few cytokines to transmit signals from human producer to porcine target cells (and vice versa) might cause a disarrangement of essential regulatory feedback mechanisms that might lead to a locally disequilibrated cytokine network under physiopathological conditions.

To date, research referring to porcine cytokines has focused mainly on the development of suitable biological reagents and assay systems, and on *in vitro* research. Current thinking suggests, however, that in the years to come we will witness the first studies employing recombinant porcine cytokines in live pigs.

ACKNOWLEDGEMENT

Work in the authors' laboratory on porcine cytokines is supported by the Belgian Ministry of Agriculture, General Management of Research and Development, Brussels, Belgium.

References for Part III

Anderson, G., Urban, O., Fedorka-Cray, P., Newell, A., Nunberg, J. and Doyle M. (1987) Interleukin-2 and protective immunity in *Haemophilus pleuropneumoniae*: Preliminary studies. In: Chanock, R.M., Lerner, R.A., Brown, F. and Ginsberg, H. (eds), *Vaccines 87. Modern Approaches to New Vaccines, Prevention of AIDS and other Viral, Bacterial, and Parasitic Diseases.* Cold Spring Harbor Laboratory, Cold Spring Harbor, New York, pp. 22–25.

Arakawa, T., Alton, N.K. and Hsu, Y.R. (1985) Preparation and characterization of recombinant DNA-derived human interferon-γ. *Journal of Biological Chemistry* 260, 14435–14439.

Bagby, G.C., Jr (1989) Interleukin-1 and hematopoiesis. *Blood Review* 3, 152–161.

Bailey, M., Clarke, C.J., Wilson, A.D., Williams, N.A. and Stokes, C.R. (1992) Depressed potential for interleukin-2 production following early weaning in piglets. *Veterinary Immunology and Immunopathology* 34, 197–207.

Bailey, M., Perry, A.C.F., Bland, P.W., Stokes, C.R. and Hall, L. (1993) Nucleotide and deduced amino acid sequence of porcine interleukin-4 cDNA derived from lamina propria lymphocytes. *Biochimica et Biophysica Acta* 1171, 328–330.

Becker, B.A. and Misfeldt, M.L. (1993) Evaluation of the mitogen-induced proliferation and cell surface differentiation antigens of lymphocytes from pigs 1 to 30 days of age. *Journal of Animal Science* 71, 2073–2078.

Bhagyam, R.C., Jarrett-Zaczek, D. and Ferguson, F.G. (1988) Activation of swine peripheral blood lymphocytes with human recombinant interleukin-2. *Immunology* 64, 607–613.

Bielefeldt-Ohmann, H. (1995) Role of cytokines in the pathogenesis and treatment of respiratory disease. In: Myers, M.J. and Murtaugh, M.P. (eds), *Cytokines in Animal Health and Disease.* Marcel Dekker, New York, pp. 291–332.

Bielefeldt-Ohmann, H. and Martinod, S.R. (1990) Interferon immunomodulation in domestic food animals. In: Blecha, F. and Charley, B. (eds), *Immunomodulation in Domestic Food Animals*. Academic Press, San Diego, 334pp.

Binns, R.M., Licence, S.T., Wooding, F.B.P. and Duffus, W.P.H. (1992) Active lymphocyte traffic induced in the periphery by cytokines and phytohemagglutinin: three different mechanisms? *European Journal of Immunology* 22, 2195–2203.

Blancho, G., Gianello, P., Germana, S., Baetscher, M., Sachs, D.H. and LeGuern, C. (1995) Molecular identification of porcine interleukin 10: regulation of expression in a kidney allograft model. *Proceedings of the National Academy of Sciences, USA* 92, 2800–2804.

Blecha, F. (1990a) *In vivo* use of interleukins in domestic food animals. In: Blecha, F. and Charley, B. (eds), *Immunomodulation in Domestic Food Animals*. Academic Press, San Diego, pp. 231–252.

Blecha, F. (1990b) Cytokine use in cattle and pigs. In: Barta, O. (ed.), *Monographs in Animal Immunology: MHC, Differentiation Antigens, and Cytokines in Animals and Birds*. Bar-Lab, Blacksburg, Virginia, pp. 113–125.

Blecha, F. (1991) Cytokines: applications in domestic food animals. *Journal of Dairy Science* 74, 328–339.

Blecha, F. and Charley, B. (1990) Rationale for using immunopotentiators in domestic food animals. In: Blecha, F. and Charley, B. (eds), *Immunomodulation in Domestic Food Animals*. Academic Press, San Diego, pp. 3–19.

Blecha, F., Reddy, D.N., Chitko-McKown, C.G., McVey, D.S., Chengappa, M.M., Goodband, R.D. and Nelssen, J.L. (1993) Interleukin-1α and interleukin-2 immunomodulation in pigs. *Veterinary Immunology and Immunopathology* 35, 22.

Blecha, F., Reddy, D.N., Chitko-McKown, C.G., McVey, D.S., Chengappa, M.M., Goodband, R.D. and Nelssen, J.L. (1995) Influence of recombinant bovine interleukin-1α and interleukin-2 in pigs vaccinated and challenged with *Streptococcus suis*. *Veterinary Immunology and Immunopathology* 44, 329–346.

Brown-Borg, H.M., Klemcke, H.G. and Blecha, F. (1993) Lymphocyte proliferative responses in neonatal pigs with high or low plasma cortisol concentration after stress induced by restraint. *American Journal of Veterinary Research* 54, 2015–2020.

Campos, M., Rossi-Campos, A., Potter, A., Harland, R., Bielefeldt-Ohmann, H. and Babiuk, L.A. (1989) Parenteral administration of interferons protects pigs from acute death caused by the gram-negative bacterium *Actinobacillus (Haemophilus) pleuropneumoniae*. *Immunobiology* 180 (Suppl. 4), 214.

Cepica, A. and Derbyshire, J.B. (1984) Antibody-dependent and spontaneous cell-mediated cytotoxicity against transmissible gastroenteritis virus infected cells by lymphocytes from sows, fetuses and neonatal piglets. *Canadian Journal of Comparative Medicine* 48, 258–261.

Chardon, P., Nunes, M., Dezeure, F., Andres-Cara, D. and Vaiman, M. (1991) Mapping and genetic organization of the TNF genes in the swine MHC. *Immunogenetics* 34, 257–260.

Charley, B. and Fradelizi, D. (1987) Differential effects on natural killing (NK) activity of newborn piglets and adult pig lymphocytes. *Annales de Recherches Vétérinaires* 18, 227–232.

Charley, B. and Laude, H. (1988) Induction of alpha interferon by transmissible gastroenteritis coronavirus: role of transmembrane glycoprotein E1. *Journal of Virology* 62, 8–11.

Charley, B., McCullough, K. and Martinod, S. (1988) Antiviral and antigenic properties of recombinant porcine interferon-γ. *Veterinary Immunology and Immunopathology* 19, 95–103.

Chrivia, J.C., Wedrychowicz, T., Young, H.A. and Hardy, K.J. (1990) A model of human cytokine regulation based on transfection of γ interferon gene fragments directly into isolated peripheral blood T lymphocytes. *Journal of Experimental Medicine* 172, 661–664.

Ciccarone, V.C., Chrivia, J., Hardy, K.J. and Young, H.A. (1990) Identification of enhancer-like elements in human IFN-γ genomic DNA. *Journal of Immunology* 1444, 725–730.

Clark, B.D., Collins, K.L., Gandy, M.S., Webb, A.C. and Auron, P.E. (1986) Genomic sequence for human prointerleukin 1 β: possible evolution from a reverse transcribed prointerleukin 1 α gene. *Nucleic Acids Research* 14, 7897–7914.

Collins, R.A., Tayton, H.K., Gelder, K.I., Britton, P. and Oldham, G. (1994) Cloning and expression of bovine and porcine interleukin-2 in baculovirus and analysis of species cross-reactivity. *Veterinary Immunology and Immunopathology* 40, 313–324.

Cummins, J.M., Mock, R.E., Shive, B.W., Krakowka, S., Richards, A.B. and Hutcheson, D.P. (1995) Oral treatment of transmissible gastroenteritis with natural human interferon alpha: a field study. *Veterinary Immunology and Immunopathology* 45, 355–360.

Czuprynski, C.J. and Brown, J.F. (1987) Recombinant murine interleukin-1 α enhancement of nonspecific antibacterial resistance. *Journal of Immunology* 55, 2061–2065.

Czuprynski, C.J., Brown, J.F., Young, K.M., Cooley, A.J. and Kurtz, R.S. (1988) Effects of murine recombinant interleukin-1 α on the host response to bacterial infection. *Journal of Immunology* 140, 962–968.

Dalsgaard, K., Hilgers, L. and Trouve, G. (1990) Classical and new approaches to adjuvant use in domestic food animals. In: Blecha, F. and Charley, B. (eds), *Immunomodulation in Domestic Food Animals.* Academic Press, San Diego, pp. 121–160.

Davidson, D.J. and Catellino, F.J. (1991) Asparagine-linked oligosaccharide processing in lepidopteran insect cells. Temporal dependence of the nature of oligosaccharides assembled on asparagine-289 of recombinant human plasminogen produced in baculovirus vector infected *Spodoptera frugiperda* (IPLB-SF-27AE cells). *Biochemistry* 30, 6165–6169.

Dijkmans, R., Vandenbroeck, K., Beuken, E. and Billiau, A. (1990) Sequence of the porcine interferon-γ (IFN-γ) gene. *Nucleic Acids Research* 18, 4259.

Drews, R.T., Coffee, B.W., Prestwood, A.K. and McGraw, R.A. (1990) Gene sequence of porcine tumor necrosis factor α. *Nucleic Acids Research* 18, 5564.

El-Awar, F.Y. and Hahn, E.C. (1991) Sources of antibody-dependent cellular cytotoxicity deficiency in young pig leukocytes. *Journal of Leukocyte Biology* 49, 227–235.

Ellegren, H., Johansson, M., Chowdhary, B.P., Marklund, S., Ruyter, D., Marklund, L., Brõuner-Nielsen, P., Edfors-Lilja, I., Gustavsson, I., Juneja, R.K. and Andersson, L. (1993a) Assignment of 20 microsatellite markers to the porcine linkage map. *Genomics* 16, 431–439.

Ellegren, H., Fredholm, M., Edfors-Lilja, I., Winteroe, A.K. and Andersson, L. (1993b) Conserved synteny between pig chromosome 8 and human chromosome 4 but rearranged and distorted linkage maps. *Genomics* 17, 599–603.

Fong, S. and Doyle, M.V. (1986) Response of bovine and porcine peripheral blood mononuclear cells to human recombinant interleukin-2125. *Veterinary Immunology and Immunopathology* 11, 91–100.

Freud, Y.R., Sgarlato, G., Jacob, C.O., Suzuki, Y. and Remington, J.S. (1992) Polymorphisms in the tumor necrosis factor (TNF-α) gene correlate with murine resistance to development of toxoplasmic encephalitits and with levels of TNF-α mRNA in infected brain tissue. *Journal of Experimental Medicine* 175, 683–688.

Gianello, P.R., Blancho, G., Fishbein, J.F., Lorf, T., Nickeleit, V., Vitiello, D. and Sachs, D.H. (1994) Mechanism of cyclosporin-induced tolerance to primarily vascularized allografts in miniature swine. *Journal of Immunology* 153, 4788–4797.

Goodall, J.C., Emery, D.C., Bailey, M., English, L.S. and Hall, L. (1991) cDNA cloning of porcine interleukin 2 by polymerase chain reaction. *Biochimica et Biophysica Acta* 1089, 257–258.

Goodman, R.B., Forstrom, J.W., Osborn, S.G., Chi, E.Y. and Martin, T.R. (1991) Identification of two neutrophil chemotactic peptides produced by porcine alveolar macrophages. *Journal of Biological Chemistry* 266, 8455–8463.

Gray, P.W. and Goeddel, D.V. (1982) Structure of the human immune interferon gene. *Nature* 298, 859–863.

Gurney, A.L., Kuang, W.J., Xie, M.H., Malloy, B.E., Eaton, D.L. and de Sauvage, F.J. (1995) Genomic structure, chromosomal location, and conserved alternative splice forms of thrombopoietin. *Blood* 85, 981–988.

Hardy, K.J., Manger, B., Newton, M. and Stoboe, J.D. (1987) Molecular events involved in regulating human interferon-γ gene expression during T-cell activation. *Journal of Immunology* 138, 2353–2358.

Heath, A.W. and Playfair, J.H.L. (1992) Cytokines as immunological adjuvants. *Vaccine* 10, 427–434.

Hennessy, K.J., Blecha, F., Fenwick, B.W., Thaler, R.C. and Nelssen, J.L. (1990) Human recombinant interleukin-2 augments porcine natural killer cell cytotoxicity *in vivo*. *Annales de Recherches Vétérinaires* 21, 101–109.

Holt, M.E., Enright, M.R. and Alexander, T.J.L. (1990) Protective effect of sera raised against different fractions of *Streptococcus suis* type 2. *Journal of Comparative Pathology* 103, 85–94.

Hosang, K., Knoke, I., Klaudiny, J., Wempe, F., Wuttke, W. and Scheit, K.H. (1994a) Porcine luteal cells express monocyte chemoattractant protein-1 (MCP-1): analysis by polymerase chain reaction and cDNA cloning. *Biochemical and Biophysical Research Communications* 199, 962–968.

Hosang, K., Knoke, I., Klaudiny, J., Wempe, F., Wuttke, W. and Scheit, K.H. (1994b) Porcine luteal cells express monocyte chemoattractant-2 (MCP-2):

analysis by cDNA cloning and Northern analysis. *Biochemical and Biophysical Research Communications* 205, 148–153.

Huether, M.J., Lin, G., Smith, D.M., Murtaugh, M.P. and Molitor, T.W. (1993) Cloning, sequencing, and regulation of an mRNA encoding porcine interleukin-1 β. *Gene* 129, 285–289.

Huh, N.D., Kim, Y.B., Koren, H.S. and Amos, D.B. (1981) Natural killing and antibody-dependent cellular cytotoxicity in specific-pathogen-free miniature swine and germ-free piglets. II. Ontogenic development of NK and ADCC. *International Journal of Cancer* 28, 175–178.

Jesmok, G., Lindsey, C., Duerr, M., Fournel, M. and Emerson, T. (1992) Efficacy of monoconal antibody against human recombinant tumor necrosis factor in *E. coli*-challenged swine. *American Journal of Pathology* 141, 1197–1207.

Johansson, M., Chowdhary, B., Ellegren, H., Gustavsson, I. and Andersson, L. (1993) Genetic analysis of the gene for porcine submaxillary gland mucin: physical assignment of the MUC and interferon-γ genes to chromosome 5. *Journal of Heredity* 84, 259–262.

Jordan, L.T. and Derbyshire, J.B. (1994) Antiviral activity of interferon against transmissible gastroenteritis virus in cell culture and ligated intestinal segments in neonatal pigs. *Veterinary Microbiology* 38, 263–276.

Kawashima, K. and Platt, K.B. (1989) The effect of human recombinant interleukin-2 on the porcine immune response to a pseudorabies virus sub-unit vaccine. *Veterinary Immunology and Immunopathology* 22, 345–353.

Keelan, E.T.M., Licence, S.T., Peters, A.M., Binns, R.M. and Haskard, D.O. (1994) Characterization of E-selectin *in vivo* with use of a radiolabeled monoclonal antibody. *American Journal of Physiology* 266, H279–H290.

Kehrli, M.E., Cullor, J.S. and Nickerson, S.C. (1991) Immunobiology of hematopoietic colony-stimulating factors: potential application to disease prevention in the bovine. *Journal of Dairy Science* 74, 4399–4412.

Knoblock, K.F. and Canning, P.C. (1992) Modulation of *in vitro* porcine natural killer cell activity by recombinant interleukin-1α, interleukin-2 and interleukin-4. *Immunology* 76, 299–304.

Kobayashi, M., Fitz, L., Ryan, M., Hewick, R.M., Clark, S.C., Chan, S., Loudon, R., Sherman, F., Perussia, B. and Trinchieri, G. (1989) Identification and purification of natural killer cell stimulatory factor (NKSF), a cytokine with multiple biologic effects on human lymphocytes. *Journal of Experimental Medicine* 170, 827–845.

Krull, D., Shalhevet, D., Clamp, P., Feltes, R., Beever, J., Atac, E., Murtaugh, M. and Schook, L.B. (1992) Linkage mapping of IL-1α, IL-1β and CRYG in the pig. *Animal Genetics* 23 (Suppl. 1), 87.

Kuhnert, P., Wüthrich, C., Peterhans, E. and Pauli, U. (1991) The porcine tumor necrosis factor-encoding genes: sequence and comparative analysis. *Gene* 102, 171–178.

Kuhnert, P., Peterhans, E. and Pauli, U. (1992) Chromatin structure and DNase I hypersensitivity in the transcriptionally active and inactive porcine tumor necrosis factor gene locus. *Nucleic Acids Research* 20, 1943–1948.

Kurtz, R.S., Young, K.M. and Czuprynski, C.J. (1989) Separate and combined effects of interleukin-1α and γ interferon on antibacterial resistance. *Infection and Immunity* 57, 553–558.

La Bonnardière, C. (1993) Nature and possible functions of interferons secreted by the preimplantation pig blastocyst. *Journal of Reproduction and Fertility* Suppl. 48, 157–170.

La Bonnardière, C. and Laude, H. (1981) High interferon titre in newborn pig intestine during experimentally induced viral enteritis. *Infection and Immunity* 32, 28–31.

La Bonnardière, C., Lefevre, F., L'Haridon, F., Bernard, S. and Charley, B. (1990) Essais préliminaires *in vitro* et *in vivo* d'un interféron-α porcin recombinant. *Journées de Recherche Porcine en France* 22, 269–274.

La Bonnardière, C., Lefèvre, F. and Charley, B. (1994) Interferon response in pigs: molecular and biological aspects. *Veterinary Immunology and Immunopathology* 43, 29–36.

Laude, H., Gelfi, J., Lavenant, L. and Charley, B. (1992) Single amino acid changes in the viral glycoprotein M affect the induction of alpha interferon by the coronavirus transmissible gastroenteritis virus. *Journal of Virology* 66, 743–749.

Lawman, M.J.P., Campos, M., Bielefeldt-Ohmann, H., Griebel, P. and Babiuk, L.A. (1989) Recombinant cytokines and their therapeutic value in veterinary medicine. In: Moo-Young, M., Babiuk, L.A. and Phillips, J.P. (eds), *Animal Biotechnology*. Pergamon Press, Oxford, pp. 63–106.

Lefevre, F., Mege, D., L'Haridon, R.L., Bernard, S., de Vaureix, C. and La Bonnardière, C. (1990) Contribution of molecular biology to the study of the porcine interferon system. *Veterinary Microbiology* 23, 245–257.

Lesnick, D.E. and Derbyshire, J.B. (1988) Activation of natural killer cells in newborn piglets by interferon induction. *Veterinary Immunology and Immunopathology* 18, 109–117.

Lin, G., Pearson, A.E., Scamurra, R.W., Zhou, Y., Baarsch, M.J., Weiss, D.J. and Murtaugh, M.P. (1994) Regulation of interleukin-8 expression in porcine alveolar macrophages by bacterial lipopolysaccharide. *Journal of Biological Chemistry* 269, 77–85.

Loewen, K.G. and Derbyshire, J.B. (1988) Interferon induction in piglets with poly I:C acid complexed with poly-L-lysine and carboxymethyl cellulose. *Research in Veterinary Science* 44, 132–133.

Maliszewski, C.R., Renshaw, B.R., Schoenborn, M.A., Urban, J.F. and Baker, P.E. (1990) Porcine IL-1 α cDNA nucleotide sequence. *Nucleic Acids Research* 18, 4282.

March, C.J., Mosley, B., Larsen, A., Cerretti, D.P., Braedt, G., Price, V., Gillis, S., Henney, C.S., Kronheim, S.R., Grabstein, K., Conlon, P.J., Hopp, T.P. and Cosman, D. (1985) Cloning, sequence and expression of two distinct human interleukin-1 complementary DNAs. *Nature* 315, 641–646.

Markowsda-Daniel, I., Pejasak, Z., Szmigielski, S., Jeljaszewicz, J. and Pulverer, G. (1992) Stimulation of granulopoiesis in pregnant swine and their offspring by *Propionibacterium avidum* KP-40. *British Veterinary Journal* 148, 133–145.

Mathialagan, N., Bixby, J.A. and Roberts, R.M. (1992) Expression of interleukin-6 in porcine, ovine and bovine preimplantation conceptuses. *Molecular Reproduction and Development* 32, 324–330.

McGuire, W., Hill, A.V., Allsop, C.E., Greenwood, B.M. and Kwiatkowski, D. (1994) Variation in the TNF-α promoter region associated with susceptibility to cerebral malaria. *Nature* 371, 508–510.

McIntyre, K.W., Unowsky, J., DeLorenzo, W. and Benjamin, W. (1989) Enhancement of antibacterial resistance of neutropenic, bone marrow-suppressed mice by interleukin-1α. *Infection and Immunity* 57, 48–54.

Megyeri, P., Abraham, C.S., Temesvari, P., Kovacs, J., Vas, T. and Speer, C.P. (1992) Recombinant human tumor necrosis factor α constricts pial arterioles and increases blood–brain barrier permeability in newborn piglets. *Neuroscience Letters* 148, 137–140.

Mellink, C., Lahbib-Mansais, Y., Yerle, M. and Gellin, J. (1994) Mapping of the regulatory type I α and catalytic β subunits of cAMP-dependent protein kinase and interleukin 1 α and β in the pig. *Mammalian Genome* 5, 298–302.

Mochizuki, D.Y., Eisenman, J.R., Conlon, P.J., Larsen, A.D. and Tushinski, R.J. (1987) Interleukin-1 regulates hematopoietic activity, a role previously ascribed to hematopoietin-1. *Proceedings of the National Academy of Sciences, USA* 84, 5267–5271.

Murtaugh, M.P. (1994) Porcine cytokines. *Veterinary Immunology and Immunopathology* 43, 37–44.

Myers, M.J. and Murtaugh, M.P. (1995) *Cytokines in Animal Health and Disease.* Marcel Dekker, New York, 465pp.

Naidoo, D. and Derbyshire, J.B. (1992) Interferon induction in porcine leukocytes with transmissible gastroenteritis virus. *Veterinary Microbiology* 30, 317–327.

Nakajima, K., Smith, C.V., Mixon, A., Sykes, M., Guzzetta, P.C., Spitzer, T.R., Eckhaus, M.A. and Sachs, D.H. (1992) *In vitro* and *in vivo* effects of recombinant human interleukin-2 in naive miniature swine. *Journal of Immunotherapy* 11, 169–175.

Nowacki, W., Cederbald, B., Renard, C., La Bonnardière, C. and Charley, B. (1993) Age-related increase of porcine natural interferon α producing cell frequency and interferon yield per cell. *Veterinary Immunology and Immunopathology* 37, 113–122.

Nunberg, J.H., Doyle, M.V., Newell, A.D., Anderson, G.A. and York, C.J. (1988) Interleukin-2 as an adjuvant to vaccination. In: Ginsberg, H., Brown, F., Lerner, R.A. and Chanock, R.M. (eds), *Vaccines 88: New Chemical and Genetic Approaches to Vaccination, Prevention of AIDS and other Viral, Bacterial, and Parasitic Diseases.* Cold Spring Harbor Laboratory, Cold Spring Harbor, New York, pp. 247–251.

Opdenakker, G., Froyen, G., Fiten, P., Proost, P. and Van Damme, J. (1993) Human monocyte chemotactic protein-3 (MCP-3): molecular cloning of the cDNA and comparison with other chemokines. *Biochemical and Biophysical Research Communications* 191, 535–542.

Pauli, U., Beutler, B. and Peterhans, E. (1989) Porcine tumor necrosis factor α: cloning with the polymerase chain reaction and determinatin of the nucleotide sequence. *Gene* 81, 185–191.

Petters, R.M. (1994) Transgenic livestock as genetic models of human disease. *Reproduction and Fertility* 6, 643–645.

Power, C.A., Proudfoot, A.E.I., Magnenat, E., Bacon, K.B. and Wells, T.N.C. (1994) Molecular cloning and characterization of a neutrophil chemotactic protein from porcine platelets. *European Journal of Biochemistry* 221, 713–719.

Proost, P., Wuyts, A., Conings, R., Lenaerts, J.P., Billiau, A., Opdenakker, G. and Van Damme, J. (1993) Human and bovine granulocyte chemotactic protein-2: complete amino acid sequence and functional characterization as chemokines. *Biochemistry* 32, 10170–10177.

Reddy, P.G., Blecha, F., Minocha, H.C., Anderson, G.A., Morrill, J.L., Fedorka-Cray, P.J. and Baker, P.E. (1989) Bovine recombinant interleukin-2 augments immunity and resistance to bovine herpesvirus infection. *Veterinary Immunology and Immunopathology* 23, 61–74.

Reddy, D.N., Reddy, P.G., Minocha, H.C., Fenwick, B.W., Baker, P.E., Davis, W.C. and Blecha, F. (1990) Adjuvanticity of recombinant bovine interleukin-1-β: Influence on immunity, infection, and latency in a bovine herpesvirus-1 infection. *Lymphokine Research* 9, 295–307.

Reddy, D.N., Reddy, P.G., Xue, W., Minocha, H.C., Daley, M.J. and Blecha, F. (1993) Immunopotentiation of bovine respiratory disease virus vaccines by interleukin-1-β and interleukin-2. *Veterinary Immunology and Immunopathology* 37, 25–38.

Richards, C.D. and Saklatvala, J. (1991) Molecular cloning and sequence of porcine interleukin 6 cDNA and expression of mRNA in synovial fibroblasts *in vitro*. *Cytokines* 3, 269–276.

Roth, J.A. and Flaming, K.P. (1990) Model systems to study immunomodulation in domestic food animals. In: Blecha, F. and Charley, B. (eds), *Immunomodulation in Domestic Food Animals*. Academic Press, San Diego, pp. 21–41.

Rucinski, B., Poggi, A., James, P., Holt, J.C. and Niewiarowski, S. (1983) Purification of two heparin-binding proteins from porcine platelets and their homology with human secreted platelet proteins. *Blood* 61, 1072–1080.

Sachs, D.H. (1994) The pig as a potential xenograft donor. *Veterinary Immunology and Immunopathology* 43, 185–191.

Saklatvala, J. and Dingle, J.T. (1980) Identification of catabolin, a protein from synovium which induces degradation of cartilage in organ culture. *Biochemical and Biophysical Research Communications* 96, 1225–1231.

Saklatvala, J., Curry, V.A. and Sarsfield, S.J. (1983) Purification to homogeneity of pig leukocyte catabolin, a protein that causes cartilage resorption *in vitro*. *Biochemical Journal* 215, 385–392.

Saklatvala, J., Pilsworth, L.M.C., Sarsfield, S.J., Gavrilovic, J. and Heath, J.K. (1984) Pig catabolin is a form of interleukin 1. Cartilage and bone resorb, fibroblast make prostaglandin and collagenase, and thymocyte proliferation is augmented in response to one protein. *Biochemical Journal* 224, 461–466.

Saklatvala, J., Sarsfield, S.J. and Townsend, Y. (1985) Pig interleukin 1. Purification of two immunologically different leukocyte proteins that cause cartilage resorption, lymphocyte activation and fever. *Journal of Experimental Medicine* 162, 1208–1222.

Saulnier, D., Martinod, S. and Charley, B. (1991) Immunomodulatory effects *in vivo* of recombinant porcine interferon γ on leukocyte functions of immunosuppressed pigs. *Annales de Recherches Vétérinaires* 22, 1–9.

Scamurra, R.W. (1995) Characterization of the Biological Activities of Porcine Interleukin-6. PhD thesis, University of Minnesota.

Shi, J., Goodband, R.D., Chengappa, M.M., Nelssen, J.L., Tokach, M.D., McVey, D.S. and Blecha, F. (1994) Influence of interleukin-1 on neutrophil function and resistance to *Streptococcus suis* in neonatal pigs. *Journal of Leukocyte Biology* 56, 88–94.

Sica, A., Tan, T.H., Rice, N., Kretzschmar, M., Ghosh, P. and Young, H.A. (1992) The c-rel protooncogene product c-Rel but not NF-kB binds to the intronic region of the human interferon-γ gene at a site related to an interferon-stimulable response element. *Proceedings of the National Academy of Sciences, USA* 89, 1740–1744.

Singer, D.S, Parent, L.J. and Ehrlich, R. (1987) Identification and DNA sequence of an interspersed repetitive DNA element in the genome of the miniature swine. *Nucleic Acids Research* 15, 2780.

Solinas, S., Hasler-Rapacz, J., Maeda, N., Rapacz, J. and Fries, R. (1992) Assignment of the porcine tumor necrosis factor α and β genes to the chromosome region 7p11–q11 by *in situ* hybridization. *Animal Genetics* 23, 267–271.

Sordillo, L.M. and Daley, M.J. (1995) Role of cytokines in the prevention and treatment of mastitis. In: Myers, M.J. and Murtaugh, M.P. (eds), *Cytokines in Animal Health and Disease*. Marcel Dekker, New York, pp. 389–419.

Stott, J.L., Fenwick, B.W. and Osburn, B.I. (1986) Human recombinant interleukin-2 augments *in vitro* blastogenesis of bovine and porcine lymphocytes. *Veterinary Immunology and Immunopathology* 13, 31–44.

Su, X., Prestwood, A.K. and McGraw, R.A. (1992) Production of recombinant porcine tumor necrosis factor α in a novel *E. coli* expression system. *Biotechniques* 13, 756–762.

Ulich, T.R., Del Castillo, J., Keys, M., Granger, G.A. and Rong-Xiang, N. (1987) Kinetics and mechanisms of recombinant human interleukin 1 and tumor necrosis factor-α-induced changes in circulating numbers of neutrophils and lymphocytes. *Journal of Immunology* 139, 3406–3415.

Vandenbroeck, K., Dijkmans, R., Van Aerschot, A. and Billiau, A. (1991) Engineering by PCR-based exon amplification and ligation of the genomic porcine interferon-γ DNA for expression in *Escherichia coli*. *Biochemical and Biophysical Research Communications* 180, 1408–1415.

Vandenbroeck, K., Martens, E., d'Andrea, S. and Billiau, A. (1993a) Refolding and single step purification of porcine interferon-γ from *Escherichia coli* inclusion bodies. Conditions for reconstitution of dimeric IFN-γ. *European Journal of Biochemistry* 215, 481–486.

Vandenbroeck, K., Fiten, P., Beuken, E., Martens, E., Jannsen, A., Van Damme, J., Opdenakker, G. and Billiau, A. (1993b) Gene sequence, cDNA construction, expression in *Escherichia coli* and genetically approached purification of porcine interleukin-1-β. *European Journal of Biochemistry* 217, 45–52.

Vandenbroeck, K., Willems, L., Billiau, A., Opdenakker, G. and Huybrechts, R. (1994a) Glycoform heterogeneity of porcine interferon-γ expressed in Sf9 cells. *Lymphokine and Cytokine Research* 13, 253–258.

Vandenbroeck, K., Fiten, P., Opdenakker, G. and Billiau, A. (1994b) Two distinct interleukin-1-β genes in the pig genome. *Immunogenetics* 40, 82.

Von Niederhäusern, B., Bertoni, G., Hertig, C., Pfister, H., Peterhans, E. and Pauli, U. (1993) Cloning and expression in mammalian cells of porcine tumor necrosis factor α: examination of biological properties. *Veterinary Immunology and Immunopathology* 38, 57–74.

Walz, A. and Baggiolini, M. (1989) A novel cleavage product of β-thromboglobulin formed in cultures of stimulated mononuclear cells activates human neutrophils. *Biochemical and Biophysical Research Communications* 159, 969–975.

Warren, E.J., Finck, B.N., Arkins, S.F., Kelley, K.W. and Johnson, R.W. (1995a) Brain derived tumor necrosis factor-α modulates food intake and behavior in the pig. *Journal of Animal Science* (Suppl. 1) 73, 147.

Warren, E.J., Finck, B.N., Scamurra, R.W., Murtaugh, M.P. and Johnson, R.W. (1995b) Behavioral and physiological responses induced by peripheral immune stimulation are mimicked in pigs by central injection of porcine tumor necrosis factor α. *Journal of Animal Science* (Suppl. 1) 73, 92.

Wood, P.R., Seow, H.-F. and Rothel, J.S. (1994a) Cytokines and regulation of the immune response. In: Wood, P.R., Willadsen, P., Vercoe, J.E., Hoskinson, R.M. and Demeyer, D. (eds) *Vaccines in Agriculture: Immunological Applications to Animal Health and Production.* CSIRO Information Services, East Melbourne, pp. 49–55.

Wood, P.R., Willadsen, P., Vercoe, J.E., Hoskinson, R.M. and Demeyer, D. (1994b) *Vaccines in Agriculture: Immunological Applications to Animal Health and Production.* CSIRO Information Services, East Melbourne, pp. 207

Wooley, S.T., Whyte, A., Licence, S.T., Haskard, D.O., Wooding, F.B. and Binns, R.M. (1995) Differences in E-selectin expression and leucocyte infiltration induced by inflammatory agents in a novel subcutaneous sponge matrix model. *Immunology* 84, 55–63.

Yan, Z., Holt, J.C., Stewart, G.J. and Niewiarowski, S. (1993) Neutrophil-activating intercrine secreted by porcine platelets is active without proteolytic processing. *American Journal of Physiology* 265, C1396–C1404.

Yang, W.C. and Schultz, R.D. (1986) Ontogeny of natural killer cell activity and antibody dependent cell mediated cytotoxicity in pigs. *Developmental and Comparative Immunology* 10, 405–418.

Yasue, H., Takahashi, H., Atawa, T. and Popescu, P.C. (1991) Uneven distribution of short interspersed repetitive sequence, PRE-1, on swine chromosomes. *Cellular Structure and Function* 16, 475–479.

IV

Equine Cytokines

Edited by D.W. Horohov

Cytokine Production in Equine Disease

11

D.W. Horohov, C.E. Swiderski, J.A. Robinson and T.R. Klei

School of Veterinary Medicine, Louisiana State University, Baton Rouge, Louisiana 70803, USA

CYTOKINES AND EQUINE MEDICINE

Horses have been used as experimental models for a variety of immunological studies beginning with some of the earliest work on natural antibodies by Glenny and Sordelli (cited in Wilson and Miles, 1948), combined immunodeficiency diseases (McGuire and Poppie, 1973), antiviral cytotoxicity responses (Greural *et al.*, 1977; Fujimiya *et al.*, 1979) and antigenic variation during lentivirus infection (Cheevers and McGuire, 1985; Rwambo *et al.*, 1990). The role of the immune system in both causing and preventing a variety of equine diseases has been long recognized (Cross, 1966; Henson and McGuire, 1971; Banks and Henson, 1972; Sellon *et al.*, 1994). As the field of immunology has focused more recently on the role of cytokines in regulating immune responses, so too have equine immunologists. While relatively few groups have been involved in such efforts, the tools of modern biotechnology have helped to accelerate the rate of advancement. Thus the cloning, sequencing and expression of a number of equine cytokines has been accomplished (Table 11.1). The availability of these reagents will help lead to a better understanding of the equine immune system and, perhaps, to the development of novel therapeutic strategies for treating diseases of the horse. As in the past, information gleaned from the equine system may be applicable to other species.

The first section of this chapter summarizes the current status of equine cytokines. Investigation of equine cytokines has focused on three major areas: inflammatory cytokines; IFNs; and T-cell-derived cytokines. Chapters

Table 11.1. Cloned equine cytokines.

Cytokine	Reference
Interleukin-1α	Kato *et al.*, 1995
Interleukin-1β	Kato *et al.*, 1995
Interleukin-2	Tavenor *et al.*, 1993; Vandergrifft and Horohov, 1993
Interleukin-4	Vandergrifft *et al.*, 1994
Interleukin-6	Swiderski *et al.*, in preparation
Interleukin-10	Swiderski and Horohov, 1995
Interferon-α	Himmler *et al.*, 1986
Interferon-β	Himmler *et al.*, 1986
Interferon-ω	Himmler *et al.*, 1986
Interferon-γ	Curran *et al.*, 1994
Tumour necrosis factor-α	Su *et al.*, 1991

12 and 13, by Drs Himmler, Grunig and May, discuss the role of the inflammatory cytokines IL-1, IL-6 and TNF-α in the pathogenesis of equine joint disease, and the potential clinical applications of the IFNs in the horse. Inflammatory joint disease remains a serious health problem in both human and veterinary medicine and novel therapeutic and diagnostic approaches based on these cytokines are now becoming available. Type I (-α, -β and -ω) and type II (-γ) IFNs of the horse have been cloned and expressed as recombinant proteins and preliminary investigations into their clinical application have begun. As additional information is gained regarding the role of cytokines in disease processes one can anticipate that other clinical applications will be forthcoming.

EQUINE CYTOKINES

Equine IL-1 exists in two forms, IL-1α and IL-1β, identifiable by chromatographic separation and isoelectric focusing (May *et al.*, 1990). Equine IL-1 drives IL-2 production by the murine EL-4 cell line (May *et al.*, 1990) and is cytocidal for D10.G4.1 cells (May *et al.*, 1990). Equine articular cells will respond to rHuIL-1 (May *et al.*, 1992a). The cDNA for equine IL-1α and IL-1β were recently cloned and sequenced (Kato *et al.*, 1995). The deduced aa sequence of equine IL-1α exhibits 72% and 60% similarity with that of human and murine IL-1α, respectively. Likewise, the aa sequence of equine IL-1β exhibits 67% and 62% similarity with that of human and murine IL-1β.

Equine IL-2 was partially characterized by Magnuson and co-workers and was estimated to have a molecular mass of approximately 14 kDa (Magnuson *et al.*, 1983). Equine IL-2 has been cloned and sequenced (Tavenor *et al.*, 1993; Vandergrifft and Horohov, 1993) yielding a protein core of predicted molecular mass of 14.9 kDa. Equine IL-2 shares 72% aa similarity with the human sequence, 70% with porcine, 62% with bovine and ovine, 56% with rat and 54% with the mouse. Recombinant equine IL-2 expressed in COS and CHO cells induces proliferation and LAK cell activity in equine PBMC, but has no effect on murine CTLL-2 cells likely due to unique aa substitutions in regions relevant to receptor binding (Vandergrifft and Horohov, 1993; Horohov *et al.*, 1996). Human IL-2 induces proliferation and LAK cell activity in equine PBMC (Fenwick *et al.*, 1988; Hormanski *et al.*, 1992), though it has lower specific activity (Fenwick *et al.*, 1988; Horohov *et al.*, 1996).

Equine IL-4 cDNA was cloned and sequenced from mitogen-stimulated equine PBMC and shown to share homology with IL-4 sequences from other species (Vandergrifft *et al.*, 1994). The equine IL-4 aa sequence shares 62% similarity with the human IL-4 and 40% sequence similarity with the mouse and rat IL-4 aa sequences. The precursor equine IL-4 protein has a predicted molecular mass of 15.3 kDa. Equine IL-4 expressed in COS cells augments the proliferative response of B cells to mitogens and inhibits the generation of lymphokine activated killer cells (Horohov *et al.*, 1996).

Equine IL-6 activity can be detected using the murine hybridoma cell line B 13.29 clone B.9 (Morris *et al.*, 1992). Equine IL-6 cross-reacts with antibodies to human IL-6 in slot-blot radioimmunoassays (Billinghurst *et al.*, 1995). The cDNA for equine IL-6 has recently been cloned and sequenced (Swiderski *et al.*, unpublished). The deduced aa sequence yields a 20.6 kDa protein which exhibits 71% homology to porcine IL-6 and 62% aa homology with human IL-6.

Equine IL-10 has recently been cloned and sequenced (Swiderski and Horohov, 1995). Regions of equine IL-10 exhibit homology to the IL-10-like sequence of equine herpesvirus type 2 (EHV-2) (Rode *et al.*, 1993). Both genes share homology with IL-10 of humans and with the Epstein–Barr virus protein BCRF1.

Equine TNF-α activity can be detected with an *in vitro* cytotoxicity assay using the murine fibrosarcoma cell line WEHI 164 clone 13 (Morris *et al.*, 1990b) or L929 cells (MacKay *et al.*, 1991). The molecular cloning and nucleotide sequence of the 2610-bp genomic sequence encoding equine TNF-α has been reported (Su *et al.*, 1991). The deduced aa sequence of equine TNF-α exhibits 84% homology with its human counterpart. Monoclonal antibodies and antisera to rEqTNF-α have been produced and a capture ELISA has been reported (Su *et al.*, 1992).

THE ROLE OF CYTOKINES IN EQUINE DISEASE

It has become clear that the induction of protective resistance or the exacerbation of disease is dependent on the pattern of cytokine gene expression during an immune response (Liles and Van Voorhis, 1995). The two major areas of emphasis for identifying the role of cytokines in disease processes are inflammatory responses where certain cytokines, notably IL-1, IL-6 and TNF-α, play a central role in the disease process and T-cell-mediated immune responses where the cytokines produced by specific Th cell subsets may either prevent or exacerbate disease.

Cytokine Production During Septicaemia

Septicaemia remains a common health problem for young foals (Hovda, 1992). Septicaemia occurs when intact bacteria or their toxins gain access to the systemic circulation. Of the potentially toxic bacterial components, cytokine responses to the lipopolysaccharide (LPS) portion of the Gram-negative bacterial cell wall have been extensively evaluated in a number of species including the horse (Burrell, 1990; MacKay *et al.*, 1991; Morris, 1991; MacKay and Lester, 1992). Upon entering circulation, LPS initiates a cascade of inflammatory mediators that target multiple organs resulting in the syndrome of clinical signs termed endotoxaemic shock (Bradley, 1979). The horse, by virtue of its profound physiologic responses to relatively low doses of endotoxin, is categorized as an endotoxin-sensitive species (Breuhaus and DeGraves, 1993). A growing body of evidence indicates that TNF-α, IL-1 and IL-6 act as early mediators of equine endotoxaemic shock (MacKay *et al.*, 1991; Morris, 1991; MacKay and Lester, 1992).

Tumour necrosis factor-α plays a pivotal role in initiating systemic toxaemic reactions during Gram-negative bacterial and endotoxaemic sepsis (Beutler *et al.*, 1985). TNF-α is released rapidly by LPS-stimulated macrophages in many species including the horse (Morris *et al.*, 1990b) and TNF-α infusion mimics many of the signs of endotoxaemic shock (Beutler *et al.*, 1985; Han *et al.*, 1990; MacKay *et al.*, 1991; Allen *et al.*, 1993). TNF-α causes fever, progressive acidosis, haemoconcentration, hypotension, and coagulation abnormalities that are characteristic of the toxaemic syndrome (Grunfeld and Palladino, 1990). Neutralization of TNF-α activity by infusion of anti-TNF-α has been shown to improve the clinical course of endotoxaemic and Gram-negative septic shock in a variety of species including horses (Beutler *et al.*, 1985; Cargile *et al.*, 1995). TNF-α activity has been detected in adult horses and foals after experimental induction of endotoxaemia, in horses with colic attributable to gastrointestinal tract disease, and in foals with presumed sepsis (Morris *et al.*, 1990a, 1991; Morris and Moore, 1991). Significantly, as demonstrated in other species, high TNF-α activity has been shown to correlate with disease severity and negative outcomes in foals with presumed sepsis and adult horses with gastrointestinal disease (Morris and Moore, 1991; Morris *et al.*, 1991). Though TNF-α appears to be the most proximal cytokine mediator of endotoxaemic and Gram-negative bacterial septic shock, TNF-α induces in a cascade manner both IL-1 and IL-6 which have demonstrated roles in the pathogenesis of toxaemic shock (Dinarello *et al.*, 1986; Shalaby *et al.*, 1989).

Equine leukocytes have been shown *in vitro* to release TNF-α followed by IL-1 in response to stimulation by endotoxin (Seethanathan *et al.*, 1990). TNF-α induces IL-1 production by monocytes (Dinarello *et al.*, 1986). Both TNF-α and IL-1 induce the production of IL-6 which appears to provide negative feedback for TNF-α and IL-1 production (Aderka *et al.*, 1989; Shalaby *et al.*, 1989). The biological activities of IL-1 make it significant in the pathogenesis of fever, depression, hypoglycaemia, negative protein balance, acidosis, oedema, and coagulation defects demonstrated in septicaemic shock (Dinarello, 1988). IL-1 release has been documented in both clinical and experimental sepsis of humans and laboratory animals (Cannon *et al.*, 1990). Though *in vivo* production of IL-1 in equine septic shock has yet to be documented, *in vitro* evidence noted previously supports its production *in vivo*.

Interleukin-6, which is produced primarily by mononuclear phagocytes, is an extremely pleotrophic cytokine whose biological effects have been extensively reviewed (Kishimoto, 1989; Hirano *et al.*, 1990). Gram-negative endotoxin is a potent stimulus for IL-6 production both directly and indirectly via TNF-α and IL-1 induction (Dinarello *et al.*, 1986; Shalaby *et al.*, 1989; Morris *et al.*, 1992). Increases in IL-6 activity have been documented following endotoxin infusion in horses and foals (MacKay and Lester, 1992; Robinson *et al.*, 1993; Durando *et al.*, 1994). The role of IL-6 in endotoxaemia has not been fully defined. Potentially beneficial effects of IL-6 production during endotoxaemia have been described. These include decreased production of TNF-α and IL-1 by monocytes (Aderka *et al.*, 1989; Shalaby *et al.*, 1989), and induction of the full range of acute phase proteins that not only augment non-specific and specific immunologic defences against microorganisms but also mediate protection and repair from tissue damage caused by inflammatory mediators such as TNF-α and IL-1 (Holsti and Raulet, 1989; Hirano *et al.*, 1990; Almawi *et al.*, 1991; Lue *et al.*, 1991). Other studies have indicated that IL-6 may contribute to the pathogenic effects of Gram-negative infections or endotoxaemia and that administration of IL-6 antagonists may actually be beneficial in these disease states (Tiao *et al.*, 1994; Meyer *et al.*, 1995).

In summary, the contribution of inflammatory cytokines in the systemic toxaemic reactions to endotoxin in the horse parallels that seen in humans and experimental models. Likewise, the continued development of equine-specific cytokine reagents can be expected to lead to better characterization, diagnosis and treatment of Gram-negative infections or endotoxaemia in the horse.

Possible Equine Models for Th1 and Th2 Responses

The existence of two defined $CD4^+$ Th cell subsets based on distinct patterns of cytokine gene expression was initialy recognized in the mouse (Mosmann *et al.*, 1986). These subsets have been designated either Th1 or Th2 where Th1 cells produce IL-2, IFN-γ and TNF-α and Th2 cells produce IL-4, IL-5 and IL-10 (Mosmann, 1991). Although both subsets of cells may be induced during an immune response, it has been shown that one or the other population often dominates a given response (Mosmann and Coffman, 1989). Protection from intracellular parasites generally requires a Th1 (type 1) response whereas helminth parasites evoke a Th2 (type 2) response. Deviation from this pattern, as has been demonstrated using different strains of mice, is associated with susceptibility to the invading organism (Heinzel *et al.*, 1989; Else *et al.*, 1994).

Less information is available regarding Th1 and Th2 responses in other species. Reports have associated $CD4^+$ T cells with resistance against gastrointestinal helminths of sheep, but the cytokine characteristics of this response have not been defined (Gill *et al.*, 1993). Further, the induction of Th2 T-cell-dependent effectors such as eosinophilia, mastocytosis and elevations of IgE has been reported in a number of species (Urban *et al.*, 1992). While cytokine responses characteristic of Th1 and Th2 cells have been identified in the bovine (Brown *et al.*, 1994), the correlation of a defined cytokine pattern with immune responses associated with either protective resistance or disease of non-murine species remains unproven.

The existence of Th1 and Th2 responses in the horse has yet to be demonstrated. Inferential evidence suggests several scenarios where such cells may play a role in protection or pathology. Since Th2 subsets have been associated with immune responses to metazoan parasites in other species (Urban *et al.*, 1992), a similar condition in the horse may provide evidence for this response. Acquired resistance to *Strongylus vulgaris* in the horse, as demonstrated by multiple or single experimental infections, is associated with an eosinophilia and other signs of immediate-type hypersensitivity (Dennis *et al.*, 1993). *In vitro* stimulation of PBMC from infected ponies yielded supernates containing an eosinophil chemotactic factor (Dennis *et al.*, 1993). While it is not known whether this chemotactic factor is IL-5, this type of response is consistent with the induction of the Th2 subset. Comparisons of ponies vaccinated with either parenteral inoculations of *S. vulgaris* antigen, or oral administration of radiation-attenuated larvae showed markedly different outcomes following challenge infections, and responses characteristic of either a type 1 or a type 2 cytokine response (Monahan *et al.*, 1995). Those ponies receiving the parenteral administration of antigen showed exacerbated signs of disease following challenge infestations without an eosinophilia, while those receiving the oral larval vacine were protected from infection and exhibited an eosinophilia characteristic of a type 2 response. In a recent study of a group of ponies naturally infected with strongyle nematodes we observed a spontaneous reduction in faecal egg counts (EPG) in one of the ponies consistent with an immune mediated expulsion (Fig. 11.1). A marked difference in luminal cyathostome burden was also noted for the same pony at necropsy when compared to an age-matched cohort that did not undergo this decrease in EPG (Table 11.2). It is interesting to note that the inverse relationship is true for mucosal cyathostome populations of these same ponies where the increase in larvae may be related to the induction of this expulsion. When PBMC collected at necropsy were analysed for cytokine-specific mRNA using a reverse transcription-polymerase chain reaction (RT-PCR), no discernible differences in cytokine gene expression were evident (Table 11.2). However, the colonic lymph node cells from the pony undergoing the spontaneous reduction in EPG exhibited increased levels of IL-4 compared

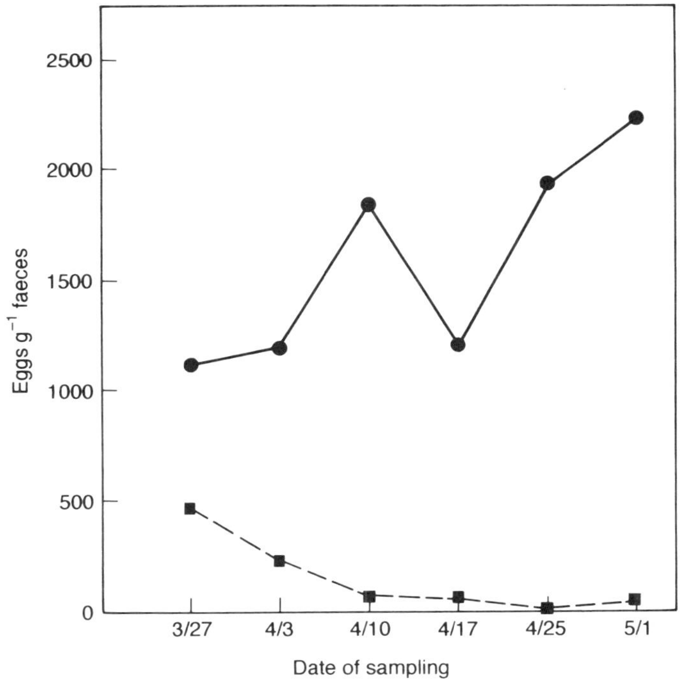

Fig. 11.1. Progression of strongyle faecal egg counts from two ponies. One pony (number 40: squares) underwent a spontaneous reduction in EPG over the 6-week-period while an age-matched cohort (number 66: circles) showed an increasing EPG. Necropsies were performed on these ponies on 5/1 and the results of cyathostome parasite counts are included in Table 11.2.

Table 11.2. Parasite burdens and cytokine production in a pony following a spontaneous reduction in faecal strongyle egg counts and a pony with a stable nematode parasite population.

	Pony 40	Pony 66
Parasite burdens[1]		
Luminal cyathostomes		
L4	3.7	2.0
Adults	4.2	82.8
Mucosal stages		
Early L3	127.0	14.0
L3–L4	8.8	4.8
Cytokine mRNA[2]		
PBMC		
IL-2	200	275
IFN-γ	60	36
IL-4	1825	1221
CLN		
IL-2	184	246
IFN-γ	53	55
IL-4	3600	439

[1]Parasite burdens in the lumen of the large intestine (fourth stage larvae, L4, and adults) were determined by a direct count in 1% of the total ingesta. Cyathostome larvae encysted in the mucosa, hypobiotic third stage (EL3) and developing third and fourth stage (L3–L4) were enumerated by digestion of 2% of the large intestinal mucosa. Numbers are expressed as counts $\times 10^{-3}$.
[2]The cytokine-specific mRNA was reverse transcribed into cDNA which was amplified using the Taq thermostable polymerase and primers based on the sequences for the equine cytokines. The amplified products were subsequently quantitated using an automated fluorescence detection system (QPCR 5000, Perkin Elmer). The results are presented as luminosity units which are proportional to the amount of amplified product present after PCR amplification.

to the unaffected pony. Similar compartmentalization of cytokine responses to the intestine of helminth-infected mice have been demonstrated (Pond *et al.*, 1992). Further characterization of this response in these ponies is underway.

Atopic individuals exhibit characteristic type 2 responses with elevated levels of IL-4 leading to production of IgE antibodies (Kapsenberg *et al.*, 1992). Chronic obstructive pulmonary disease is a respiratory disease of horses which has long been suspected to be immunologic in nature. The demonstration of fungal antigen-specific degranulation in basophils from diseased horses (Dirscherl *et al.*, 1993) and the presence of allergen-specific IgE in the bronchoalveolar lavage (BAL) fluids of affected horses (Halliwell *et al.*, 1993) support the contention that some forms are immune mediated.

Since IgE production is under the control of IL-4, it is possible that affected horses exhibit increased Th2 activity. The finding that PBMC from affected horses exhibited no blastogenic activity to the allergens is consistent with the absence of a DTH (type 1) response (Dirscherl *et al.*, 1993).

Infection with influenza viruses leads to the induction of a Th1 response in humans (Romagnani, 1991). This is not surprising considering the well-recognized role CMI responses play in the recovery from influenza virus infections (Mitchell *et al.*, 1985). It is likely that the protective response to equine influenza virus would also be a Type I response since immunized horses exhibit CMI responses to viral antigens *in vitro* (Hannant and Mumford, 1989). Differences in the antibody responses generated by vaccination versus infection have also been demonstrated in both humans and horses (Bender and Small, 1992; Nelson *et al.*, 1995). Thus the immune response to a primary infection with equine influenza virus is characterized by serum and nasal IgGa, IgGb, and IgA antibody responses while vaccination induced only IgGc and IgG(T) responses (Nelson *et al.*, 1995). Since isotype switching is under the control of T cell cytokines (Schultz and Coffman, 1991), it is likely that these differences in isotype production between infected and vaccinated horses are the result of differential expression of those cytokines. Comparison of RLN and PBMC cytokine production between a vaccinated and an infected pony demonstrated differences in the cytokine production profiles. The infected pony demonstrated a Type I response in its influenza-virus-stimulated PBMC and respiratory lymph node cultures while the vaccinate's PBMC cytokine response was Th0-like producing both IL-2 and IL-4 and its virus-stimulated RLN cells exhibited minimal production of any of the cytokine mRNA (Table 11.3).

A somewhat different situation may exist in horses infected with equine infectious anaemia virus. Peripheral blood mononuclear cells from infected ponies exhibit marked *in vitro* lymphoproliferative responses to viral antigens early in the infection, however, the intensity of this response decreases with time (Kono *et al.*, 1976; Issel *et al.*, 1992). While CMI responses can be

Table 11.3. Cytokine mRNA levels in influenza-vaccinated and infected ponies.[1]

	Vaccinated pony		Infected pony	
Cytokine	PBMC	RLN	PBMC	RLN
IL-2	287	57	558	494
IFN-γ	65	39	238	328
IL-4	366	65	45	30

[1]Peripheral blood mononuclear cells and respiratory lymph node cells from two ponies (vaccinated and infected) were stimulated *in vitro* with equine influenza virus antigen and the cells harvested 48 h later for QPCR analysis of cytokine mRNA levels. Results are represented as luminosity units (see Table 11.2).

significantly depressed with disease progression, the infected ponies continue to mount a strong antibody response to the virus (Newman *et al.*, 1991). These results mimic the TH1 to TH2 switch which is thought to occur in AIDS patients (Clerici and Shearer, 1994; Mosmann, 1994). Whether this is characteristic of lentivirus infections remains to be determined.

CONCLUSIONS

While research into the role of cytokines in equine immunity and disease is in its infancy, considerable advances have been made over the past several years. The horse provides a number of well-characterized scenarios which lend themselves to the study of cytokine interactions in either disease progression or prevention. The emphasis of the earliest work in these areas has been on the characterization of the cytokine response itself, first through the use of conventional bioassays and later, with the growing availability of recombinant cytokines and mAbs, through immunoassays and molecular analyses. One can expect a continuation of these types of investigation leading, perhaps, to a better understanding of the immunological basis for a number of important equine diseases. This avenue of investigation may not only lead to improved diagnostic and therapeutic approaches for use by equine clinicians, but also provide important information regarding the role of cytokines in diseases of other species as well – thus continuing a trend which was begun nearly a century ago.

The Cloning of Equine Type I and Type II Interferons

12

A. Himmler[1] and G. Grünig[2]

[1]*Boehringer Ingelheim R&D Vienna, Bender + Co GmbH, Dr Boehringer-Gasse 5–11, A-1121 Vienna, Austria;* [2]*DNAX, 901 California Avenue, Palo Alto, California 93404, USA*

INTRODUCTION

Structure and Function of Interferons

In 1957, Isaacs and Lindenmann discovered a substance termed 'interferon' which induced virus-infected cells to produce substances which would prevent other invading viruses from replicating. Since then it has become clear that there are various forms of IFNs, broadly classified as type I or type II IFNs (reviewed by Johnson *et al.*, 1994). Type I IFNs are single chains of aa transcribed from intron-less genes, some of them can be made by virtually every cell. They bind to a single type of receptor complex (Flores *et al.*, 1991). Based on the aa sequence of the proteins, type I IFNs are subdivided into three classes, namely IFN-α, IFN-β and IFN-ω, and a subclass of the latter, IFN-τ. These IFNs are encoded by multigene families of different sizes, depending on the species. IFN-γ is the only member of the type II IFNs and binds to its own receptor. The active form of IFN-γ is a dimer of two intertwining molecules which are orientated upside down to each other. IFN-γ is produced by T lymphocytes and NK cells.

The main function of type I IFNs is to induce cells to produce substances which interfere with virus and cell proliferation (reviewed by Johnson *et al.*, 1994). This is not the only function though, IFN-α and -β also have immunomodulating activities, for example by enhancing destruction of tumour cells by NK cells, phagocytosis of macrophages and ADCC of neutrophil granulocytes. Furthermore, IFN-α is an inhibitor of angiogenesis

 Cytokines in Veterinary Medicine (eds V.E.C.J. Schijns and M.C. Horzinek)

(Gresser, 1991; Folkman and Ingber, 1992). IFN-τ secreted by trophoblast cells is the signal of pregnancy which prevents the premature degradation of the corpus luteum in sheep, goats and cows (reviewed by Roberts *et al.*, 1992). Generally speaking IFN-γ is not as good as the type I IFNs in inhibiting virus replication but it is an important immunomodulator. It is a major cytokine of the type-1 T lymphocyte response which also includes TNF-β and IL-2 (Mosmann and Coffman, 1989) reviewed by Paul and Seder (1994). Collectively these cytokines promote CMI by enhancing the ability of macrophages to kill intracellular and extracellular microorganisms or tumour cells. For example mice genetically rendered unable to produce IFN-γ or to express the IFN-γ receptors cannot combat infection with mycobacteria (Dalton *et al.*, 1993; Kamijo *et al.*, 1993). The immunologic counter-player of IFN-γ is IL-4, one of the major cytokines of the type-2 T lymphocyte response. Type-2 T lymphocyte response greatly enhances the production of antibodies (reviewed by Paul and Seder, 1994; Horohov, this volume Chapter 11).

Virus-infected cells themselves are able to make type I IFNs (reviewed by Johnson *et al.*, 1994). In contrast, IFN-γ production is under the control of a feedback loop because it is induced by two molecules secreted by macrophages, IL-12 (Hsieh *et al.*, 1993) and IFN-γ-inducing factor (Okamura *et al.*, 1995). Through the binding to their receptors IFNs signal the cell to transcribe certain genes into messenger RNA which are then translated into novel proteins. IFN binding to their receptors activates tyrosine kinases of the Janus kinase (Jak) family (reviewed by Darnell *et al.*, 1994). The activated Jak proteins phosphorylate Stat proteins (signal transducers and activators of transcription) which then translocate to the nucleus and bind to specific DNA sequences and stimulate transcription. Type I and type II IFN receptors interact with different members of the Jak and Stat protein families which bind to different DNA sequence motives.

Interferons as Therapeutics

Although early hopes that IFNs would serve as a panacea for treating virus infections and cancer have proved to be unrealistic, IFNs have been approved throughout the world for treatment of various diseases. IFN-α became the first interferon to be approved by the Food and Drug Administration of the USA in 1986 (reviewed by Gutterman, 1994). Then it was only approved for therapy of hairy-cell-leukaemia, a rare B cell malignancy. Since then the approved therapeutic use encompasses papillomavirus- induced genital warts, chronic hepatitis B and C and Kaposi's sarcoma (an angioproliferative disease commonly found in patients suffering from AIDS). IFN-β has been approved for therapy of multiple sclerosis. IFN-β is thought to suppress the misdirected immune response to one or more myelin antigens (Goodkin, 1994). Clinical trials are under way to test the efficacy of type I IFNs in other virus-mediated diseases (e.g. AIDS) and tumours (lymphomas, renal cell carcinoma, or haemangioma).

Interferon-γ has been approved for therapy of granulomatous disease which is an inherited immunodeficiency disease characterized by recurrent bacterial infections due to deficient bactericidal activity of neutrophils and mononuclear phagocytes. IFN-γ is in clinical trials for treatment of patients infected with various intracellular microorganisms such as *M. leprae*, *M. avium*, *Leishmania* species and *T. gondii* (reviewed by Galin, 1995).

Under most physiologic conditions IFNs are produced in relatively small amounts to act locally. Therefore it is not surprising that when used systemically, IFNs can produce toxic side-effects (reviewed by Johnson *et al.*, 1994). Toxicity differs among the IFNs used and appears to be highest in IFN-α. Most commonly influenza-like symptoms like fever and fatigue are observed. Other side-effects include decreased haematopoiesis and mental depression.

Interferons in Equine Medicine

Research has been prompted by potential therapeutic applications and by the need for reagents to study normal physiology and disease pathogenesis. Type I IFNs could prove valuable in limiting the spreading of influenza virus infections in horse stocks or in treatment of equine sarcoid (caused by infection with bovine papilloma virus). Reagents for IFN-γ will be very valuable to study diseases caused by dysfunction of the immune system, e.g. chronic pulmonary disease. The results from these studies are summarized in Table 12.1.

EQUINE TYPE I INTERFERONS

Seven members of the equine type I IFN family have been cloned to date. All were isolated from the same genomic DNA library which has been constructed from horse liver DNA in the bacteriophage vector λEMBL3A (Himmler *et al.*, 1986). Recombinant bacteriophages containing equine IFN genes were identified by hybridization with radiolabelled human IFN-α2c and IFN-β cDNA probes, using conditions of low stringency. As with all other type I IFNs isolated from many different species, equine IFN-α, -β and -ω genes are devoid of introns. This fact allowed the construction of bacterial expression plasmids for the production of the recombinant proteins directly from genomic DNA fragments, obviating the need for the isolation of cDNAs for removal of intervening sequences.

Table 12.1. Characterization of equine interferons.

Interferons	Gene family	Antiviral activity of recombinant molecule	Available reagents for detection	Biological activity	Therapeutic applications
Type I					
IFN-α	≥ 7 members; 4 members cloned	Present	Gene probes	Not characterized yet	Not tested
IFN-β	At least 2 members; 1 member cloned	Present, about 5×10^8 IU mg^{-1} on equine dermal fibroblasts challenged with VSV	Neutralizing antibodies, antibodies for ELISA assay	Blood leukocytes show increased cytotoxicity, decreased proliferation and a variety of proteins are induced	Blood cells from treated horses completely inhibit the growth of a VSV test virus and have increased cytotoxicity
IFN-ω	≥ 8 members; 2 members cloned	Not characterized yet	Gene probes	Apparently no physiologic role in pregnancy	Not tested
Type II					
IFN-γ	1 member; gene and cDNA cloned	Present	Primers, probes and conditions for mRNA detection by polymerase chain reaction; attempts to make antibodies under way	mRNA present in blood lymphocytes and in the endometrium but not in trophoblast cells	Not tested

Equine Interferon-α

Nucleotide Sequence

Four highly similar IFN-α genes have been isolated and sequenced (Fig. 12.1). They encode polypeptides with 184 aa. The amino-terminal 23 residues constitute a typical hydrophobic signal sequence, when removed giving rise to mature proteins of 161 aa with a predicted mass of 18.5 kDa. *eqIFNA1* and *eqIFNA2* are very similar (98% identity at the nucleotide and 99% at the aa level) and might have evolved recently by gene duplication. The deduced aa sequence of mature eqIFN-α1 is identical to *eqIFNA3*. The low frequency of nucleotide changes between the eqIFN-α1 and eqIFN-α3 genes, even in non-coding regions, suggested the possibility of allelic variations of the same gene. The same could be true for the *eqIFNA2* and *eqIFNA4* genes. Southern blot analysis of the horse genome led to a minimum estimate of seven members of the IFN-α class.

The mature equine IFN-α proteins contain cysteine residues at the positions 1, 29, 99 and 139, which are highly conserved in almost all IFNs of this family in man, cattle, mouse and rat. In human IFNs these cysteine residues have been shown to form two intramolecular disulphide bridges from cysteine 1 to 99, and from 29 to 139, important for the correct folding and activity of the proteins (Wetzel *et al.*, 1981). eqIFN-α2 contains in addition a fifth, therefore most likely unpaired, cysteine residue at position 151. Like most IFN-αs of other species, none of the isolated equine IFN-αs possess potential *N*-linked glycosylation sites of the consensus sequence Asn-X-Ser/Thr, with X being any aa residue except proline. The described equine IFN-α proteins are about 5 aa shorter compared to most other IFN-α species.

Surprisingly, equine IFN-α proteins are most highly related to human IFN-α (71–77% aa identity), followed by bovine (65–67%), rat (61%) and murine (54–59%). The comparison with human IFN-α genes suggested a single bp deletion within codon 162 of the mature IFN of a common equine IFN-α ancestor gene causing a premature translation stop codon.

Antiviral Activity

Recombinant eqIFN-α1 and eqIFN-α2 present in lysates of *E. coli* bacteria transformed with the corresponding expression plasmids was stable to acid treatment (pH 2) and showed antiviral activity in cytopathic effect reduction assays on equine skin epidermis cells (NBL-6) challenged with VSV. Both IFNs also had considerable activity on human lung carcinoma cells (A549) challenged with encephalomyocarditis virus (EMCV), being only about one order of magnitude less active than in the homologous system. No larger quantities of rEqIFN-α proteins have been produced or purified yet. Therefore, no detailed studies could be performed on the biological and biochemical properties of this family of equine IFNs.

Natural equine IFN-α preparations obtained by induction of PBMC with Newcastle disease virus (NDV) showed greater activity on bovine and ovine cells than on equine cells (Yilma *et al.*, 1982). Natural IFN-α induced

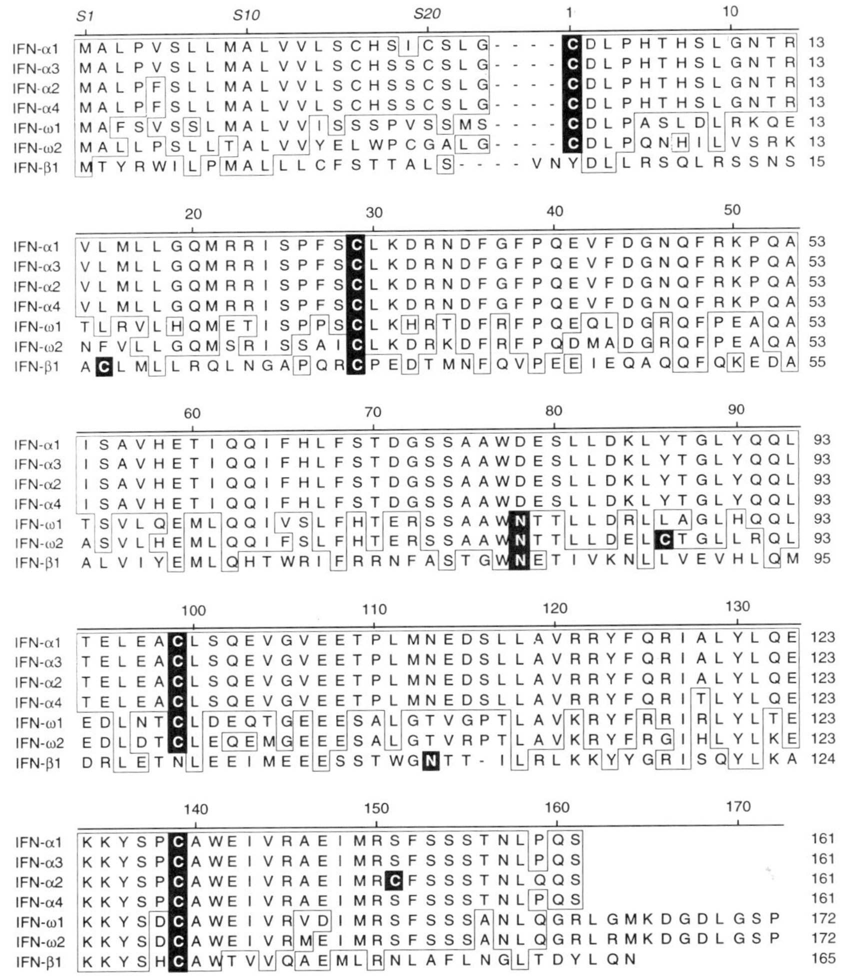

Fig. 12.1. Alignment of equine type I IFN aa sequences deduced from the gene structure. Residues in the signal sequences are indicated with an S prefix. Numbering was performed on the basis of the mature interferons in analogy to the human proteins. Dashes represent gaps introduced for optimal alignment. Amino acids conserved in most known equine type I IFNs are boxed. Cysteine residues are underlaid in black, **C**, as are potential *N*-linked glycosylation sites, **N**

from equine embryonic kidney cells by equine influenza virus had antiviral activity on the same cell type challenged with Semliki Forest virus (Edington *et al.*, 1989).

Equine Interferon-β

Nucleotide Sequence

One of at least two members of the equine IFN-β gene family has been isolated and characterized. The *eqIFNB1* gene encodes a polypeptide precursor of 186 aa, predicting – in analogy to IFN-β of other species – a 21 aa signal peptide followed by a mature IFN protein of 165 aa (Fig. 12.1). Of the known IFN-β primary structures, the mature equine protein shows most similarity to porcine (66%), followed by human (57%) and bovine (50–56%), and least to murine IFN-β (46%). Mature eqIFN-β1 contains three cysteine residues at positions 17, 31 and 140, which are conserved in man, cattle and pig. Cysteine residues 31 and 140 are likely to form a disulphide bridge stabilizing the protein conformation as shown in the human homologue. In contrast to the equine IFN-αs, IFN-β is most likely glycosylated, since it contains two potential *N*-linked glycosylation sites, the first of which (position 80–82 Asn-Glu-Thr) is conserved in the IFN-βs of man, pig and mouse and corresponds to the *N*-glycosylation site of the IFN-ωs, as well as that of most murine IFN-αs.

Antiviral Activity

The *eqIFNB1* gene expressed in *E. coli* yielded three orders of magnitude higher antiviral activity on equine dermal fibroblasts than *eqIFNA1* or *eqIFNA2* employing identical expression vectors and growth conditions (Himmler *et al.*, 1986). For this reason eqIFN-β1 was chosen as first equine type I IFN for production in larger quantities. eqIFN-β1 was prepared in highly purified form (> 98%) from recombinant *E. coli* extracts and subsequently further characterized by amino-terminal protein sequence analysis and determination of the aa composition. These analyses verified the predicted sequence deduced from the nucleotide sequence of the gene (Adolf *et al.*, 1990), except that the molecule contained an additional amino-terminal (initiation) methionine residue due to the cytoplasmic expression strategy employed. Bacterially produced unglycosylated eqIFN-β1 had a specific antiviral activity of about 5×10^8 antiviral units per milligram protein, determined on equine dermal fibroblasts challenged with VSV. Under *in vitro* conditions, 0.1 pg eqIFN-β1 ml^{-1} was already sufficient to completely inhibit the replication of a VSV test virus for a short period of time (Marquardt *et al.*, 1994). Increasing the eqIFN-β1 concentration to 1 ng ml^{-1} blocked the VSV replication for 2–3 days. eqIFN-β1 acts in a very species-specific way, showing very little cross-reactivity on cells derived from other species. The activity observed on human, murine, ovine or bovine cells challenged with VSV or EMCV was more than five orders of magnitude lower than in the homologous system (Adolf *et al.*, 1990). The cross-species reactivity of the total equine 'fibroblast' IFN reported by Yilma *et al.* (1982) may be due to an included minor fraction of IFN-α, which showed higher activity on bovine and ovine than on homologous cells.

Monoclonal Antibodies

A series of murine IgG mAbs were developed which neutralize the antiviral activity of eqIFN-β1 (Adolf *et al.*, 1990). None of these antibodies, nor polyclonal rabbit antiserum directed against eqIFN-β1, neutralized human IFN-β; neither was antiserum to human IFN-β able to neutralize eqIFN-β1. Two of the mAbs against eqIFN-β1 were used to develop a sensitive and rapid one-step solid-phase ELISA allowing the determination of eqIFN-β1 with a detection limit of approximately 3 ng ml^{-1}. The concentrations determined using the ELISA correlated very well with the biological assay measuring antiviral activity. eqIFN-β1 is acid stable, but heat labile. Incubation of the protein at 50°C for 4 h reduced the activity of the IFN by more than 95%, as determined both in the antiviral assay and the ELISA.

Equine Interferon-ω

This class of type I IFNs found first in man and cattle (Capon *et al.*, 1985; Hauptmann and Swetly, 1985) is characterized by an extended carboxy terminus. IFN-ω is more closely related to IFN-α than to IFN-β, but has antigenically distinct properties and does (at least in the human system) not cross-react with antibodies to other IFN classes (Adolf, 1987).

Nucleotide Sequence

Genomic analysis led to a minimal estimate of eight members of the *IFNω* gene family in the horse. Two of these genes, *eqIFNω1* and *eqIFNω2*, have been cloned. *eqIFNω2* was found to be closely linked to the *eqIFNA2* gene in the genome, separated by only 9 kb. This indicated that the equine type I IFN genes might also be clustered on a single chromosome as was shown in man and mouse (Coulombel *et al.*, 1991; Diaz *et al.*, 1994). Both eqIFN-ω genes encode mature IFNs of 172 aa preceded by a 23 aa long signal peptide. Starting with codon 29 of the mature protein these two genes are very similar. The deduced aa sequences revealed in both mature IFN-ω proteins conserved cysteine residues at positions 1, 29, 99 and 139, as in IFN-α. eqIFN-ω2 contains a fifth, probably unpaired cysteine residue at position 86, which corresponds in position to the fifth cysteine residue of human IFN-αD and IFN-α in mouse and rat. Both eqIFN-ωs contain a single potential *N*-linked glycosylation site (Asn-Thr-Thr) at positions 78–80, characteristic for IFN-β and IFN-ω of man and cattle.

Antiviral Activity

So far, no expression of recombinant eqIFN-ω or characterization of the proteins for their antiviral or other activities has been reported.

EQUINE TYPE II INTERFERON

Interferon-γ

Nucleotide Sequence

As in all other species analysed to date, horses contain a single copy gene for IFN-γ. The gene covers over about 4 kb. The coding region is distributed over four exons, interrupted by three introns (A. Himmler, unpublished results), as in the homologous genes of man, mouse, rat, dog and pig. IFN-γ is the only equine IFN the cDNA of which has been cloned additionally to the gene (Curran *et al.*, 1994; Grünig *et al.*, 1994). The cDNA was isolated from peripheral blood lymphocytes stimulated with Con A and phorbol myristate acetate by reverse transcription followed by PCR amplification using oligonucleotide primers from conserved regions of other species. The nucleotide sequence predicted a mature IFN-γ protein with 143 aa preceded by a 23 aa-long signal peptide (Fig. 12.2). The deduced aa sequence for mature horse IFN-γ is most similar to IFN-γ of sheep and dog (82%), cattle (81%), cervus (79%), cat (76%), pig (73%), followed by man and baboon (68%), rabbit (66%) and gerbil (55%). Much less similarity is found with mouse, rat (46%) and chicken (32%) sequences. eqIFN-γ contains two potential *N*-linked glycosylation sites. That close to the amino terminus (position 16–18) is conserved among all species, except for man and monkeys where the position of the glycosylation is shifted nine residues towards the carboxy terminus. The second *N*-glycosylation site in the middle of the polypeptide is conserved among the sequences of domestic animals. Except for chicken, all presently characterized IFN-γ proteins contain an additional potential *N*-glycosylation site in the region between residues 69 and 99. The regions between residues 51–90 and 126–145 are the most conserved between species. These regions contain helixes B, C or F in humans and rabbits and make up the core of the IFN-γ dimer (Ealick *et al.*, 1991; Samudzi *et al.*, 1991). Much less conserved are the amino and carboxy termini of the mature molecule which are thought to be critical for receptor binding or triggering of biologic responses (reviewed by Ealick *et al.*, 1991).

Antiviral Activity

Expression of the cDNA in *E. coli* resulted in active eqIFN-γ, which showed antiviral activity on equine cells but no cross-reactivity on human cells (G.R. Adolf, R. Hauptmann, A. Himmler, unpublished results).

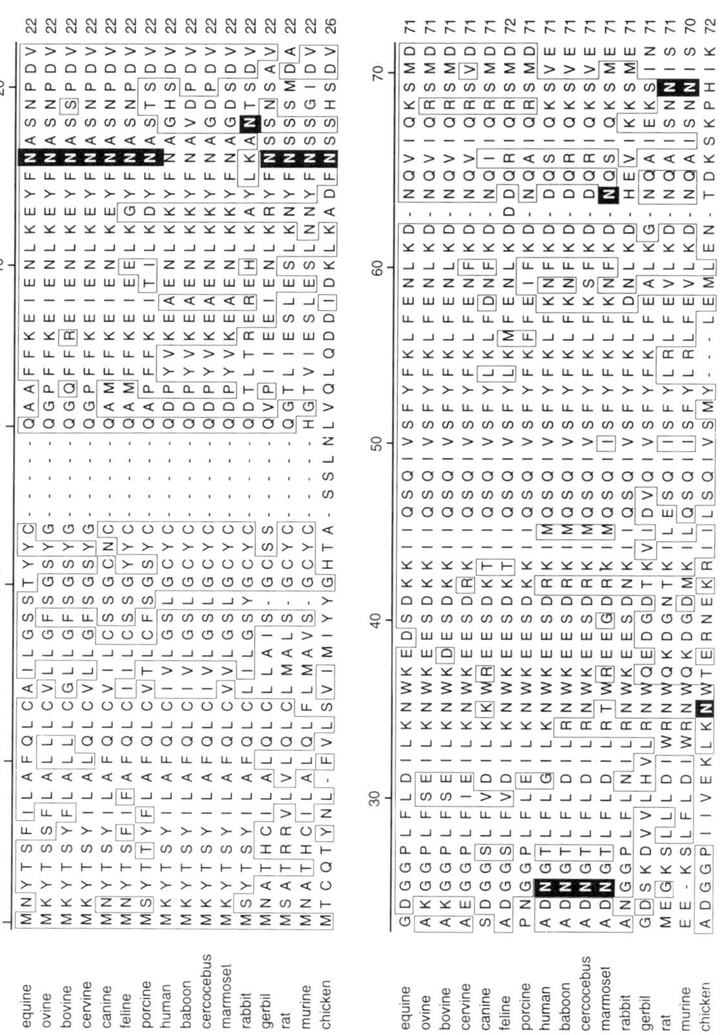

80 90 100 110

equine	T I K E D L F V - - K F F N S S T S K L E D F Q K L I Q I P V N D L K V Q R K A I S E L I K V M N D	119
ovine	I I K Q D M F Q - - K F L N G S S E K L E D F K R L I Q I P V D D L Q I Q R K A I N E L I K V M N D	119
bovine	I I K Q D M F Q - - K F L N G S S E K L E D F K K L I Q I P V D D L Q I Q R K A I N E L I K V M N D	119
cervine	I I K Q D M F Q - - K F L N G S S E K L E D F K K L I Q I S V D D M Q I Q R K A I N E L I K V M N D	119
canine	T I K E D M L G - - K F L N S S T S K R E D F L K L I Q I P V N D L Q V Q R K A I N E L I K V M N D	119
feline	T I K E D M L D - - K L L N T S S S K R D D F L K L I Q I P V N D L Q V Q R K A I N E L F K V M N D	120
porcine	V I K Q D M F Q - - R F L N G S S G K L N D F E K L I K I P V D N L Q I Q R K A I S E L I K V M N D	119
human	T I K E D M N V - - K F F N S N K K K R D D F E K L T N Y S V T D L N V Q R K A I H E L I Q V M A E	119
baboon	T I K E D I N V - - K F F N S N K K K W D D F E K L T N Y S V T D L N V Q R K A V H E L I Q V M A E	119
cercocebus	T I K E D I N V - - K F F N S N K K K R D D F E K L T N Y S V T D L N V Q R K A V H E L I Q V M A E	142
marmoset	T I K E D M N V - - K F F N S N K R K Q D D F E R L T N Y S V N D L N V Q R K A I H E L I Q V M A E	119
rabbit	S I K E D I F V - - K F F N S N L T K M D D F Q N L T R I S V D D R L V Q R K A V S E L S N V L N F	119
gerbil	A I R A D L I A - - N F F N N S E A K Y D G F M S I M K I E V N D P Q I Q S K A I N E L V K V M G H	119
rat	V I E S H L I T - - N F F S N S K A K K D A F M S I A K F E V N N P Q I Q H K A V N E L I R V I H Q	119
murine	V I E S H L I T - - T F F S N S K A K K D A F M S I A K F E V N N P Q V Q R Q A F N E L I R V V H Q	118
chicken	H I S E E L Y T L K N N L P D G V K K V K D I M D L A K L P M N D L R I Q R K A A N E L F S I L Q K	122

120 130 140 150

equine	L S P K A N L R K R K R S Q N P F R G R R A L Q	143
ovine	L S P K S N L R K R K R S Q N L F R G R R A S M	143
bovine	L S P K S N L R K R K R S Q N L F R G R R A S M	143
cervine	L S P K S N L I K R K R S Q N L F R G R R A S M	143
canine	L S P R S N L R K R K R S Q N L F R G R R A S K	143
feline	L S P R S N L R K R K R S Q N L F R G R R A S K	144
porcine	L S P R S N L R K R K R S Q T M F Q G Q R A S K	143
human	L S P A A K T G K R K R S Q M L F R G R R A S Q	143
baboon	L S P A A K I G K R K R S Q - M F R G R R A S Q	142
cercocebus	L S P A A K I G K R K R S Q - T F R G R R A S Q	142
marmoset	L S P A P K I G K R R R S Q T L F R G R R A S Q	143
rabbit	L S P K S N L K K R K R S Q T L F R G R R A S K Y	144
gerbil	L S P R V T L R K R K R S R C C F G G G N R L N K N N P A S T I	151
rat	L S P E S S L R K R K R S R C	133
murine	L L P E S S L R K R K R S R C	133
chicken	L V D P P S F - K R K R S Q S - - - Q R R C N C	142

Fig. 12.2. The primary structure of interferon-γ of various species as derived from cDNA or gene sequences present in the GenBank/EMBL databases. Dashes represent gaps introduced for better alignment. Residues conserved in the majority of IFN-γ proteins are boxed. Putative N-glycosylation sites are underlaid in black, **N**
The numbering was done according to the mature proteins, determined experimentally or in analogy to it.

BIOLOGICAL ROLE AND CLINICAL APPLICATIONS

Equine Interferon-β1

To assess the feasibility of IFN-β therapy, highly purified recombinant eqIFN-β1 made in the laboratory of G.R. Adolf was used in a series of horse studies performed *in vitro* and *in vivo* by the group around J. Marquardt, W. Leibold and E. Deegen in Hannover.

In Vivo *Studies*

Initially performed pharmacokinetic studies showed no difference between i.v. and i.m. eqIFN-β1 application. For this reason the animal studies were performed by i.m. injection of the IFN (Marquardt *et al.*, 1994). Addition of 1% (w/v) highly pure equine serum albumin with a very low endotoxin content was found advantageous for *in vivo* applications. A transient, dose-dependent increase in rectal temperature up to 39°C was observed with doses of 0.1–1 μg kg^{-1} body weight of eqIFN-β1, rectal temperatures peaked 8 h post-injection and were back to baseline 24 h post-injection. Also, slight leukopenia was observed 6 and 24 h after administration of eqIFN-β1 with full recovery at 48 h suggestive of a transiently altered homing of leukocytes rather than of leukodepression. No other side-effects were observed (Marquardt *et al.*, 1994).

The growth of a VSV test virus was completely inhibited in PBMC prepared from horses 6 h after treatment with eqIFN-β1 at 1 μg kg^{-1} body weight (Marquardt *et al.*, 1994). Blood mononuclear cells already obtained 3 h after treatment were 99% protected. Multiple administrations of IFN-β1 did not increase the duration of the antiviral state after each administration. But importantly, Marquardt *et al.* (1994) showed that it was possible to prolong the inhibition of VSV growth in PBMC for 2 weeks when horses were dosed with an initial injection of 1 μg kg^{-1} body weight of eqIFN-β1 followed by injections of 0.3 μg kg^{-1} body weight every other day.

Concomitant with the antiviral effect, neutrophil granulocytes obtained from those horses showed increased cytotoxic activity towards the human Jurkat T cell line (Heymer *et al.*, 1992). Also, blood cells prepared from horses treated with 1 μg eqIFN-β1 kg^{-1} body weight had reduced mitogen-induced proliferation. The antiproliferative effect was maximal after 6–24 h and lasted for up to 72 h (Marquardt *et al.*, unpublished). Together these data suggest that eqIFN-β1 induces an increased defence ability against virus-infected or tumour cells in the horse.

In Vitro *Studies*

EqIFN-β1 was also able to reduce the proliferation of mitogen-induced PBMC *in vitro* when used at high concentrations (Marquardt *et al.*, 1992). EqIFN-β1 also increased cytotoxicity of PBMC *in vitro* (Heymer *et al.*,

1992). Because eqIFN-β1 did not induce cytotoxicity in purified granulocytes *in vitro*, optimal induction of cytotoxicity most probably requires additional interactions with mononuclear cells (Heymer *et al.*, 1992). Culture of PBMC or neutrophil granulocytes with eqIFN-β1 influenced the expression and/or modification of a number of different proteins (Heinz *et al.*, 1994). In mononuclear cells, the expression of the equine analogue of the human Mx protein was upregulated. This might be functionally important, because of the well-studied inhibiting action of Mx proteins on influenza virus replication, most impressively shown in transgenic mouse models (Arnheiter *et al.*, 1990). In neutrophil granulocytes four proteins in the size range of 25 to 27 kDa were induced. These proteins could be cytokines such as IL-1, IL-6 or TNF-α (Heinz *et al.*, 1994).

Conclusions

To date, eqIFN-β1 has only been studied for its inhibiting activity against VSV. Therefore, no solid statement about the efficacy against other viral infections can be made. However, the quickly achieved protective effect suggests eqIFN-β1 to be suitable for the prophylaxis or treatment of infections with IFN-sensitive viruses. *In vivo* application of eqIFN-β1 in healthy horses indicated that this cytokine is well tolerated by the animals, showing no side-effects which would restrict the therapeutic use of eqIFN-β1.

Equine Interferon-ω

In ruminants (Roberts, 1989) and possibly rabbit (Mege *et al.*, 1991) a class of IFNs formerly classified as IFN-ω, now termed IFN-τ, represents the signal for maternal recognition of pregnancy. IFN-τ is produced by trophoblast cells and extends the life span of the corpus luteum. In contrast to sheep and cattle, conceptus membranes from mares pregnant for 11–16 days did not release any antiviral activity into the culture supernatant (Sharp *et al.*, 1989). Furthermore it was impossible to detect proteins of expected isoelectric point or molecular mass range (Sharp *et al.*, 1989). In a further study, Baker *et al.* (1991) were unable to detect trophoblast mRNA which would hybridize with a labelled eqIFN-ω gene probe.

Maintenance of the corpus luteum in early horse pregnancy does not appear to be due to the secretion of an IFN.

Equine Interferon-γ

Studies to elucidate the immunoregulatory role of eqIFN-γ have only been reported for pregnancy physiology thus far. Grünig and Antczak (1995) were looking for the role of cytokines that might mediate the interaction between the semi-allogeneic fetus and the maternal immune system. In the horse the maternal immune response to the invading trophoblast cells which form the endometrial cups is particularly striking (reviewed by Antczak and

Allen, 1989). The mare mounts a strong antibody response to paternal molecules expressed by the invading trophoblast cells and a dense infiltrate of maternal lymphocytes surrounds the endometrial cups. Therefore Grünig and Antczak (1995) determined whether mRNA for several cytokines including eqIFN-γ was expressed in the gravid endometrium or by horse trophoblasts. Preliminary results using Northern hybridization showed that horse peripheral blood lymphocytes stimulated with a mixture of mitogens (Con A and phorbol myristate acetate) express mRNA for IFN-γ from 2 to 24 h post-stimulation, peaking at 6 h post-stimulation (unpublished results). Using the sensitive RT-PCR, mRNA for IFN-γ could even be detected in unstimulated peripheral blood lymphocytes as well as in uterine tissue from the gravid and non-gravid horn. Trophoblasts from conceptuses recovered from mares pregnant for 30–55 days neither expressed mRNA for IFN-γ nor for IL-4, the immunologic counter-player of IFN-γ. Thus, trophoblasts of the horse do not directly communicate with maternal T cells through these classical T lymphocyte cytokines. However, the authors were unable to assess the role of IFN-γ in the endometrium because of the lack of a quantitative assay for eqIFN-γ.

The cloning of eqIFN-γ is a first step in examining the immunoregulatory role of eqIFN-γ in various physiologic and pathologic conditions. Further research will be greatly facilitated when antibodies to this protein become available for detection by ELISA or for neutralization. It can be hoped that this goal could be achieved in the near future because the cDNA of IFN-γ has been subcloned into a bacterial expression vector which is available for research use.

ACKNOWLEDGEMENTS

The authors thank their collaborators for all their support. The cloning and characterization of the equine type I IFNs was a collaborative effort between Drs Adolf, Hauptmann, Himmler and Swetly (Boehringer Ingelheim R&D Vienna). Equine IFN-γ was cloned and characterized in a collaboration between Drs Himmler, Hauptmann and Adolf (Boehringer Ingelheim R&D Vienna) and Drs Grünig and Antzcak (Cornell University, Ithaca, New York). The work described from Dr Antczak's laboratory was supported by USPHS grant HD-15799 and the Dorothy Russell Havemeyer Foundation Inc. The authors thank Dr Donna Rennick (DNAX Research Institute, Palo Alto) and Dr Günther R. Adolf (Boehringer Ingelheim R&D Vienna) for support and the critical reading of the manuscript.

Cytokines in the Pathogenesis of Equine Joint Disease

13

S.A. May

Royal Veterinary College, University of London, Hawkshead Lane, North Mymms, Hatfield, Hertfordshire AL9 7TA, UK

THE EARLY EVIDENCE

Two parallel, but as it turns out complementary, research paths resulted in the implication of cytokines, in particular IL-1, in joint disease in the late 1970s and early 1980s. Dame Honor Fell, working in the Strangeways Institute, Cambridge, UK, observed, *in vitro*, an intriguing interaction between porcine synovial tissue and articular cartilage (Fell and Jubb, 1977; Jubb and Fell, 1980). In the presence of minced synovium, proteoglycan loss from articular cartilage was accelerated, but, crucially, this depended on the presence of living chondrocytes. Thus it appeared that the synovial cells were producing a factor which stimulated proteoglycan-degrading enzyme production by chondrocytes. Named catabolin, from its effect on cartilage, this factor was subsequently identified as the pI 5 form of porcine (po) IL-1, namely poIL-1α (Pilsworth and Saklatvala, 1983; Saklatvala *et al.*, 1984).

Independently, in research in Harvard Medical School, USA, a mononuclear cell factor (MCF) was identified which stimulated increased production of collagenase and prostaglandin E2 (PGE2) by human synovial cells obtained from patients with rheumatoid arthritis (Dayer *et al.*, 1979). Like catabolin, this was shown to be IL-1 (Mizel *et al.*, 1981), in this case the human (hu) form, and subsequently recombinant huIL-1β has been shown to stimulate collagenase, stromelysin and PGE2 production by human synovial cells (Dayer *et al.*, 1986; Yaron *et al.*, 1987; Saus *et al.*, 1988).

OTHER 'INFLAMMATORY CYTOKINES'

Biological activity related to TNF has been recognized for much longer than IL-1, in the context of bacterial infections and endotoxin-related events (Coley, 1893; Carswell *et al.*, 1975). Originally, the tumour necrosis was thought to be due directly to material derived from Gram-negative bacteria. However, in 1975, a macrophage-derived factor, TNF, was identified which caused haemorrhagic necrosis of transplanted tumours in mice, as well as death of neoplastic cells *in vitro*. Similarities between IL-1 and TNF, in terms of their systemic inflammatory properties, led to a comparison of their local effects. Like IL-1, recombinant huTNF-α inhibits proteoglycan synthesis and stimulates proteoglycan release from porcine articular cartilage (Saklatvala, 1986). TNF also has similar effects to IL-1 on isolated cells (Bunning and Russell, 1989). Recombinant huTNF-α stimulates the production of stromelysin and PGE2 by human articular chondrocytes *in vitro*.

The other 'inflammatory cytokine' produced by resident articular cells is IL-6 (Guerne *et al.*, 1989, 1990). In inflammatory arthropathies in man, the concentrations of IL-6 in synovial fluids often exceed the concentrations of IL-1 and TNF. Systemically, IL-6 makes an important contribution to the acute phase response, by inducing the production of various acute phase proteins by hepatocytes (Geiger *et al.*, 1988; Castell *et al.*, 1989). The local role for IL-6 is less obvious (Mihara *et al.*, 1995). However, this may relate to early *in vitro* comparisons with IL-1 and TNF, in which cells, such as synovial fibroblasts, which responded to the latter cytokines in a variety of ways, including increased multiplication and increased production of enzymes and prostaglandins, failed to respond to IL-6. If the experiment is repeated with both IL-6 and soluble IL-6 receptors, rheumatoid synovial fibroblasts are stimulated to multiply, emphasizing how complex *in vivo* cytokine and cytokine receptor interactions may lead to effects not predicted in isolated cell systems.

CYTOKINE INHIBITORS AND BINDING PROTEINS

The identification and characterization of the cytokines has been paralleled by the discovery of various naturally occurring inhibitors of their activity, including the IL-1 receptor antagonist (ra), non-signalling cell surface receptors and soluble receptors (binding proteins). The protein sequence of huIL-1ra is sufficiently similar to huIL-1α (69% homology) and huIL-1β (72% homology) to have led to the suggestion that, in the case of IL-1, biological control of activity is, at least in part, exerted by the third member of an IL-1 family of proteins, IL-1ra, which binds to the IL-1 receptors without cell stimulation (Carter *et al.*, 1990; Eisenberg *et al.*, 1990; Hannum *et al.*, 1990).

The more common mechanism for biological control of inflammatory cytokine activity appears to be by soluble receptors. The type II IL-1 receptor, which selectively binds IL-1β but does not produce cell activation (Slack *et al.*, 1992), may be released by proteolytic cleavage to form a soluble (s) IL-1 binding protein (Symons *et al.*, 1991). Thus cell surface IL-1 receptor, sIL-1 receptor and IL-1ra may act together to control IL-1β activity *in vivo*. Similarly, two TNF binding proteins (BP), now known to be soluble receptors, which inhibit the activity of TNF, have been identified (Cope *et al.*, 1992; Roux-Lombard *et al.*, 1993). In contrast, as indicated above, a soluble IL-6 receptor both aids and enhances IL-6 activity (Mihara *et al.*, 1995), further complicating the cytokine network.

CYTOKINES IN HUMAN JOINT DISEASE

There is a marked difference between the so-called 'inflammatory cytokines' and the T cell cytokines in terms of their occurrence in human joint disease. IL-1 (Wood *et al.*, 1983; Hopkins *et al.*, 1988; Westacott *et al.*, 1990), TNF-α (Di Giovine *et al.*, 1988; Hopkins and Meager, 1988; Saxne *et al.*, 1988; Westacott *et al.*, 1990; Holt *et al.*, 1992; Yoshida *et al.*, 1992) and IL-6 (Houssiau *et al.*, 1988; Swaak *et al.*, 1988; Holt *et al.*, 1992; Yoshida *et al.*, 1992) have all been detected in relative abundance in synovial fluid samples from a variety of arthritides in man, whereas IL-2, IL-4, IFN-γ and TNF-β are present at much lower levels (Saxne *et al.*, 1988; Firestein and Zvaifler, 1990). In general, IL-1 and IL-6 concentrations were highest in the inflammatory arthritides such as rheumatoid arthritis, and lower in osteoarthritis. However, the results for TNF-α are less clear cut. In at least two studies, mean TNF-α concentrations were higher in samples from patients with osteoarthritis than those from patients with rheumatoid arthritis, and the presence of TNF-α appeared to correlate with disease chronicity (Di Giovine *et al.*, 1988; Westacott *et al.*, 1990).

Immunohistochemistry using mAbs has identified IL-1α, IL-1β, IL-6 and TNF-α, as well as other cytokines, in synovial membrane samples from patients with rheumatoid arthritis and osteoarthritis (Farahat *et al.*, 1993). In rheumatoid arthritis, increased levels of IL-1α together with IL-1 receptors (Deleuran *et al.*, 1992a) and TNF-α together with its receptors (Chu *et al.*, 1991; Deleuran *et al.*, 1992b) have been identified at the cartilage–pannus junction, where most of the cartilage destruction is taking place.

CYTOKINES *IN VIVO* – LABORATORY ANIMAL STUDIES

The recognition of the potential involvement of cytokines in naturally occurring joint diseases led to attempts to develop cytokine-based arthritis models in laboratory animals. A single dose of either purified huIL-1 or recombinant huIL-1β injected into the rabbit femorotibial joint reduced the articular cartilage proteoglycan concentration by nearly 20% at 24 h (Pettipher *et al.*, 1986). By 72 h, the cartilage proteoglycan concentration had returned to normal. The change in proteoglycan content was mirrored by an increase and then a decrease in PMN count in synovial washes. However, the leukocytes did not explain the loss of the proteoglycans from cartilage, as a similar increase in leukocyte numbers stimulated by the injection of bacterial LPS had no effect on the articular cartilage. Further work revealed that a single injection of recombinant huIL-1β into rabbit joints increased the concentrations of stromelysin and proteoglycans in synovial fluid (McDonnell *et al.*, 1992). The increased concentration of stromelysin closely correlated with increased mRNA production in synovial tissue and cartilage, supporting *in vitro* conclusions that resident articular cells might be the source of cartilage degrading enzymes in a variety of arthritides.

The intra-articular IL-1 model of joint disease was developed further by repeated daily injection for 7 days (Dingle *et al.*, 1987), and subsequently continuous infusion over a 14 day period (Feige *et al.*, 1989). These treatments resulted in marked loss of proteoglycans from articular cartilage, together with chronic synovitis including fibrosis and pannus formation.

In a slightly different approach to the question of cytokine involvement in arthritis, the production of cytokines by articular cells has been measured in other laboratory animal models of joint disease. A single impact to the canine patella results in fissuring of the articular cartilage in the region of the impact, and loss of safranin O staining of proteoglycans around these cracks (Pickvance *et al.*, 1993). Immunohistochemistry revealed increased levels of IL-1β, TNF-α and stromelysin around the cartilage fissures 14 days after injury.

Transection of the cranial cruciate ligament in the canine femorotibial joint is a well-established instability model of degenerative joint disease (Pond and Nuki, 1973; McDevitt *et al.*, 1977). Elevated concentrations of TNF and IL-6, but not IL-1, have been measured by bioassay in synovial fluid from this model, 12 weeks after surgery (Venn *et al.*, 1993). However, immunohistochemistry has identified increased levels of IL-1α and IL-1β in both cartilage and synovial tissue 12 weeks after cranial cruciate desmotomy (Pelletier *et al.*, 1993).

PROBLEMS WITH THE CYTOKINE STORY IN ARTHRITIS

A number of problems, in reconciling *in vitro* and *in vivo* experimental data with the cellular and molecular events seen in naturally occurring joint diseases, indicate that the sequence suggesting an inciting factor stimulates IL-1 and TNF production, which in turn induce collagenase and stromelysin production, leading to cartilage degradation is too simplistic. In general, in both early naturally occurring osteoarthritis (Mankin *et al.*, 1971) and early experimental osteoarthritis (Sandy *et al.*, 1984), an increase in proteoglycan synthesis is detected despite the progressive degeneration of articular cartilage. However, both IL-1 (Tyler, 1985) and TNF-α (Saklatvala, 1986) induce a reduction in proteoglycan synthesis at the same time as cartilage degradation. This apparent contradiction would appear to be explained by the more complex cytokine environment present *in vivo*. In cranial cruciate desmotomy-induced arthritis, despite the increase in TNF-α in synovial fluid, proteoglycan synthesis is increased (Venn *et al.*, 1993), suggesting the presence of other factors which overcome at least part of the action of TNF and stimulate proteoglycan synthesis.

The original research which characterized IL-1 used juvenile porcine cartilage, bovine nasal cartilage, or isolated articular cells. However, as scientists attempted *in vitro* to mimic human disease, by replacing animal cartilage with adult human cartilage, it became clear that additional factors played a role in the degradation of mature cartilage. Neither MCF (Huber-Bruning *et al.*, 1986) nor the recombinant cytokines IL-1α, IL-1β and TNF-α, by themselves or in combination (Hollander *et al.*, 1991), stimulated proteoglycan loss from adult human cartilage, although such loss was produced by synovial fluid from patients with rheumatoid and osteoarthritis (Hollander *et al.*, 1991). In a repeat of Dame Honor Fell's experiment, synovial fluid caused significantly less degradation of dead cartilage than living cartilage, indicating the likely involvement of both cytokines and at least one other catabolic cofactor.

Although IL-1 can stimulate the degradation of juvenile cartilage (Saklatvala *et al.*, 1984), and this process is associated with increased metalloproteinase production and inhibited by metalloproteinase inhibitors (Mort *et al.*, 1993), there is increasing evidence that the enzymatic cleavage of the core protein of the proteoglycan molecule, in pathological joints, is not due to the originally characterized metalloproteinases, collagenase, stromelysin and gelatinase (Sandy *et al.*, 1992; Hughes *et al.*, 1995). An as yet uncharacterized enzyme, 'aggrecanase', presumably a metalloproteinase, producing core protein cleavage at a different site from the other metalloproteinases, would appear to be responsible. Increased activity of this enzyme is also stimulated by IL-1 (Witt *et al.*, 1995).

INTERLEUKIN-1 IN EQUINE JOINT DISEASE

Equine Articular Cells *in Vitro*

The majority of work with equine cartilage has involved equine MCF or recombinant huIL-1α and -β, although the recent cloning of equine IL-1α and -β (Kato *et al.*, 1995) should allow the recombinant equine molecules to be investigated in the near future. Original characterization of equine MCF indicated that the main form of IL-1 present was IL-1β (May *et al.*, 1990) and this has been confirmed recently by dot blot analysis of total RNA preparations from stimulated equine PBMC using equine IL-1α and -β probes (Kato *et al.*, 1995). Equine PBMC supernatants contain both IL-1 and TNF activity when assayed in EL4.NOB-1/CTLL or D10.(N4)M and L929 bioassays, respectively. Inhibition of IL-1 bioactivity down to background levels has been demonstrated when excess recombinant huIL-1ra is incorporated in the D10 assay, and the similar stimulation curves generated by recombinant huIL-1β and equine IL-1, with varying concentrations of huIL-1ra, suggest that the IL-1 bioactivity is mediated solely through the IL-1 receptor (A.K. Austin and S.A. May, unpublished results). Similar inhibition of TNF activity is seen following the incorporation of either recombinant huTNF-BPI or II in the L929 assay (A.K. Austin and S.A. May, unpublished results). This has helped confirm the nature of the active factors, present in equine biological fluids, in these bioassays.

Both equine MCF and recombinant huIL-1, α or β, stimulated the production of PGE2 by equine chondrocytes and synovial cells *in vitro* (May *et al.*, 1992a, b). However, although a given concentration of MCF produced a broadly similar increase in PGE2 production by equine chondrocytes and synovial cells, in a series of experiments both recombinant huIL-1α and β consistently produced greater stimulation of equine synovial cells than chondrocytes. In addition, equine MCF stimulated stromelysin production by virtually all adult and foal articular cell lines tested, but, again in a series of experiments, only a single synovial cell line from a foal ever responded to huIL-1, in that case huIL-1β. This apparent partial restriction in the ability of recombinant huIL-1 to cross the species barrier was also investigated in an equine thymocyte assay (May *et al.*, 1992c). Recombinant huIL-1α and β failed to stimulate thymocyte proliferation in either a direct or a phytohaemagglutinin comitogenic assay. Similar results have been obtained in attempts to stimulate bovine thymocytes with recombinant murine IL-1α and huIL-1α (Lederer and Czuprynski, 1989).

As in other species, recombinant huIL-1 inhibits proteoglycan synthesis in equine articular cartilage (May, 1989; MacDonald *et al.*, 1992; Morris and Treadwell, 1994; Platt and Bayliss, 1994). Like other effects of IL-1 on equine tissues, this inhibition was more rapid in onset and more profound at a given concentration of huIL-1β than the same concentration of huIL-1α (May, 1989; Platt and Bayliss, 1994). The reduced number of newly synthesized proteoglycan molecules had the same hydrodynamic size and electrophoretic heterogeneity as the proteoglycans synthesized by unstimulated control cartilage, but their sulphation pattern was altered with a decrease in chondroitin-6-sulphate relative to chondroitin-4-sulphate

(Platt and Bayliss, 1994). A similar change in the chondroitin-6-sulphate : chondroitin-4-sulphate ratio is seen in osteoarthritis in man, and this has led to the suggestion that chondrocytes in degenerate cartilage are reverting to a more immature phenotype (Mankin and Lipiello, 1971).

As with human cartilage, a distinct difference is seen between equine cartilage from young and old horses in terms of the ability of huIL-1 to stimulate cartilage degradation. Concentrations of huIL-1 which significantly reduced proteoglycan synthesis failed to produce significant degradation of cartilage from older horses, although in some individuals a trend towards proteoglycan loss was seen (May, 1989; Morris and Treadwell, 1994). In one study which looked in more detail at the age effect, this change in responsiveness appeared to occur at about one year of age, with cartilage from animals younger than this showing both a reduction in proteoglycan synthesis and increased proteoglycan loss (MacDonald *et al.*, 1992). The difference between old and young cartilage may be quantitative rather than qualitative. Although, following stimulation with recombinant huIL-1β, adult equine chondrocytes failed to produce stromelysin which could be measured in a bioassay (May *et al.*, 1992a), immunoprecipitation using an antibovine stromelysin antibody revealed an increase in production of stromelysin in adult cartilage medium following stimulation with huIL-1α (Morris and Treadwell, 1994). However, these observations are of importance for those interested in conditions such as osteoarthritis. Inferences made on the actions of cytokines *in vivo* from work on juvenile cartilage may not apply to joint disease in what tends to be an older, adult population.

In addition to a paracrine action on neighbouring cells, the possibility of an autocrine action, amplifying chondrocyte stimulation, exists for IL-1 (May *et al.*, 1992b). Equine chondrocytes can be stimulated with bacterial LPS to produce IL-1 in a dose-dependent fashion. IL-1 production has also been seen with bovine and rat chondrocytes (Towle *et al.*, 1987; Rath *et al.*, 1988), and IL-1α and β mRNA has been detected in human chondrocytes (Ollivierre *et al.*, 1986).

Interleukin-1 *in Vivo*

Interleukin-1-like activity has been identified in the synovial fluids of horses with a variety of arthritides, including osteoarthritis, osteochondrosis, traumatic arthritis and infective arthritis (E.A. Morris *et al.*, 1990; Alwan *et al.*, 1991; Billinghurst *et al.*, 1995). Levels of IL-1 tended to be highest in animals with severe synovitis (Bertone *et al.*, 1993), such as those with joint infections (Alwan *et al.*, 1991). In addition, synovial fluid from normal horses, and horses with arthritis, inhibits both huIL-1α- and β-mediated PGE2 production by equine synovial cells *in vitro* (May *et al.*, 1992d). However, in more chronic joint disease there appeared to be a selective loss of a specific IL-1β inhibitor. A similar, specific IL-1β inhibitor, corresponding to the sIL-1R, has been identified in human synovial fluid (Symons *et al.*, 1991). As indicated above, this may play an important role in the control of IL-1 activity *in vivo*, with either an absolute increase in IL-1 concentration or a decrease in sIL-1R concentration resulting in an increase in IL-1 bioactivity.

The effects of the intra-articular injection of IL-1 have been investigated in three ponies (May, 1989): 24 h after the injection of 25 ng recombinant huIL-1β into the left radiocarpal joints, there was a marked increase in total nucleated cell count in synovial fluid ($95 \pm 15 \times 10^9\ l^{-1}$) compared to the opposite saline-treated joints ($12 \pm 5 \times 10^9\ l^{-1}$). A similar, though not statistically significant, increase was seen in the PGE2 concentration in synovial fluid (IL-1 treated: 588 ± 324 pg ml^{-1}; opposite control: 41 ± 16 pg ml^{-1}), although the ponies exhibited surprisingly little lameness at that stage. In the single animal which was sampled at 72 h, the total nucleated cell count had returned to $1.7 \times 10^9\ l^{-1}$ (opposite control $1.8 \times 10^9\ l^{-1}$), and the PGE2 concentration was also considerably reduced. The relative lack of swelling and pain in the IL-1-treated joints, despite the increase in total nucleated cell count in synovial fluid, was similar to previously reported findings following a single injection of IL-1 into the rabbit femorotibial joint (Pettipher *et al.*, 1986).

TUMOUR NECROSIS FACTOR AND INTERLEUKIN-6 IN EQUINE JOINT DISEASE

Both equine chondrocytes and synovial cells produce IL-6 *in vitro*. In one study, both early and late passage chondrocytes produced relatively high levels of IL-6 without stimulation (Kelly, 1994). Unstimulated synovial cells produced much lower levels of IL-6, but recombinant huTNF-α, huIL-1α and huIL-1β all stimulated IL-6 production. IL-1 appeared to be more potent than TNF as an inducer of IL-6, an observation which has also been made on human chondrocytes (Guerne *et al.*, 1990).

Intra-articular injection of bacterial endotoxin has been advocated as a model of acute synovitis and even infective arthritis in the horse (Firth *et al.*, 1987; Palmer and Bertone, 1994). The ability of bacterial LPS to stimulate cytokine production *in vitro*, and the likely involvement of cytokines in equine arthritis, led to the measurement of TNF and IL-6 in this model (Hawkins *et al.*, 1993). The TNF concentration in synovial fluid peaked 2 h after injection and returned to baseline levels by 8 h. IL-6 peaked later at 8 h, and was still significantly elevated at 6 days when the experiment concluded. Evidence from other species would suggest that this delayed IL-6 peak is stimulated by IL-1 and/or TNF (Shalaby *et al.*, 1989).

Tumour necrosis factor and IL-6 have also been measured in synovial fluids from clinical cases of equine arthritis (Bertone *et al.*, 1993; Billinghurst *et al.*, 1995). As with IL-1, concentrations of TNF and IL-6 were found to be highest in cases with acute synovitis, and, in these animals, IL-6 concentration correlated significantly with total nucleated cell count in synovial fluid (Bertone *et al.*, 1993). TNF concentrations were also measured in the synovial fluid of five racehorses, with arthritis secondary to intra-articular chip fractures, at 2, 12 and 26 h after racing (Billinghurst *et al.*, 1995). This revealed a significant increase in TNF concentration between 2 and 12 h, which was sustained at 26 h, suggesting direct stimulation of articular cells in diseased joints by routine competitive exercise.

THE FUTURE

In the horse, as in all domestic species, investigation of the role of cytokines in disease has been hampered by limited resources and the relatively small number of research groups working in this area. Even in a structure as well defined as the equine joint, the cytokine network is likely to be complex with similar interactions of the inflammatory cytokines, with each other, their naturally occurring inhibitors and other growth factors, to those seen in other species. However, the size of the horse means that large amounts of cartilage and synovial fluid can be obtained from both normal and abnormal joints, and even the joints of exercised and non-exercised horses. Therefore, direct comparisons are possible, potentially allowing elucidation of the different physiological and pathological processes involved in cartilage matrix turnover, with the volume of material allowing cytokine and growth factor profiles to be established in each situation.

The importance of joint disease, in particular osteoarthritis, as a limiting factor in the use of the horse in all equestrian pursuits, makes the unravelling of the puzzle of arthritis a key diagnostic and therapeutic target. In addition, the parallels between osteoarthritis in man and the horse mean that any advance in our understanding of equine arthritis is likely to be of direct benefit to our own species!

References for Part IV

Aderka, D., Le, J. and Vilcek, J. (1989) IL-6 inhibits lipopolysaccharide-induced tumor necrosis factor production in cultured human monocytes U937 cells and in mice. *Journal of Immunology* 143, 3517–3523.

Adolf, G.R. (1987) Antigenic structure of human interferon w1 (interferon $a_{II}1$): Comparison with other human interferons. *Journal of General Virology* 68, 1669–1676.

Adolf, G.R., Traxler, E. and Maurer-Fogy, I. (1990) Recombinant equine interferon-b_1: purification and preliminary characterization. *Journal of Interferon Research* 10, 255–267.

Allen, G., Green, E., Robinson, J., Garner, H., Loch, W. and Walsh, D. (1993) Serum tumor necrosis factor α concentrations and clinical abnormalities in colostrum-fed and colostrum-deprived neonatal foals given endotoxin. *American Journal of Veterinary Research* 54, 1404–1410.

Almawi, W.Y., Lipman, M.L., Stevens, A.C., Zanker, B., Hadro, E.T. and Strom, T.B. (1991) Abrogation of gluccorticoid-mediated inhibition of T cell proliferation by the synergistic action of IL-1, IL-6, and IFN γ. *Journal of Immunology* 146, 3523–3527.

Alwan, W.H., Carter, S.D., Dixon, J.B., Bennett, D., May, S.A. and Edwards, G.B. (1991) Interleukin-1-like activity in synovial fluids and sera of horses with arthritis. *Research in Veterinary Science* 51, 72–77.

Antczak, D.F. and Allen, W.R. (1989) Maternal immunological recognition of pregnancy in equids. *Journal of Reproduction and Fertility* Suppl. 37, 69–78.

Arnheiter, H., Skuntz, S., Noteburn, M., Chang, S. and Meier, E. (1990) Transgenic mice with intracellular immunity to influenza virus. *Cell* 62, 51–61.

Baker, C.B., Adams, M.H. and McDowell, K.J. (1991) Lack of expression of alpha or omega interferons by the horse conceptus. *Journal of Reproduction and Fertility* Suppl. 44, 439–443.

Banks, K. and Henson, J. (1972) Immunologically mediated glomerulitis of horses. II. Antiglomerular basement membrane antibody and other mechanisms of spontaneous disease. *Laboratory Investigation* 26, 708–715.

Bender, B. and Small, P. (1992) Influenza: pathogenesis and host defense. *Seminars in Respiratory Infection* 7, 38–45.

Bertone, A.L., Palmer, J.L. and Jones, J. (1993) Synovial fluid inflammatory mediators as markers of equine synovitis. *Veterinary Surgery* 22, 372.

Beutler, B.I., Milsark, W. and Cerami, A. (1985) Passive immunization against cachectin/ tumor necrosis factor (TNF) protects mice from the lethal effect of endotoxin. *Science* 229, 869–871.

Billinghurst, R.C., Fretz, P.B. and Gordon, J.R. (1995) Induction of intra-articular tumour necrosis factor during acute inflammatory responses in equine arthritis. *Equine Veterinary Journal* 27, 208–216.

Bradley, S.G. (1979) Cellular and molecular mechanisms of action of bacterial endotoxins. *Annual Review of Microbiology* 33, 67–94.

Breuhaus, B.A. and DeGraves, F.J. (1993) Plasma endotoxin concentrations in clinically normal and potentially septic equine neonates. *Journal of Veterinary Internal Medicine* 7, 296–302.

Brown, W.C., Woods, V.M., Chitko-McKown, C.G., Hash, S.M. and Rice-Ficht, A.C. (1994) Interleukin-10 is expressed by bovine type 1 helper, type 2 helper, and unrestricted parasite-specific T-cell clones and inhibits proliferation of all three subsets in an accessory-cell-dependent manner. *Infection and Immunity* 62, 4697–4708.

Bunning, R.A. and Russell, R.G.G. (1989) The effect of tumor necrosis factor α and γ-interferon on the resorption of human articular cartilage and on the production of prostaglandin E and of caseinase activity by human articular chondrocytes. *Arthritis and Rheumatism* 32, 780–784.

Burrell, R. (1990) Immunomodulation by bacterial endotoxin. *Critical Reviews in Microbiology* 17, 189–208.

Cannon, J., Tompkins, R., Gelfand, J., Michie, H., Stanford, G., Van der Meer, J., Endres, S., Lonneman, G., Corseti, J., Chernow, B., Wilmore, D., Wolf, S., Burke, J. and Dinarello, C. (1990) Circulating interleukin-1 and tumor necrosis factor in septic shock and experimental endotoxin fever. *Journal of Infectious Disease* 161, 79–84.

Capon, D.J., Shepard, H.M. and Goeddel, D.V. (1985) Two distinct families of human and bovine interferon-a genes are coordinately expressed and encode functional polypeptides. *Molecular and Cellular Biology* 5, 768–779.

Cargile, J.L., MacKay, R.J., Dankert, J.R. and Skelley, L. (1995) Effect of treatment with a monoclonal antibody against equine tumor necrosis factor (TNF) on clinical, hematologic, and circulating TNF responses of Miniature Horses given endotoxin. *American Journal of Veterinary Research* 56, 1451–1459.

Carswell, E.A., Old, L.J., Kassel, R.L., Green, S., Fiore, N. and Williamson, B. (1975) An endotoxin-induced serum factor that causes necrosis of tumors. *Proceedings of the National Academy of Sciences* 72, 3666–3670.

Carter, D.B., Deibel, M.R., Dunn, C.J., Tomich, C.-S.C., Laborde, A.L., Slightom, J.L., Berger, A.E., Bienkowski, M.J., Sun, F.F., McEwan, R.N., Harris, P.K.W., Yem, A.W., Waszak, G.A., Chosay, J.G., Sieu, L.C., Hardee, M.M., Zurcher-Neely, H.A., Reardon, I.M., Heinrikson, R.L., Truesdell, S.E., Shelly, J.A., Eessalu, T.E., Taylor, B.M. and Tracey, D.E. (1990) Purification, cloning, expression and biological characterization of an interleukin-1 receptor antagonist protein. *Nature* 344, 633–638.

Castell, J.V., Gomez-Lechon, M.J., David, M., Andus, T., Geiger, T., Trullenque, R., Fabra, R. and Heinrich, P.C. (1989) Interleukin-6 is the major regulator of acute phase protein synthesis in adult human hepatocytes. *Federation of European Biochemical Societies Letters* 242, 237–239.

Cheevers, W.P. and McGuire, T.C. (1985) Equine infectious anemia virus: Immunopathogenesis and persistence. *Reviews in Infectious Disease* 7, 83–88.

Chu, C.Q., Field, M., Feldmann, M. and Maini, R.N. (1991) Localization of tumor necrosis factor α in synovial tissues and at the cartilage/pannus junction in patients with rheumatoid arthritis. *Arthritis and Rheumatism* 34, 1125–1132.

Clerici, M. and Shearer, G.M. (1994) The TH1-TH2 hypothesis of HIV infection: new insights. *Immunology Today* 15, 575–581.

Coley, W.B. (1893) The treatment of inoperable sarcoma by the mixed toxins of erysipelas: with a report on ten original cases. *American Journal of Medical Science* 105, 487–511.

Cope, A.P., Aderka, D., Doherty, M., Engelmann, H., Gibbons, D., Jones, A.C., Brennan, F.M., Maini, R.N., Wallach, D. and Feldmann, M. (1992) Increased levels of soluble tumor necrosis factor receptors in the sera and synovial fluid of patients with rheumatic diseases. *Arthritis and Rheumatism* 35, 1160–1169.

Coulombel, C., Vodjdani, G. and Doly, J. (1991) Isolation and characterization of a novel interferon-a-encoding gene, IFN-a11, within a murine IFN cluster. *Gene* 104, 187–195.

Cross, R. (1966) Equine periodic ophthalmia. *Veterinary Record* 78, 8–13.

Curran, J.A., Argyle, D.J., Cox, P., Onions, D.E. and Nicolson, L. (1994) Nucleotide sequence of the equine interferon gamma cDNA. *DNA Sequence – The Journal of Sequencing and Mapping* 4, 405–407.

Dalton, D.K., Pitts-Meek, S., Keshav, S., Figari, I.S., Bradley, A. and Stewart, T.A. (1993) Multiple defects of immune cell function in mice with disrupted interferon-γ genes. *Science* 259, 1739–1742.

Darnell, J.E., Kerr, I.M. and Stark, G.R. (1994) Jak-STAT pathways and transcriptional activation in response to IFNs and other extracellular signaling proteins. *Science* 264, 1415–1421.

Dayer, J.-M., Breard, J., Chess, L. and Krane, S.M. (1979) Participation of monocyte-macrophages and lymphocytes in the production of a factor that stimulates collagenase and prostaglandin release by rheumatoid synovial cells. *Journal of Clinical Investigation* 64, 1386–1392.

Dayer, J.-M., de Rochemonteix, B., Burrus, B., Demczuk, S. and Dinarello, C.A. (1986) Human recombinant interleukin 1 stimulates collagenase and prostaglandin E2 production by human synovial cells. *Journal of Clinical Investigation* 77, 645–648.

Deleuran, B.W., Chu, C.Q., Field, M., Brennan, F.M., Katsikis, P., Feldmann, M. and Maini, R.N. (1992a) Localization of interleukin-1α, Type I interleukin-1 receptor and interleukin-1 receptor antagonist in the synovial membrane and cartilage/pannus junction in rheumatoid arthritis. *British Journal of Rheumatology* 31, 801–809.

Deleuran, B.W., Chu, C.Q., Field, M., Brennan, F.M., Mitchell, T., Feldmann, M. and Maini, R.N. (1992b) Localization of tumor necrosis factor receptors in the synovial tissue and cartilage/pannus junction in patients with rheumatoid arthritis. *Arthritis and Rheumatism* 35, 1170–1178.

Dennis, V.A., Klei, T.R. and Chapman, M.R. (1993) Generation and partial characterization of an eosinophil chemotactic cytokine produced by sensitized equine mononuclear cells stimulated with *Strongylus vulgaris* antigen. *Veterinary Immunology and Immunopathology* 37, 135–149.

Diaz, M.O., Pomykala, H.M., Bohlander, S.K., Maltepe, E., Malik, K., Brownstein, B. and Olopade, O.I. (1994) Structure of the human type-I interferon gene cluster determined from a YAC clone contig. *Genomics* 22, 540–552.

Di Giovine, F., Nuki, G. and Duff, G.W. (1988) Tumour necrosis factor in synovial exudates. *Annals of the Rheumatic Diseases* 47, 768–772.

Dinarello, C.A. (1988) Biology of interleukin 1. *FASEB Journal* 2, 108–115.

Dinarello, C., Cannon, J., Wolf, S., Bernheim, H., Cerami, A., Figari, I., Palladino, M., Jr and O'Connor, J. (1986) Tumor necrosis factor (cachetin) is an endogenous pyrogen and induces production of interleukin-1. *Journal of Experimental Medicine* 163, 1433–1450.

Dingle, J.T., Thomas, D.P.P., King, B. and Bard, D.R. (1987) *In vivo* studies of articular tissue damage mediated by catabolin/interleukin 1. *Annals of the Rheumatic Diseases* 46, 527–533.

Dirscherl, P., Grabner, A. and Buschmann, H. (1993) Responsiveness of basophil granulocytes of horses suffering from chronic obstructive pulmonary disease to various allergens. *Veterinary Immunology Immunopathology* 38, 217–227.

Durando, M.M., MacKay, R.J., Linda, S. and Skelley, L.A. (1994) Effects of polymyxin B and *Salmonella typhimurium* antiserum on horses given endotoxin intravenously. *American Journal of Veterinary Research* 55, 921–927.

Ealick, S., Cook, W.J., Vijay-Kumar, S., Carson, M., Nagabhushan, T.L., Trotta, P.P. and Bugg, C.E. (1991) Three-dimensional structure of recombinant human interferon-γ. *Science* 252, 698–702.

Edington, N., Bridges, C.G. and Griffiths, L. (1989) Further investigations of equine interferons *in vitro*. *Journal of Interferon Research* 9, 385–387.

Eisenberg, S.P., Evans, R.J., Arend, W.P., Verderber, E., Brewer, M.T., Hannum, C.H. and Thompson, R.C. (1990) Primary structure and functional expression from complementary DNA of a human interleukin-1 receptor antagonist. *Nature* 343, 341–346.

Else, K.J., Finkelman, F.D., Maliszewski, C.R. and Grencis, R.K. (1994) Cytokine mediated regulation of chronic intestinal helminth infection. *Journal Experimental Medicine* 179, 347–356.

Farahat, M.N., Yanni, G., Poston, R. and Panayi, G.S. (1993) Cytokine expression in synovial membranes of patients with rheumatoid arthritis and osteoarthritis. *Annals of the Rheumatic Diseases* 52, 870–875.

Feige, U., Karbowski, A., Rordorf-Adam, C. and Pataki, A. (1989) Arthritis induced by continuous infusion of hr-interleukin-1 α into the rabbit knee joint. *International Journal of Tissue Reactions* 11, 225–238.

Fell, H.B. and Jubb, R.W. (1977) The effect of synovial tissue on the breakdown of articular cartilage in organ culture. *Arthritis and Rheumatism* 20, 1359–1371.

Fenwick, B.W., Schore, C.E. and Osburn, B.I. (1988) Human Recombinant Interleukin-2(125) induced *in vitro* proliferation of equine caprine, ovine, canine and feline peripheral blood lymphocytes. *Immunology and Microbiology of Infectious Disease* 11, 51–60.

Firestein, G.S. and Zvaifler, N.J. (1990) How important are T cells in chronic rheumatoid synovitis? *Arthritis and Rheumatism* 33, 768–773.

Firth, E.C., Wensing, T. and Seuren, F. (1987) An induced synovitis disease model in ponies. *Cornell Veterinarian* 77, 107–118.

Flores, I., Mariano, T.M. and Pestka, S. (1991) Human interferon omega (w) binds to the a/b receptor. *Journal of Biological Chemistry* 266, 19875–19877.

Folkman, J. and Ingber, D. (1992) Inhibition of angiogenesis. *Seminars in Cancer Biology* 3, 89–96.

Fujimiya, Y., Perryman, L.E. and Crawford, T.B. (1979) Leukocyte cytotoxicity in a persistent virus infection: Presence of direct cytotoxicity but absence of antibody-dependent cellular cytotoxicity in horses infected with equine infectious anemia virus. *Infection and Immunity* 24, 618–626.

Galin, J.I. (1995) Interferon-γ in the management of infectious diseases. *Annales of Internal Medicine* 123, 216–224.

Geiger, T., Andus, T., Klapproth, J., Hirano, T., Kishimoto, T. and Heinrich, P.C. (1988) Induction of rat acute-phase proteins by interleukin 6 *in vivo*. *European Journal of Immunology* 18, 717–721.

Gill, H.S., Watson, D.L. and Brandon, M.R. (1993) Monoclonal antibody to $CD4^+$ T-cells abrogates genetic resistance to *Haemonchus contortus* in sheep. *Immunology* 78, 43–49.

Goodkin, D.E. (1994) Interferon beta-1b. *Lancet* 344, 1057–1060.

Gresser, I. (1991) Antitumour effects of interferons: past present and future. *British Journal of Haematology* 79, 1–5.

Greural, A., Rouse, B. and Babiuk, L. (1977) Mechanism of resistance to herpesvirus: comparison of resistance of the effectiveness of different cell types in mediating antibody-dependent cytotoxicity. *Infection and Immunity* 15, 698–705.

Grunfeld, C. and Palladino, M., Jr (1990) Tumor necrosis factor: immunologic, antitumor, metabolic, and cardiovascular activities. *Advances in Internal Medicine* 35, 45–72.

Grünig, G. and Antczak, D.F. (1995) Horse trophoblasts produce tumor necrosis factor alpha but not interleukin 2, interleukin 4, or interferon gamma. *Biology of Reproduction* 52, 531–539.

Grünig, G., Himmler, A. and Antczak, D.F. (1994) Cloning and sequencing of equine interferon-γ cDNA. *Immunogenetics* 39, 448–449.

Guerne, P.-A., Zuraw, B.L., Vaughan, J.H., Carson, D.A. and Lotz, M. (1989) Synovium as a source of interleukin 6 *in vitro*: Contribution to local and systemic manifestations of arthritis. *Journal of Clinical Investigation* 83, 585–592.

Guerne, P.-A., Carson, D.A. and Lotz, M. (1990) IL-6 production by human articular chondrocytes: Modulation of its synthesis by cytokines, growth factors and hormones *in vitro*. *Journal of Immunology* 144, 499–505.

Gutterman, J.U. (1994) Cytokine therapeutics: Lessons from interferon a. *Proceedings of the National Academy of Science, USA* 91, 1198–1205.

Halliwell, R.E.W., McGorum, B.C., Irving, P. and Dixon, P.C. (1993) Local and systemic antibody production in horses affected with chronic obstructive pulmonary disease. *Veterinary Immunology and Immunopathology* 38, 201–215.

Han, J., Brown, T. and Beutler, B. (1990) Endotoxin-responsive sequences control cachectin/tumor necrosis factor biosynthesis at the translational level. *Journal of Experimental Medicine* 171, 465–475.

Hannant, D. and Mumford, J.A. (1989) Cell mediated immune responses in ponies following infection with equine influenza virus (H3N8): the influence of induction culture conditions on the properties of cytotoxic effector cells. *Veterinary Immunology and Immunopathology* 21, 327–337.

Hannum, C.H., Wilcox, C.J., Arend, W.P., Joslin, F.G., Dripps, D.J., Heimdal, P.L., Armes, L.G., Sommer, A., Eisenberg, S.P. and Thompson, R.C. (1990) Interleukin-1 receptor antagonist activity of a human interleukin-1 inhibitor. *Nature* 343, 336–340.

Hauptmann, R. and Swetly, P. (1985) A novel class of human type I interferons. *Nucleic Acids Research* 13, 4739–4749.

Hawkins, D.L., MacKay, R.J., Gum, G.G., Colahan, P.T. and Meyer, J.C. (1993) Effects of intra-articularly administered endotoxin on clinical signs of disease and synovial fluid tumor necrosis factor, interleukin 6, and prostaglandin E2 values in horses. *American Journal of Veterinary Research* 54, 379–386.

Heinz, H., Marquardt, J., Schuberth, H.J., Adolf, G.R. and Leibold, W. (1994) Proteins induced by recombinant equine interferon-b_1 within equine peripheral blood mononuclear cells and polymorphonuclear neutrophilic granulocytes. *Veterinary Immunology and Immunopathology* 42, 221–235.

Heinzel, F.P., Sadick, M.D., Holaday, B.J., Coffman, R.L. and Locksley, R.M. (1989) Reciprocal expression of interferon-γ or IL-4 during the resolution or progression of murine leishmaniasis. Evidence for expansion of distinct helper T cell subsets. *Journal of Experimental Medicine* 169, 59–65.

Henson, J. and McGuire, T. (1971) Immunopathology of equine infectious anemia. *American Journal of Clinical Patholology* 56, 306–313.

Heymer, J., Marquardt, J., Heinz, H., Adolf, G.R. and Leibold, W. (1992) 'Discordant' influence of equine recombinant interferon-b_1 on the cytotoxic capacity of equine polymorphonuclear neutrophils and peripheral blood mononuclear cells *in vitro* and *in vivo*. *Journal of Interferon Research* 12, 415–423.

Himmler, A., Hauptmann, R., Adolf, G.R. and Swetly, P. (1986) Molecular cloning and expression in *Escherichia coli* of equine type I interferons. *DNA* 5, 345–356.

Hirano, T., Akira, S., Taga, T. and Kishimoto, T. (1990) Biological and clinical aspects of interleukin-6. *Immunology Today* 12, 443–449.

Hollander, A.P., Atkins, R.M., Eastwood, D.M., Dieppe, P.A. and Elson, C.J. (1991) Degradation of human cartilage by synovial fluid but not cytokines *in vitro*. *Annals of the Rheumatic Diseases* 50, 57–58.

Holsti, M.A. and Raulet, D.H. (1989) IL-6 and IL-1 synergize to stimulate IL-2 production and proliferation of peripheral T cells. *Journal of Immunology* 143, 2514–2519.

Holt, I., Cooper, R.J., Denton, J., Meager, A. and Hopkins, S.J. (1992) Cytokine interrelationships and their association with disease activity in arthritis. *British Journal of Rheumatology* 31, 725–733.

Hopkins, S.J. and Meager, A. (1988) Cytokines in synovial fluid: II. The presence of tumour necrosis factor and interferon. *Clinical and Experimental Immunology* 73, 88–92.

Hopkins, S.J., Humphreys, M. and Jayson, M.I.V. (1988) Cytokines in synovial fluid. I. The presence of biologically active and immunoreactive IL-1. *Clinical and Experimental Immunology* 72, 422–427.

Hormanski, C.E., Truax, R., Pourciau, S.S., Folsom, R.W. and Horohov, D.W. (1992) Induction of lymphokine-activated killer cells of equine origin: specificity for equine target cells. *Veterinary Immunology and Immunopathology* 32, 25–36.

Horohov, D.W., Pourciau, S.S., Swiderski, C.E. and E.V. Vandergrifft. (1996) Biological characterization of recombinant equine interleukin-2 (IL-2) and interleukin-4 (IL-4). *Veterinary Immunology and Immunopathology* (in press).

Houssiau, F.A., Devogelaer, J.P., Van Damme, J. and Dedeuxchaisnes, C.N. (1988) Interleukin-6 in synovial fluid and serum of patients with rheumatoid arthritis and other inflammatory arthritides. *Arthritis and Rheumatism* 31, 784–788.

Hovda, L. (1992) Neonatal septicemia. In: Robinson, N. (ed.), *Current Therapy in Equine Medicine*. W.B. Saunders Co., Philadelphia, Volume 3, pp. 435–442.

Hsieh, C.S., Macatonia, S.E., Tripp, C.S., Wolf, S.F., OGarra, A. and Murphy, K.M. (1993) Development of Th1 $CD4^+$ T cells through IL-12 produced by *Listeria*-induced macrophages. *Science* 260, 547–549.

Huber-Bruning, O., Wilbrink, B., Vernooij, J.E., Bijlsma, J.W.J., Den Otter, W. and Huber, J. (1986) Contrasting *in vitro* effects of retinol and mononuclear cell factor on young and old human cartilage. *Journal of Pathology* 150, 21–27.

Hughes, C.E., Caterson, B., Fosang, A.J., Roughley, P.J. and Mort, J.S. (1995) Monoclonal antibodies that specifically recognise neoepitope sequences generated by aggrecanase and matrix metalloproteinase cleavage of aggrecan: Application to catabolism *in situ* and *in vitro*. *Biochemical Journal* 305, 799–804.

Isaacs, A. and Lindemann, J. (1957) Virus interference. I: The interferons. *Proceedings of the Royal Society, London (Biology)* 147, 258–267.

Issel, C.J., Horohov, D.W., Lea, D.F., Adams, W.V., Hagius, S.D., McManus, J.M., Allison, A.C. and Montelaro, R.C. (1992) Efficacy of inactivated whole-virus and subunit vaccines in preventing infection and disease caused by equine infectious anemia virus. *Journal of Virology* 66, 3398–3408.

Johnson, H.M., Bazer, F.W., Szente, B.E. and Jarpe, M.A. (1994) How interferons fight disease. *Scientific American* 270, 68–75.

Jubb, R.W. and Fell, H.B. (1980) The effect of synovial tissue on the synthesis of proteoglycan by the articular cartilage of young pigs. *Arthritis and Rheumatism* 23, 545–555.

Kamijo, R., Le, J., Shapiro, D., Havell, E.A., Huang, S., Aguet, M., Bosland, M. and Vilcek, J. (1993) Mice that lack the interferon-γ receptor have profoundly altered responses to infection with Bacillus Calmette-Guerin and to subsequent challenge with lipopolysaccharide. *Journal of Experimental Medicine* 178, 1435–1440.

Kapsenberg, M.L., Jansen, H.M., Bos, J.D. and Wierenga, E.A. (1992) Role of type 1 and type 2 T helper cells in allergic diseases. *Current Opinion in Immunology* 4, 788–793.

Kato, H., Ohashi, T., Nakamura, N., Nishimura, Y., Watari, T., Goitsuka, R., Tsujimoto, H. and Hasegawa, A. (1995) Molecular cloning of equine interleukin-1 α and β cDNAs. *Veterinary Immunology and Immunopathology* 48, 221–231.

Kelly, L. (1994) The Role of Cytokines in Equine Joint Diseases. PhD thesis. University of Liverpool, UK.

Kishimoto, T. (1989) The biology of interleukin-6. *Journal of The American Society of Hematology* 74, 1–10.

Kono, Y., Sentsui, H. and Murakami, Y. (1976) Development of specific *in vitro* lymphocyte stimulation responses in horses infected with equine infectious anemia virus. *Veterinary Microbiology* 1, 31–43.

Lederer, J.A. and Czuprynski, C.J. (1989) Species preference of bovine thymocytes and fibroblasts for bovine interleukin 1. *Veterinary Immunology and Immunopathology* 23, 213–222.

Liles, W.C. and Van Voorhis, W.C. (1995) Review: nomenclature and biologic significance of cytokines involved in inflammation and the host immune response. *Journal of Infectious Disease* 172, 1573–1580.

Lue, C., Kiyono, H., Mcghee, J.R., Fujihashi, K., Kishimoto, T., Hirano, T. and Mestecky, J. (1991) Recombinant human interleukin-6 (rhIL-6) promotes the terminal differentiation of *in vivo*-activated human B-cells into antibody-secreting cells. *Cellular Immunology* 132, 423–432.

MacDonald, M.H., Stover, S.M., Willits, N.H. and Benton, H.P. (1992) Regulation of matrix metabolism in equine cartilage explant cultures by interleukin 1. *American Journal of Veterinary Research* 53, 2278–2285.

MacKay, R. and Lester, G. (1992) Induction of the acute phase cytokine, hepatocyte-stimulating factor/interleukin 6, in the circulation of horses treated with endotoxin. *American Journal of Veterinary Research* 53, 1285–1289.

MacKay, R.J., Merritt, A.M., Zertuche, J.M., Whittington, M. and Skelley, L.A. (1991) Tumor necrosis factor activity in the circulation of horses given endotoxin. *American Journal of Veterinary Research* 52, 533–538.

Magnuson, N.S., Namen, A.E., Perryman, L.E. and Magnuson, J.A. (1983). Studies on interleukin-2 from horses. *Federation Proceedings* 42, 446.

Mankin, H.J., Dorfmann, H., Lipiello, L. and Zarins, A. (1971) Biochemical and metabolic abnormalities in articular cartilage from osteoarthritic human hips – II. Correlation of morphology with biochemical and metabolic data. *Journal of Bone and Joint Surgery (American)* 53, 523–537.

Mankin, H.J. and Lipiello, L. (1971) The glycosaminoglycans of normal and arthritic cartilage. *Journal of Clinical Investigation* 50, 1712–1719.

Marquardt, J., Heymer, J., Heinz, H., Deegen, E., Adolf, G.R. and Leibold, W. (1992) Monitoring of effects induced by recombinant equine interferon-b1 in whole blood and separated fractions of peripheral blood of horses. *Zentralblatt für Veterinärmedizin Reihe B* 39, 327–336.

Marquardt, J., Heymer, J., Heinz, H., Adolf, G.R. and Deegen, E. (1994) Kinetics of inhibition of replication of vesicular stomatitis virus in blood mononuclear cells of horses after *in vitro* and *in vivo* treatment with recombinant equine interferon-b1. *American Journal of Veterinary Research* 55, 1093–1100.

May, S.A. (1989) Mediators of Equine Joint Disease. PhD thesis. The Royal Veterinary College, University of London, UK.

May, S.A., Hooke, R.E. and Lees, P. (1990) The characterisation of equine interleukin-1. *Veterinary Immunology and Immunopathology* 24, 169–175.

May, S.A., Hooke, R.E. and Lees, P. (1992a) Interleukin-1 stimulation of equine articular cells. *Research in Veterinary Science* 52, 342–348.

May, S.A., Hooke, R.E. and Lees, P. (1992b) Equine chondrocyte activation by a variety of stimuli. *British Veterinary Journal* 148, 389–397.

May, S.A., Hooke, R.E. and Lees, P. (1992c) Species restrictions demonstrated by the stimulation of equine cells with recombinant human interleukin-1. *Veterinary Immunology and Immunopathology* 30, 373–384.

May, S.A., Hooke, R.E. and Lees, P. (1992d) Inhibition of interleukin-1 activity by equine synovial fluid. *Equine Veterinary Journal* 24, 99–102.

McDevitt, C., Gilbertson, E. and Muir, H. (1977) An experimental model of osteoarthritis: Early morphological and biological changes. *Journal of Bone and Joint Surgery (British)* 59, 24–35.

McDonnell, J., Hoerrner, L.A., Lark, M.W., Harper, C., Dey, T., Lobner, J., Eiermann, G., Kazazis, D., Singer, I.I. and Moore, V.L. (1992) Recombinant human interleukin-1 β-induced increase in levels of proteoglycans, stromelysin, and leukocytes in rabbit synovial fluid. *Arthritis and Rheumatism* 35, 799–805.

McGuire, T. and Poppie, M. (1973) Hypogammaglobulinemia and thymic hypoplasia in horses: a primary combined immunodeficiency disorder. *Infection and Immunity* 8, 272–277.

Mege, D., Lefevre, F. and Labonnardiere, C. (1991) The porcine family of interferon-ω: cloning, structural analysis, and functional studies of five related genes. *Journal of Interferon Research* 11, 341-350.

Meyer, T.A., Wang, J., Tiao, G.M., Ogle, C.K., Fischer, J.E. and Hasselgren, P.O. (1995) Sepsis and endotoxemia stimulate intestinal interleukin-6 production. *Surgery* 118, 336–342.

Mihara, M., Moriya, Y., Kishimoto, T. and Ohsugi, Y. (1995) Interleukin-6 (IL-6) induces the proliferation of synovial fibroblastic cells in the presence of soluble IL-6 receptor. *British Journal of Rheumatology* 34, 321–325.

Mitchell, D.M., McMichael, A.J. and Lamb, J.R. (1985) The immunology of influenza. *British Medical Bulletin* 41, 80–85.

Mizel, S.B., Dayer, J.-M., Krane, S.M. and Mergenhagen, S.E. (1981) Stimulation of rheumatoid synovial cell collagenase and prostaglandin production by partially purified lymphocyte-activating factor (interleukin 1). *Proceedings of the National Academy of Sciences, USA* 78, 2474–2477.

Monahan, C., Taylor, H., Chapman, M. and Klei, T. (1995) Experimental immunization of ponies with *Strongylus vulgaris* radiation-attenuated larvae or crude soluble somatic extracts from larvae or adult stages. *Journal of Parasitology* 80, 911–923.

Morris, D. (1991) Endotoxemia in horses. A review of cellular and humoral mediators involved in its pathogenesis. *Journal of Veterinary Internal Medicine* 5, 167–181.

Morris, D. and Moore, J. (1991) Tumor necrosis factor activity in the serum of neonatal foals with presumed septicemia. *Journal of the American Veterinary Medical Association* 199, 1584–1589.

Morris, D., Crowe, N. and Moore, J. (1990a) Correlation of clinical and laboratory data with serum tumor necrosis factor activity in horses with experimentally induced endotoxemia. *American Journal of Veterinary Research* 51, 1935–1940.

Morris, D., Moore, J., Fischer, K. and Tarleton, R.L. (1990b) Endotoxin-induced tumor necrosis factor activity production by equine peritoneal macrophages. *Circulatory Shock* 30, 229–236.

Morris, D., Moore, J. and Crowe, N. (1991) Serum tumor necrosis factor activity in the circulation of horses with colic attributable to gastrointestinal tract disease. *American Journal of Veterinary Research* 52, 1565–1569.

Morris, D.D., Crowe, N., Moore, J.N. and Moldawer, L.L. (1992) Endotoxin-induced production of interleukin 6 by equine peritoneal macrophages *in vitro*. *American Journal of Veterinary Research* 53, 1298–1301.

Morris, E.A. and Treadwell, B.V. (1994) Effect of interleukin 1 on articular cartilage from young and aged horses and comparison with metabolism of osteoarthritic cartilage. *American Journal of Veterinary Research* 55, 138–146.

Morris, E.A., McDonald, B.S., Webb, A.C. and Rosenwasser, L.J. (1990) Identification of interleukin-1 in equine osteoarthritic joint effusions. *American Journal of Veterinary Research* 51, 59–64.

Mort, J.S., Dodge, G.R., Roughley, P.J., Liu, J., Finch, S., DiPasquale, G. and Poole, A.R. (1993) Direct evidence for active metalloproteinases mediating matrix degradation in interleukin 1-stimulated human articular cartilage. *Matrix* 13, 95–102.

Mosmann, T.R. (1991) Regulation of immune responses by T cells with different cytokine secretion phenotypes: role of a new cytokine, cytokine synthesis inhibitory factory (IL10). *International Archives of Allergy and Applied Immunology* 94, 110–115.

Mosmann, T.R. (1994) Cytokine patterns during the progression to AIDS. *Science* 265, 193–194.

Mosmann, T.R. and Coffman, R.L. (1989) TH1 and TH2 cells: Different patterns of lymphokine secretion lead to different functional properties. *Annual Review of Immunology* 7, 145–173.

Mosmann, T.R., Cherwinski, H., Bond, M.W., Giedlin, M.A. and Coffman, R.L. (1986) Two types of murine helper T cell clone. I. Definition according to profiles of lymphokine activities and secreted proteins. *Journal of Immunology* 136, 2348–2357.

Nelson, K., Olsen, C., Schram, B., McGregor, M., Hinshaw, V. and Lunn, D. (1995) Local and circulating antibody isotype responses to equine influenza virus infection or vaccination. *International Veterinary Immunology Symposium Proceedings* IV, 261.

Newman, M.J., Issel, C.J., Truax, R.E., Powell, M.D., Horohov, D.W. and Montelaro, R.C. (1991) Transient suppression of equine immune responses by equine infectious anemia virus. *Virology* 184, 55–66.

Okamura, H., Tsutsul, H., Komatsu, T., Yutsudo, M., Hakura, A., Tanimoto, T., Torigoe, K., Okura, T., Nukada, Y., Hattori, K., Akita, K., Namba, M.,

Tanabe, F., Konishi, K., Fukuda, S. and Kurimoto, M. (1995) Cloning of a new cytokine that induces IFN-γ production by T cells. *Nature* 378, 88–91.

Ollivierre, F., Gubler, U., Towle, C.A., Laurencin, C. and Treadwell, B.V. (1986) Expression of IL-1 genes in human and bovine chondrocytes: A mechanism for autocrine control of cartilage matrix degradation. *Biochemical and Biophysical Research Communications* 141, 904–911.

Palmer, J.L. and Bertone, A.L. (1994) Experimentally-induced synovitis as a model for acute synovitis in the horse. *Equine Veterinary Journal* 26, 492–495.

Paul, W.E. and Seder, R.A. (1994) Lymphocyte responses and cytokines. *Cell* 76, 241-251.

Pelletier, J.-P., Faure, M.-P., DiBattista, J.A., Wilhelm, S., Visco, D. and Martel-Pelletier, J. (1993) Coordinate synthesis of stromelysin, interleukin-1, and oncogene proteins in experimental osteoarthritis. *American Journal of Pathology* 142, 95–105.

Pettipher, E.R., Higgs, G.A. and Henderson, B. (1986) Interleukin-1 induces leucocyte infiltration and cartilage proteoglycan degradation in the synovial joint. *Proceedings of the National Academy of Sciences, USA* 83, 8749–8753.

Pickvance, E.A., Oegema, T.R. and Thompson, R.C. (1993) Immunolocalization of selected cytokines and proteases in canine articular cartilage after transarticular loading. *Journal of Orthopaedic Research* 11, 313–323.

Pilsworth, L.M.C. and Saklatvala, J. (1983) The cartilage-resorbing protein catabolin is made by synovial fibroblasts and its production is increased by phorbol myristate acetate. *Biochemical Journal* 216, 481–489.

Platt, D. and Bayliss, M.T. (1994) An investigation of the proteoglycan metabolism of mature equine articular cartilage and its regulation by interleukin-1. *Equine Veterinary Journal* 26, 297–303.

Pond, M.J. and Nuki, G.A. (1973) Experimentally induced osteoarthritis in the dog. *Annals of the Rheumatic Diseases* 32, 387–388.

Pond, L., Wassom, D.L. and Hayes, C.E. (1992) Influence of resistant and susceptible genotype, IL-1, and lymphoid organ on *Trichinella spiralis*-induced cytokine secretion. *Journal of Immunology* 149, 957–965.

Rath, N.C., Oronsky, A.L. and Kerwar, S.S. (1988) Synthesis of interleukin-1-like activity by normal rat chondrocytes in culture. *Clinical Immunology and Immunopathology* 47, 39–46.

Roberts, R.M. (1989) A novel group of interferons associated with the early ovine and bovine embryo. *Journal of Interferon Research* 9, 373–378.

Roberts, R.M., Cross, J.C. and Leaman, D.W. (1992) Interferons as hormones of pregnancy. *Endocrinology Review* 13, 432–452.

Robinson, J., Allen, G., Green, E., Garner, H., Loch, W. and Walsh, D. (1993) Serum interleukin-6 concentrations in endotoxin-infused neonatal foals. *American Journal of Veterinary Research* 54, 1411–1414.

Rode, H.J., Janssen, W., Rosen-Wolff, A., Bugert, J.J., Thein, P., Becker, Y. and Darai, G. (1993) The genome of equine herpesvirus type 2 harbors an interleukin 10 (IL10)-like gene. *Virus Genes* 7, 111–116.

Romagnani, S. (1991) Human TH1 and TH2 subsets: doubt no more. *Immunology Today* 12, 256–257.

Roux-Lombard, P., Punzi, L., Hasler, F., Bas, S., Todesco, S., Gallati, H., Guerne, P.A. and Dayer, J.-M. (1993) Soluble tumor necrosis factor receptors in human inflammatory synovial fluids. *Arthritis and Rheumatism* 36, 485–489.

Rwambo, P.M., Issel, C.J., Adams, W.V., Hussain, K.A., Miller, M. and Montelaro, R.C. (1990) Equine infectious anemia virus (EIAV) humoral response of recipient ponies and antigenic variation during persistent infection. *Archives in Virology* 111, 119–126.

Saklatvala, J. (1986) Tumour necrosis factor α stimulates resorption and inhibits synthesis of proteoglycan in cartilage. *Nature* 322, 547–549.

Saklatvala, J., Pilsworth, L.M.C., Sarsfield, S.J., Gavrilovic, J. and Heath, J.K. (1984) Pig catabolin is a form of interleukin 1. *Biochemical Journal* 224, 461–466.

Samudzi, C.T., Burton, L.E. and Rubin, J.R. (1991) Crystal structure of recombinant rabbit interferon-γ at 2.7 Å resolution. *Journal of Biological Chemistry* 266, 21791–21797.

Sandy, J.D., Adams, M.E., Billingham, M.E.J., Plaas, A. and Muir, H. (1984) *In vivo* and *in vitro* stimulation of chondrocyte biosynthetic activity in early experimental osteoarthritis. *Arthritis and Rheumatism* 27, 388–397.

Sandy, J.D., Flannery, C.R., Neame, P.J. and Lohmander, L.S. (1992) The structure of aggrecan fragments in human synovial fluid: Evidence for the involvement in osteoarthritis of a novel proteinase which cleaves the Glu373–Ala374 bond of the interglobular domain. *Journal of Clinical Investigation* 89, 1512–1516.

Saus, J., Quinones, S., Otani, Y., Nagasi, H., Harris, E.D. and Kurkinen, M. (1988) The complete primary structure of human matrix metalloproteinase-3: Identity with stromelysin. *Journal of Biological Chemistry* 263, 6742–6745.

Saxne, T., Palladino, M.A., Heinegard, D., Talal, N. and Wollheim, F.A. (1988) Detection of tumor necrosis factor α but not tumor necrosis factor β in rheumatoid arthritis synovial fluid and serum. *Arthritis and Rheumatism* 31, 1041–1045.

Schultz, C.L. and Coffman, R.L. (1991) Control of isotype switching by T cells and cytokines. *Current Opinion in Immunology* 3, 350–354.

Seethanathan, P., Bottoms, G.D. and Schafer, K. (1990) Characterization of release of tumor necrosis factor, interleukin-1, and superoxide anion from equine white blood cells in response to endotoxin. *American Journal of Veterinary Research* 51, 1221–1225.

Sellon, D.C., Fuller, F.J. and McGuire, T.C. (1994) The immunopathogenesis of equine infectious anemia virus. *Virus Research* 32, 111–138.

Shalaby, M.R., Waage, A., Aarden, L. and Espevik, T. (1989) Endotoxin, tumor necrosis factor-α and interleukin 1 induce interleukin 6 production *in vivo*. *Clinical Immunology and Immunopathology* 53, 488–498.

Sharp, D.C., McDowell, K.J., Weithenauer, J., Franklin, K., Mirando, M. and Bazer, F. (1989) Is an interferon-like protein involved in the maternal recognition of pregnacy in mares? *Equine Veterinary Journal* Suppl. 8, 7–9.

Slack, J., McMahan, C.J., Waugh, S., Schooley, K., Spriggs, M.K., Sims, J.E. and Dower, S.K. (1992) Independent binding of interleukin-1α and interleukin-1β to type I and type II interleukin-1 receptors. *Journal of Biological Chemistry* 268, 2513–2524.

Su, X., Morris, D. and McGraw, R. (1991) Cloning and characterization of gene TNF-α encoding equine tumor necrosis factor α. *Gene* 107, 319–321.

Su, X., Morris, D.D., Crowe, N.A., Moore, J.N., Fischer, K.J. and McGraw, R. A. (1992) Equine tumor necrosis factor α: cloning and expression in *Escherichia coli*, generation of monoclonal antibodies, and development of a sensitive enzyme-linked immunosorbent assay. *Hybridoma* 11, 715–727.

Swaak, A.J., Van Rooye, A., Niewenhuis, E. and Aarden, L.A. (1988) Interleukin-6 (IL-6) in synovial fluid and serum of patients with rheumatic diseases. *Scandinavian Journal of Rheumatology* 17, 469–474.

Swiderski, C. and Horohov, D. (1995) Molecular cloning and sequencing of equine interleukin 10 (IL-10) and β-actin. *Conference of Research Workers in Animal Diseases Abstracts,* Chicago, Illinois.

Symons, J.A., Eastgate, J.A. and Duff, G.W. (1991) Purification and characterization of a novel soluble receptor for interleukin-1. *Journal of Experimental Medicine* 174, 1251–1254.

Tavenor, A.S., Allen, W.R. and Butcher, G.W. (1993) cDNA cloning of equine interleukin-2 by polymerase chain reaction. *Equine Veterinary Journal* 25, 242–243.

Tiao, G., Rafferty, J., Ogle, C., Fischer, J.E. and Hasselgren, P.O. (1994) Detrimental effect of nitric oxide synthase inhibition during endotoxemia may be caused by high levels of tumor necrosis factor and interleukin-6. *Surgery* 116, 332–337.

Towle, C.A., Trice, M.E., Ollivierre, F., Awbrey, B.J. and Treadwell, B.V. (1987) Regulation of cartilage remodelling by IL-1: Evidence for autocrine synthesis of IL-1 by chondrocytes. *Journal of Rheumatology* Suppl. 14, 1–13.

Tyler, J.A. (1985) Articular cartilage cultured with catabolin (pig interleukin 1) synthesizes a decreased number of normal proteoglycan molecules. *Biochemical Journal* 227, 869–878.

Urban, J.F., Madden, K.B., Svetic, A., Cheever, A., Trotta, P.P., Gause, W.C., Katona, I.M. and Finkelman, F.D. (1992) Importance of TH2 cytokines in protective immunity to nematodes. *Immunology Reviews,* 127, 205–220.

Vandergrifft, E.V. and Horohov, D.W. (1993) Molecular cloning and expression of equine interleukin 2. *Veterinary Immunology and Immunopathology,* 39, 395–406.

Vandergrifft, E.V., Swiderski, C.E. and Horohov, D.W. (1994) Molecular cloning and sequencing of equine interleukin 4. *Veterinary Immunology and Immunopathology,* 40, 379–384.

Venn, G., Nietfeld, J.J., Duits, A.J., Brennan, F.M., Arner, E., Covington, M., Billingham, M.E.J. and Hardingham, T.E. (1993) Elevated synovial fluid levels of interleukin-6 and tumor necrosis factor associated with early experimental canine osteoarthritis. *Arthritis and Rheumatism* 36, 819–826.

Westacott, C.I., Whicher, J.T., Barnes, I.C., Thompson, D., Swan, A.J. and Dieppe, P.A. (1990) Synovial fluid concentration of five different cytokines in rheumatic diseases. *Annals of the Rheumatic Diseases* 49, 676–681.

Wetzel, R., Perry, L.J., Estell, D.A., Lin, N., Levine, H.L., Slinker, B., Fields, F., Ross, M.J. and Shively, J. (1981) Properties of a human alpha-interferon purified from *E. coli* extracts. *Journal of Interferon Research* 1, 381–390.

Wilson, G. and Miles, A. (1948) *Topley and Wilson's Principles of Bacteriology and Immunity.* Edward Arnold, London.

Witt, M., Fosang, A.J., Hughes, C.E. and Hardingham, T.E. (1995) Changes in the patterns of aggrecan cleavage in cartilage explants following stimulation with

IL-1 α or retinoate and inhibition of cleavage by a metalloproteinase inhibitor. *Transactions of the 41st Annual Meeting of the Orthopaedic Research Society* 20, 122.

Wood, D.D., Ihrie, E.J., Dinarello, C.A. and Cohen, P.L. (1983) Isolation of an interleukin-1-like factor from human joint effusions. *Arthritis and Rheumatism* 26, 975–983.

Yaron, I., Meyer, F.A., Dayer, J.-M. and Yaron, M. (1987) Human recombinant interleukin-1β stimulates glycosaminoglycan production in human synovial fibroblasts. *Arthritis and Rheumatism* 30, 424–430.

Yilma, T., McGuire, T.C. and Perryman, L.E. (1982) Preliminary characterization of equine interferons and their antiviral activities on bovine, ovine, and human cells. *Journal of Interferon Research* 2, 363–370.

Yoshida, K., Kobayashi, K., Yamagata, N., Iwabuchi, H., Katsura, T., Sugihara, S., Negishi, M., Ide, H., Mori, Y. and Takahashi, T. (1992) Inflammatory cytokines and enzymes in synovial fluid of patients with rheumatoid arthritis and other arthritides. *International Archives of Allergy and Immunology* 98, 286–292.

V

Canine Cytokines

Edited by E. Pinelli

Cytokines in Canine Visceral Leishmaniasis

14

E. Pinelli

Department of Immunology, Faculty of Veterinary Medicine, Yalelaan 1, 3584 CL Utrecht, The Netherlands

INTRODUCTION

Visceral leishmaniasis (VL) is a parasitic disease caused by protozooan parasites of the *Leishmania donovani* complex. It is estimated that 200 million people are at risk of contracting the disease with approximately 10,000 new cases annually (Ashford *et al.*, 1992).The disease is generally zoonotic, dogs being the main parasite reservoir and humans only infected secondarily. In Mediterranean countries VL affects both man and dogs and is caused by *Leishmania infantum* (Laison and Shaw, 1987). In man the disease is characterized by irregular fever, anaemia, hypergammaglobulinaemia, wasting and enlargement of the liver and spleen. Dogs develop similar clinical signs of the disease in addition to enlargement of lymph nodes, dermatitis around the eyes and nose, alopecia and increased nail growth. In man and dogs the disease may have a fatal outcome.

Studies in mouse models have demonstrated that the generation of protective immunity against leishmaniasis is T cell and cytokine mediated (Scott *et al.*, 1989; Liew, 1990; Cox and Liew, 1992). Thus, the absence of such immune responses has been associated with chronicity of the disease and high mortality in untreated cases (Sacks *et al.*, 1987; Carvallo *et al.*, 1989). However, studies on the protective role of T cells (Bradonisio *et al.*, 1988; Abranches *et al.*, 1991; Cabral *et al.*, 1992) and cytokines in natural host models of VL, such as the dog, are limited. Studies carried out by our group have focused on the identification and characterization of the factors and mechanisms involved in protective immunity in dogs infected with

Leishmania infantum. We observed that at 3 years post-challenge, a group of experimentally infected beagles (Nieto *et al.*, 1991) comprised asymptomatic, apparently resistant dogs and symptomatic animals, susceptible to the experimental infection. The availability of dogs either resistant or susceptible to *Leishmania* prompted us to examine the roles of antibodies, T cells and cytokines in *Leishmania* infections in these dogs.

DEVELOPMENT OF PROTECTIVE IMMUNITY TO *LEISHMANIA* PARASITES IN EXPERIMENTALLY INFECTED DOGS

For the experimental infection of beagle dogs, infective doses of *L. infantum* from experimentally infected sandflies were prepared (Killick-Kendrick *et al.*, 1994). The infective dose included the addition of sandfly salivary gland extracts which have been suggested to improve parasite infectivity (Titus and Ribiero, 1988). In this study, dogs were infected by the i.d. inoculation of $5–8 \times 10^3$ metacyclic promastigotes harvested from the midguts of sandflies that were experimentally infected by allowing them to feed on a dog with clinical VL. Evidence that the dogs became infected after inoculation of metacyclic promastigotes was obtained by isolation of parasites from these animals.

An interesting finding from the experimentally infected outbred beagle population was that although parasites were isolated from most of these animals (Killick-Kendrick *et al.*, 1994), some of them developed clinical signs of VL whereas others did not. Results from these studies demonstrate that experimental infections of dogs with *L. infantum* may result in canine VL, associated with impaired cellular immune responses. Thus, we observed that PBMC from symptomatic animals failed to respond to parasite antigens both *in vitro* as indicated by lymphocyte stimulation tests and *in vivo* as indicated by absence of DTH reactions (Table 14.1; Pinelli *et al.*, 1994). Similar observations have been reported in humans (Haldar *et al.*, 1983; Sacks *et al.*, 1987). Carvalho and colleagues described that also in American VL there is an impaired proliferative response of lymphocytes from patients to parasite antigens and that this is not due to reduced numbers of circulating T cells (Carvalho *et al.*, 1981).

An important immunological feature of VL, also in the dog, is a marked humoral response. We found a considerable increase in serum Ig levels, in particular of the IgG and IgM classes (unpublished results). This increase in serum Ig levels leads to a decrease of the albumin/globulin ratio often found in dogs with VL (Slappendel, 1988). The increase in serum Igs appears to be mainly due to an increased production of Ig as a result of polyclonal B cell activation (Galvão-Castro *et al.*, 1984). High antileishmanial antibody titres

Table 14.1. Immunological parameters of asymptomatic and symptomatic, experimentally *L. infantum*-infected dogs.

	Asymptomatic	Symptomatic
Lymphocyte proliferation to LSA	++	–
Lymphocyte proliferation to Con A	+++	+++
Lymphocyte proliferation to Dohyvac	+	+
DTH reaction to leishmanin	++	±
Serum anti-leishmanial antibodies	±	+++
Parasite isolation	+	+
Symptoms of visceral leishmaniasis	±	+++

+ and ± refer to relative values for each parameter studied; LSA: *Leishmania* soluble antigen; DTH: delayed-type hypersensitivity.

could be detected using an immunofluorescence technique in all experimentally infected dogs with symptoms of leishmaniasis. Serum antileishmanial titres in asymptomatic dogs were either absent or low. Although some of these animals developed serum antileishmanial titres during the 5 year observation period, they were transient and low compared to those found in symptomatic dogs (Table 14.1).

On the basis of the results obtained in these studies it is now apparent that a correlation exists between disease and marked humoral responses whereas the development of a strong parasite-specific cellular immune response seems to prevail in apparently resistant dogs. However, these immunological parameters are not necessarily predictive of the state of resistance of these animals to *Leishmania* parasites. They may, nevertheless, be useful parameters when evaluating a candidate vaccine in dogs.

The approach for the development of a successful vaccine against leishmaniasis includes the identification of the factors that would induce a protective cellular immune response. Stimulation of the appropriate T cell subpopulations must be considered, since studies in murine models of leishmaniasis have demonstrated that T cells are involved in both protection and exacerbation of the disease.

Protective Th1 Cells Versus Disease-promoting Th2 Cells

Infection of different strains of mice with *L. major* and studies of T cell transfer and depletion in these mice have led to the discovery of protective and disease-promoting $CD4^+$ Th cell subpopulations (Liew, 1989; Müller *et al.*, 1989). Studies by Mosmann *et al.* demonstrated that $CD4^+$ T cell clones could be divided into two groups based on their cytokine secretion pattern in response to specific antigen (Mosmann *et al.*, 1986). Th1 clones were found to produce IL-2 and IFN-γ whereas Th2 clones secreted IL-4, IL-5 and IL-10. Assessment of the cytokine mRNA levels in lymph nodes and spleens isolated from resistant *L. major*-infected C57BL/6 and healed BALB/c mice indicated higher levels of IFN-γ mRNA, whereas IL-4 and IL-10 mRNA were hardly detected. In susceptible BALB/c mice that developed lesions upon infection, considerably less IFN-γ but higher levels of IL-4 and IL-10 mRNA were found (Bogdan *et al.*, 1993; Liew and O'Donell, 1993).

Analysis of the cytokine secretion pattern of PBMC from asymptomatic apparently resistant dogs indicated that protective immunity to *L. infantum* is associated with a Th1-like response. This conclusion is based on results indicating that stimulation of PBMC from asymptomatic animals with parasite antigens or mitogens resulted in higher production of IL-2, TNF and IFN-γ compared to symptomatic dogs (Fig. 14.1) (Pinelli *et al.*, 1994, 1995). PBMC from symptomatic animals failed to respond *in vitro* to parasite antigens both by proliferation and IFN-γ production. These cells secreted TNF and IL-2 but considerably lower amounts compared to secretion of PBMC derived from asymptomatic dogs. Similar observation on the pattern

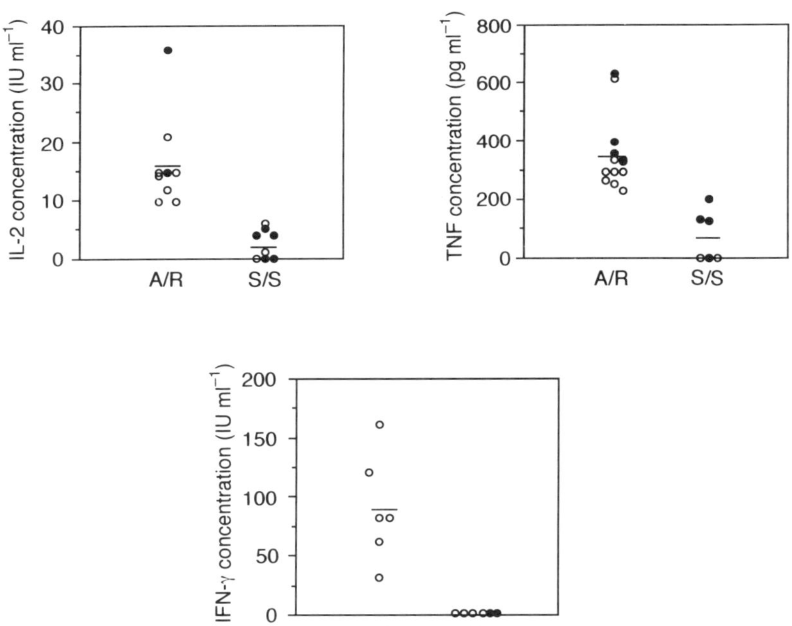

Fig. 14.1. Cytokine production by canine PBMC from *L. infantum*-infected asymptomatic/resistant (A/R) and symptomatic/susceptible (S/S) dogs. Cytokines were tested from 48-h supernatants of stimulated PBMC derived from experimentally (empty circles) and naturally infected (closed circles) mixed breed dogs. Horizontal line indicates mean value of cytokine production.

of cytokine secretion was made with dogs exposed to natural infections with this parasite. Whether susceptibility to *L. infantum* infections in the dog is associated with a Th2-like response is a question still to be addressed. At present our group is interested in developing tools to determine canine cytokines such as IL-10 and IL-4. So far, we have focused on the characterization of the protective immune response associated with resistance to VL in the dog. The identification as well as the understanding of mechanisms involved in protection are essential for designing a candidate vaccine. In addition to the protective role of cytokines produced by Th1 $CD4^+$ cells we have studied the role for $CD8^+$ T cells in protective immunity against VL in the dog.

Role of $CD8^+$ T Cells in Protective Immunity

Although most of the attention has been given to $CD4^+$ T cells, several studies show that also $CD8^+$ T cells may play a role in protective immunity against leishmaniasis. In addition to the segregation of murine $CD4^+$ T cells into Th1 and Th2 cells, it was recently shown that murine $CD8^+$ T cells also segregate (Ferrick *et al.*, 1995). Although $CD8^+$ T cells alone could not confer protection (Moll *et al.*, 1988) depletion studies of $CD8^+$ T cells in BALB/c mice have shown that these cells contribute to the resolution of the disease and resistance against reinfection (Bogdan *et al.*, 1993). The beneficial effect of $CD8^+$ T cells has been suggested to be due to the secretion of IFN-γ which is required for the activation of macrophages and the suppression of Th2 cells (Müller *et al.*, 1989, 1993; Chan, 1993).

Our studies indicate that $CD8^+$ T cells are also involved in protective immunity to VL in the dog (Pinelli *et al.*, 1995). Using depletion studies we found that these cells when derived from asymptomatic but not from symptomatic animals were not only capable of producing IFN-γ but could also lyse infected macrophages in a MHC-restricted manner. *Leishmania* parasites are intracellular pathogens residing in the endosomal compartment of macrophages, and whether parasite antigens could be presented in the context of MHC class I molecules was not clear. For other intracellular pathogens (e.g. *Listeria monocytogenes*), which are capable of egressing into the cytosol, there is no conceptual problem concerning the delivery of parasite antigen into the MHC class I pathway of presentation. Recently, lysis of *Leishmania*-infected macrophages by MHC class I-restricted $CD8^+$ CTL was also shown in mice that had resolved a primary infection (Da Conceicão-Silva *et al.*, 1994). In the past years, it has been clarified that antigens derived from phagosomes may also enter the MHC class I pathway (Rock *et al.*, 1990; Pfeifer *et al.*, 1993) by an alternative pathway responsible for protein/peptide transport from the endosome into the cytosol (Kovacsovics-Bankowsk and Rock, 1994; Rock, 1996).

Our studies indicate a role for parasite-specific CTL in protective immunity against *L. infantum*. Although these parasite-specific CTL are capable of lysing *Leishmania*-infected macrophages, we did not observe killing of intracellular parasites. Similar results have been obtained by others using alloimmune cytolytic T cells, which showed that *L. major*-infected macrophages could be lysed, however, releasing live amastigotes (Pham and Mauël, 1987). In an indirect assay system using a $CD8^+$ *Plasmodium*-specific CTL clone, it was shown that extended incubation of the cytotoxic cells with infected macrophages that were sensitized with *Plasmodium*-specific peptide, resulted in macrophage lysis (Smith *et al.*, 1991). Released parasites were found to be subsequently killed by bystander macrophages that were activated through IFN-γ released by the CTL.

A proposed model for the role of *Leishmania*-specific CTL in protective immunity against VL in dogs harbouring a chronic infection is shown in Fig. 14.2. Parasite-specific CTL may encounter infected target cells that would stimulate the cytolytic activity of these cells. The parasites would be released in a viable state from incapacitated host cells which facilitates subsequent uptake by surrounding macrophages that have been activated by IFN-γ

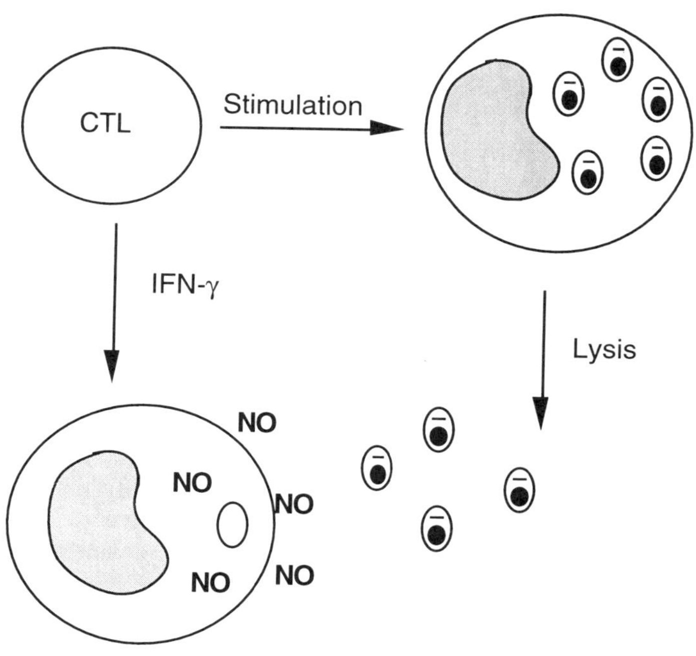

Fig. 14.2. Proposed effector mechanism involving parasite-specific CTL in protective immunity against visceral leishmaniasis in dogs harbouring a chronic *Leishmania infantum* infection.

released by parasite-specific CTL following initial stimulation. In this model, the cytolytic activity of parasite-specific CTL contributes to the elimination of the intracellular parasite which is finally killed by canine macrophages activated by T-cell-derived cytokines.

Our results concerning the cytolytic activity and production of IFN-γ by *Leishmania*-specific T cell lines derived only from asymptomatic dogs, suggest that both of these effector mechanisms are involved in resistance to VL in dogs that harbour a chronic *L. infantum* infection (Pinelli *et al.*, 1995). The identification of parasite T cell epitopes that induce IFN-γ production and cytolytic activity will be relevant to the development of a vaccine against VL.

TOWARDS A VACCINE AGAINST CANINE VISCERAL LEISHMANIASIS

Several factors influence the development of Th cell subsets, but one of the most important factors may be the early exposure to cytokines. Here it is of interest to mention the innate immune response because its condition may determine the outcome of infection. Recently, it was shown that NK cells producing IFN-γ contribute to the early resistance against *L. major* (Scharton and Scott, 1993). These findings suggest that NK cells serve as an early source of IFN-γ, which is known to promote the development of Th1 cells (Scott, 1991). IFN-γ is produced by three major types of cells: $CD4^+$ T cells, $CD8^+$ T cells and NK cells. This cytokine induces several elements of host defence, such as activation of NK cells and neutrophils (Klebanoff *et al.*, 1992) as well as stimulation of macrophage microbicidal activity (Belosevic *et al.*, 1988). In addition, IFN-γ stimulates the formation of granulomas, which are of particular importance in host defence against intracellular pathogens. Recently, IFN-γ has been shown to be an important adjuvant for treatment of certain intracellular parasites (Murray, 1994). In human leishmaniasis, both cutaneous and visceral, IFN-γ has been added to treatment using antileishmanial drugs with very encouraging results (Arana *et al.*, 1994). Furthermore, immunization of mice with parasite antigens together with the administration of IFN-γ leads to a Th1 type of response and significant protection against challenge infection in otherwise susceptible mice (Scott, 1991).

Another cytokine, secreted at the onset of infection, which regulates Th1 development is IL-12 (Hsieh *et al.*, 1993). IL-12 is produced by macrophages and B lymphocytes, and stimulates the production of IFN-γ by T cells and NK cells (Scott, 1993). However, studies with IFN-γ 'knock-out' mice show that IL-12 may exacerbate Th2-dependent pathology in the absence of IFN-γ (Wynn *et al.*, 1995).

Few studies on the immune responses in canine VL have been carried out, probably due to the limited number of tools available. Recently, more tools have been developed for research in this field (Cobbold *et al.*, 1994). At present we are interested to study the possible use of IFN-γ as modulator of the immune response, resulting in the induction of a Th1-like response required for protection against VL in the dog. This will be tested using an experimental challenge model (Killick-Kendrick *et al.*, 1994) which is presently being simplified.

The Role of Canine IFN-γ in Graft Rejection Evaluated in the Canine Transplant Model

15

L. Fuller, K. Zucker, D. Asthana, S. Zhen, M. Carreno, V. Esquenazi and J. Miller

Department of Surgery, Microbiology and Immunology at the University of Miami School of Medicine and the Miami Veterans Administration Hospital, PO Box 016310, Miami, Florida 33101, USA

INTRODUCTION

The immunogenic stimuli for allograft rejections are intimately linked to MHC class I and class II molecules. Allograft rejection is a complex immune process that results from T lymphocyte recognition of allogeneic MHC antigens on grafted tissue. We have advanced the hypothesis that the primary autolymphoproliferative response of T lymphocytes to kidney cells (termed MLKC) and purified islets of Langerhans (MLIC) results from their recognition of 'tissue associated' antigen(s) presented in conjunction with MHC class II molecules (Esquenazi *et al.*, 1987). The increased expression of MHC antigens in allografts has been well documented (Hanafusa *et al.*, 1983; Bottazzo *et al.*, 1985). However, the mechanism by which their changes enhance graft immunogenicity is still in dispute. The principal cytokine that induces MHC class II molecules in various cell types is the species-specific IFN-γ (Lindahl *et al.*, 1974).

This chapter focuses on our studies in isolating native canine IFN-γ (Fuller *et al.*, 1992), production of rCaIFN-γ (Zucker *et al.*, 1993) and polyclonal antibodies specific for these molecules (Fuller *et al.*, 1994). These studies were driven by the desire to determine the role of canine IFN-γ in autologous and allogeneic recognition of 'tissue associated' antigen present on kidney cells, and by inference, its effect on allograft rejection. To examine directly the role of canine IFN-γ on allograft rejection, we determined the effect of anti-IFN-γ on graft prolongation (Fuller *et al.*, 1994).

ISOLATION OF NATIVE CANINE INTERFERON-γ, PRODUCTION OF RECOMBINANT INTERFERON-γ AND ANTI-INTERFERON-γ ANTIBODIES

Native (n) canine IFN-γ can be readily induced by activating dog spleen and lymph node cells with A-23187 and phorbol myristate (Fuller *et al.*, 1992), subjecting the cell culture supernatant to S-Sepharose (fast flow) in 0.05 M HEPES with 0.2 M NaCl pH 8.0 bind virtually all of the nIFN-γ but less than 1% of the protein. The nIFN-γ is concentrated by elution with the above buffer containing 1 M NaCl. The rCaIFN-γ was further purified stepwise by mono-S-cation exchange column chromatography, Superose 6 HR column using the FPLC system (Fuller *et al.*, 1992). To clone canine rIFN-γ we used the PCR and primers capable of amplifying the protein coding region of mRNA, then used the resultant cloned cDNA for canine rIFN-γ to engineer bacteria to produce large quantities of rCaIFN-γ (Zucker *et al.*, 1992, 1993). Western blot analysis of purified nIFN-γ revealed bands of 31 and 42 kDa proteins. In contrast, canine rIFN-γ exhibited a 16 kDa protein. This suggests that the nIFN-γ is a dimer or exists in various states of aggregation and/or glycosylation (Kelker *et al.*, 1984).

Two mAbs designated ADI-1 and ADI-23 (both IgG1 kappa) were generated in BALB/c mice using the purified nIFN-γ for immunization. Rabbits were immunized with the canine rIFN-γ to engender polyclonal anti-rIFN-γ antibodies. In ELISA the ADI-1 specifically bound canine rIFN-γ, nIFN-γ and human rIFN-γ, although with less avidity (Fig. 15.1a). In contrast, ADI-23 was bound to the canine and human IFN-γs with equal avidity (Fig. 15.1b). Analysis by a two-site ELISA using ADI-1 and ADI-23 indicated that the two mAbs detected non-overlapping epitopes on the canine IFN-γ molecule (Fuller *et al.*, 1994). Neither of the anticanine IFN-γ mAbs bonds murine IFN-γ. This may be explained by the observation that canine rIFN-γ showed closer sequence homology to human than to murine; 65% vs. 45% (Zucker *et al.*, 1992).

EFFECTS OF INTERFERON-γ ON EXPRESSION OF MHC CLASS II MOLECULES

Our working hypothesis is that both MHC class II molecules and nominal (tissue-associated) antigens play a role in lymphocyte activation and graft rejection (Esquenazi *et al.*, 1987). Additionally, cytokine release by the activated T lymphocytes determined the magnitude of the response. As an *in vitro* correlate to kidney allograft rejection, the reaction of canine lymphocytes responding to kidney cortical cells in assay termed mlKC has been used (Roth *et al.*, 1985). In this reaction both MHC class II molecules and 'kidney

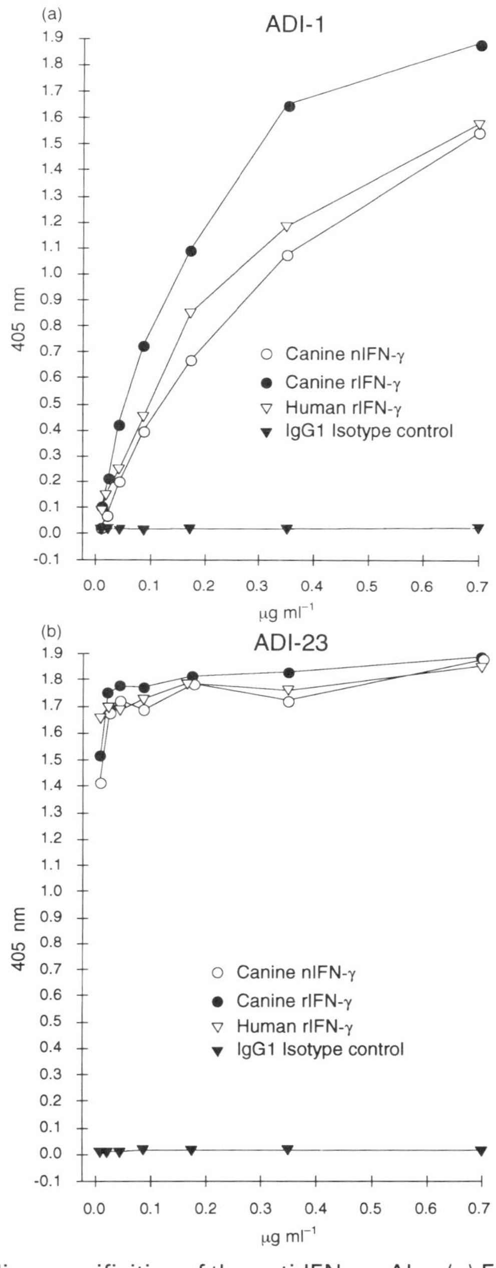

Fig. 15.1. Binding specificities of the anti-IFN-γ mAbs. (a) ELISA of ADI-1 mAb binding to canine rIFN-γ (closed circles), canine nIFN-γ (open circles), human rIFN-γ (open triangles), and the IgG1 isotype control (closed triangles). (b) ELISA (E1) of ADI-23 binding (same symbols as in (a)). From Fuller *et al.* (1994).

cell associated' antigen (KT1) serve as stimulatory elements. The KT1 antigen(s) (72 and 150 kDa) is recognized by a mAb I1F6 (Ranjan *et al.*, 1991; Asthana *et al.*, 1993). Canine IFN-γ causes increased expression of class II MHC antigens and KT1 on kidney cells (Fuller *et al.*, 1992; Asthana *et al.*, 1993; Figs 15.2, 15.3). We proposed that the enhanced expression of class II MHC antigens and KT1 allows the kidney cell to serve as surrogate APC. This enhanced expression of MHC class II and KT1 molecules after treatment with canine IFN-γ led to an increase in MLKC. However, there was a differential effect of IFN-γ on MLKC, MLC or lymphocyte-islet cell cocultures (MLIC) depending upon the type of cells used as stimulators. Fresh kidney cortical tubular cells (KC), Madin–Darby canine kidney (MDCK), and dog kidney (6247) (DK) cell lines were used in MLKC reactions, and islets of Langerhans were used in the MLIC as putative (APC) (Table 15.1). The MDCK cell line constitutively expressed low levels of MHC class II molecules and KT1. The steady-state level of the MHC class II mRNA transcription was virtually unaltered by treatment with IFN-γ (400 IU) for 48 h; however, the MHC cell surface protein expression was enhanced. The KC and DK cell lines constitutively expressed KT1, but not MHC class II

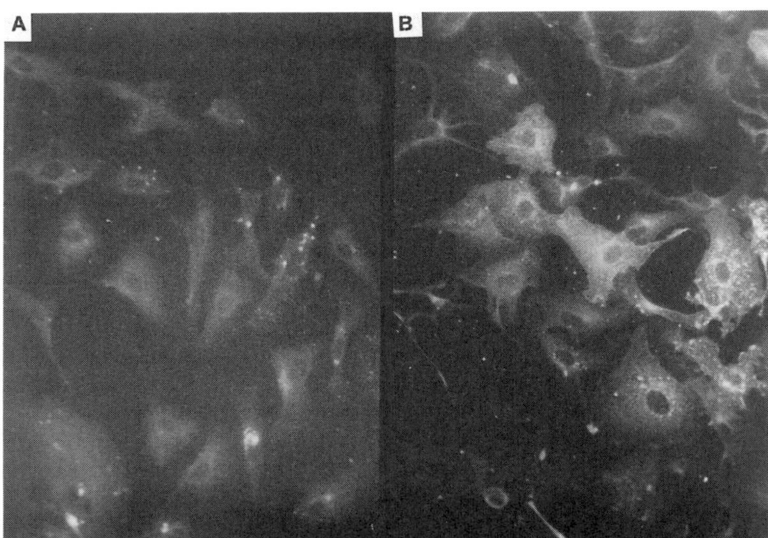

Fig. 15.2. Induction of class II MHC antigens of canine kidney cortical cells. (A) Untreated KC culture stained with anti-class II MHC (B1F6) and FITC-labelled goat anti-mouse $(Fab')_2$. (B) Kidney cortical cells incubated with 1500 IU of canine nIFN-γ for 48 h at 37°C, followed by FITC-labelled goat anti-mouse $(Fab')_2$ antibody exhibited increased MHC class II expression. From Fuller *et al.* (1992).

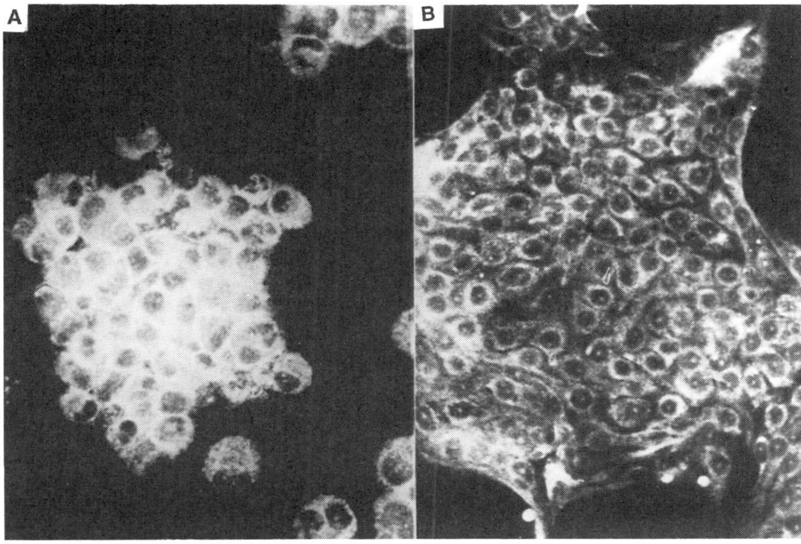

Fig. 15.3. Detection of KT1 epitope. Fresh KC cells (A) and DK cells (B) were stained by indirect immunofluorescence using I1F6 (25 mg ml^{-1}). Antibody binding was detected using FITC-labelled goat anti-mouse IgG. From Asthana *et al.* (1993).

Table 15.1. Effect of recombinant canine IFN-γ on *in vitro* lymphoproliferative responses.

Responder/stimulator combinations	Dog rIFN-γ total protein (μg well^{-1}) (0.2 ml)				
	0	0.1	1.0	2.0	4.0
MLC					
A + Ax[a]	475 ± 155[b]	783 ± 43	1430 ± 283	1947 ± 335	1889 ± 140
A + Ix	9220 ± 359	8039 ± 166	12,508 ± 1677	11,602 ± 292	9567 ± 219
I + Ix	267 ± 36	931 ± 122	902 ± 397	841 ± 324	1582 ± 85
I + Ax	1205 ± 106	2044 ± 343	2824 ± 459	4068 ± 452	3836 ± 453
MLKC					
A + MDCKx	1541 ± 250	1490 ± 108	4344 ± 1220	4732 ± 1353	8681 ± 1748
A + DKx	1170 ± 61	1284 ± 431	6182 ± 1370	12,860 ± 1972	23,622 ± 1280

[a]x, Irradiation with 3000 R for lymphocytes, and 9000 R for kidney cells; A, beagle; I, indifferent unrelated mongrel.

[b]Mean ± SD [^{3}H]thymidine uptake (CPM) of triplicate samples after 9 days in culture. This experiment is representative of three similar ones for each combination.

Table 15.2. Inhibition of MLKC response using MDCK and DK cell lines by I1F6 and anti-dog IFN-γ.

	mAb (10 μg well^{-1})			
Cell combinations	No treatment	I1F6	Anti-IFNγ	IgG1
A + Ax[a]	439 ± 371	321 ± 197	749 ± 676	366 ± 142
A + MDCKx	4655 ± 1268	2674 ± 942[b]	3912 ± 755	4184 ± 182
A + DK6247x	1158 ± 403	585 ± 132[b]	964 ± 142	1000 ± 112

[a]A, beagle dog PBL; x, irradiation with 3000 R for lymphocytes and 9000 R for kidney cells. Mean [^{3}H]thymidine uptake of triplicate samples after 9 days in culture ± SD.
[b]$P = < 0.04$ comparing I1F6 to no treatment or I1F6 to IgG1 treatment.

molecules; these cells required a minimum of 4000 IU, and a 62-h incubation with IFN-γ was needed to upregulate both surface MHC class II molecules and the transcription of corresponding specific mRNA (Asthana *et al.*, 1993). In these studies we demonstrated that in the mlKC reaction MDCK, DK and fresh KC could serve as lymphocyte activators which could be amplified by both native and rCaIFN-γ. The increased lymphoproliferative activity seen in allogeneic MLKC could be blocked by I1F6 (Table 15.2), but not by ADI-1 when kidney cell lines were used. In contrast, when fresh KC was used anti-IFN-γ inhibited both allogeneic and autologous MLKC (Asthana *et al.*, 1993).

EFFECT OF ANTI-INTERFERON-γ ON RENAL ALLOGRAFTS

The *in vitro* studies demonstrated that both purified (Fuller *et al.*, 1992) and rCaIFN-γ (Zucker *et al.*, 1993) upregulated the expression of canine class II MHC molecules on canine renal tubular epithelial cells and MDCK. Moreover, the ADI-1 mAb blocked T cell proliferation to kidney cells, but had no effect on the allogeneic mlC. This observation suggested that IFN-γ might also be a unique key substance amplifying the immune response of T cells to tissue-associated antigens on the surrogate APC, perhaps on kidney tubular cells. We also demonstrated that the anti-IFN-γ mAb bound to canine activated lymphoblast (Fuller *et al.*, 1994). We questioned if the *in vivo* administration of anti-IFN-γ would have an effect on renal allograft survival. In these experiments, one haplotype-matched mlC-reactive donor–recipient littermate pair was used (Fig. 15.4a). After engraftment, the recipients were treated with a constant infusion (1 mg day^{-1} and 3 mg day^{-1}), delivered by a Medtronic® pump into the abdominal aorta above the renal artery anastomoses.

The animals were monitored for the creatinine concentration in serum, the level of ADI-1 and absolute lymphocyte counts. In control experiments using 1 mg and 3 mg day^{-1} infusions, respectively, of the IgG isotype control, the survival time (serum creatinine concentration reaching 10 mg 100 ml^{-1} or death occurring; Fig. 15.4a) in recipients with similar genetic disparities was 9 and 13 days, respectively. Prolongation of allograft

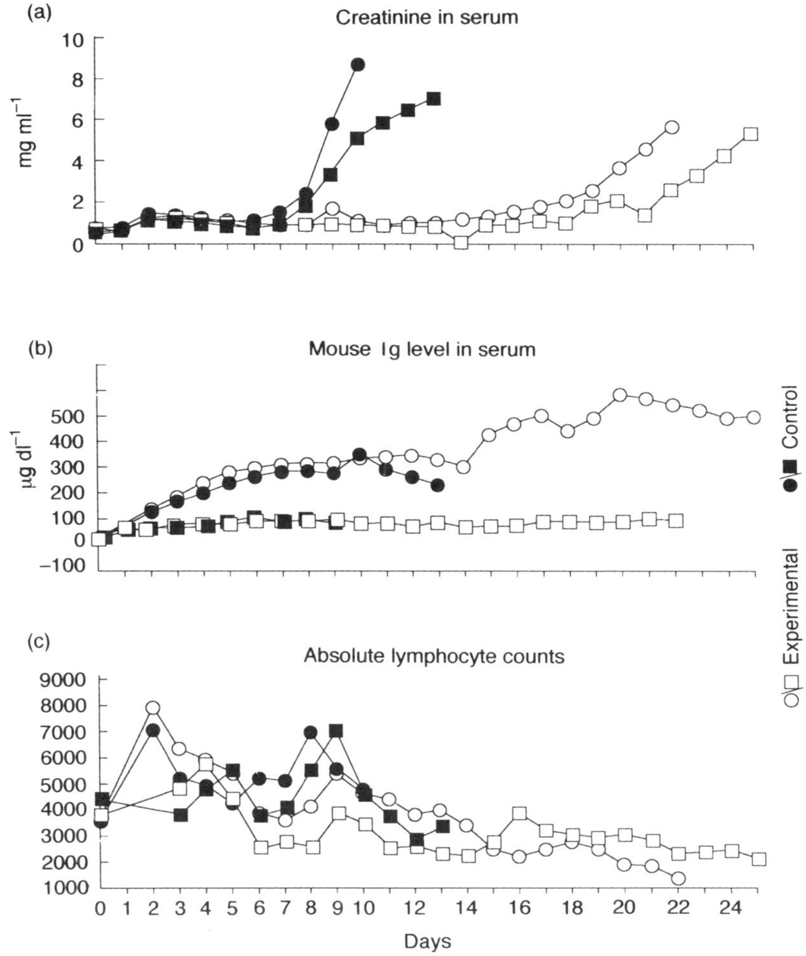

Fig. 15.4. The course of a recipient of a kidney transplant from a one haplotype-matched high MLC reactive littermate donor. (a) Serum creatinine concentration with ADI-1, 1 mg day^{-1} (open squares); 3 mg day^{-1} (open circles); IgG1 control, 1 mg day^{-1} (closed squares); 3 mg day^{-1} (closed circles). (b) ADI-1 serum level in µg dl^{-1} (same symbols as in (a)). (c) Absolute lymphocyte counts with ADI-1 (same symbols as in (a)). From Fuller *et al.* (1994).

survival occurred in the animals receiving ADI-1 (1 and 3 mg day^{-1}), in which the survival times were 22 and 25 days, respectively. In the animal receiving ADI-1 (3 mg day^{-1}), the mouse Ig in serum ranged from a minimum of *c.* 60 μg 100 ml^{-1} to 450 μg 100 ml^{-1}. In contrast, in the animal given ADI-1 (1 mg day^{-1}) the serum Ig level remained at *c.* 30 μg 100 ml^{-1} throughout the course (Fig. 15.4b). As depicted in Fig. 15.4c, the absolute lymphocyte counts in the experimental and control groups did not significantly differ.

CONCLUSIONS

We purified and characterized native canine IFN-γ and produced rCaIFN-γ. The molecules differed with respect to relative molecular mass but not biological effects. Two anticanine IFN-γ mAbs, ADI-1 and ADI-23 and a rabbit polyclonal anti-IFN-γ were produced. Although produced against native IFN-γ, ADI-1 exhibited higher avidity for rIFN-γ while ADI-23 showed equal avidity for canine and human IFN-γ. In the MlKC reaction both the MDCK and DK cell lines, as well as fresh KC cells, could serve as direct lymphocyte activators and the reactions could be amplified by exogenous IFN-γ. The requirements for IFN-γ to increase this amplification, was markedly greater in the MLKC reactions with kidney cells than in the MLC reactions. The cell types differed with respect to the abilities of anti-IFN-γ and I1F6 (anti-KT1) mAbs to inhibit lymphocyte proliferation. The anti-IFN-γ mAb did not inhibit the (allogeneic) MDCK-driven MLKC in which class II MHC molecules could be upregulated by low levels of endogenous IFN-γ, and neither I1F6 nor anti-IFN-γ had any effect on the purely allogeneic MLC reaction. These findings suggest that the paracrine effects of endogenous IFN-γ levels, at critical concentrations and durations, might play a crucial role in T lymphocyte activation by graft cells acting as surrogate APC in presenting tissue-associated nominal antigens *in vivo*. In support of this the ADI-1 mAb given *in vivo* caused a prolonged graft survival, indicating that the above-mentioned paracrine effect of IFN-γ on the allograft response may have been blocked.

Cytokines in Canine Septic Shock 16

A.G. Coran and R.A. Drongowski

CS Mott Children's Hospital, University of Michigan Medical Center, Ann Arbor, Michigan 48109-0245, USA

INTRODUCTION

Septic shock remains a major clinical problem with significant associated morbidity and mortality. Endotoxin secreted by select bacteria can induce septic shock, but it is primarily Gram-negative bacteria themselves, specifically their cell wall LPS component, which are primarily responsible for the development of septic shock. The clinical manifestations of sepsis consist of the following: tachypena, tachycardia, hyperthermia or hypothermia, and evidence of inadequate organ perfusion including hypoxaemia, elevated blood lactate levels and oliguria. A definitive diagnosis of septic shock couples the above clinical manifestations with hypotension (sustained decrease in systolic blood pressure to < 90 mmHg, or a drop of > 40 mmHg for at least 1 h) in the face of adequate intravascular volume and in the absence of other known causes of shock, i.e. myocardial infarction, hypovolaemia and pulmonary embolism (Bone, 1991; Glauser *et al.*, 1991). The major haemodynamic and metabolic changes associated with septic shock include the following: decreased cardiac output, decreased mean blood pressure, low peripheral vascular resistance (early, increased resistance, late), hyperventilation, respiratory alkalosis, and significant increases in blood lactic acid levels, oxygen free radical production, catecholamines, prostaglandins and cytokines (Cook *et al.*, 1982; Freeman and Crapo, 1982; McCord, 1985; Slotman *et al.*, 1985; Yellin *et al.*, 1986; Beck and Abel, 1987; Oettinger *et al.*, 1987). Oxygen-derived free radicals, which are generated during tissue ischaemia via the enzyme xanthine oxidase or from activated neutrophils, have been implicated as mediators of tissue injury via peroxidation of lipid membranes (Freeman and Crapo, 1982; McCord, 1985; Simpson and Lucchesi, 1987).

PATHOGENESIS

Arachidonic acid metabolism and concomitant eicosanoid generation via the cyclooxygenase and lipoxygenase pathways has been documented in numerous studies on septic and endotoxic shock (Lee *et al.*, 1989). The metabolism of arachidonic acid by cyclooxygenase yields oxygen free radicals, thromboxane A2, prostacyclin (PGI2), 6-keto-PGE1 and PGH2, which additionally may yield PGD2, PGE2 and PGF2-α. The eicosanoids, thromboxane A2 and prostacyclin, are measured as the stable metabolites thromboxane B2 and 6-keto-PGF2-α and are elevated in septic shock and endotoxic shock animal models. Arachidonic acid peroxidation via the lipoxygenase pathway in neutrophils yields 5-hydroperoxyeicosatetraenoic acid (5-HPETE), which can undergo metabolism to either 5-hydroeicosatetraenoic acid (5-HETE) or leukotriene A4 (LTA4). LTA4 can be hydrolysed to the LTB4 series of compounds, which can stimulate adhesion, migration, aggregation, enzyme release and generation of superoxide radicals by PMN. LTA4 may also be metabolized to LTD4, LTDE4 and LTF4. These leukotrienes are powerful vasodilators and bronchoconstrictors and have been shown to be identical with the 'slow-reacting substance of anaphylaxis'. Additionally, stimulated human leukocytes can convert 15-hydroperoxyeicosatetraenoic acid (15-HPETE) to lipoxin A and B, which are potent superoxide radical generators as well as inhibitors of NK cell cytotoxicity (Cook *et al.*, 1980, 1985; Butler *et al.*, 1983). Inhibition of cyclooxygenase has been achieved by many investigators using non-steroidal anti-inflammatory drugs (NSAID), primarily ibuprofen, salicylates, indomethacin and phenylbutazone (Simon and Mills, 1980; Carlin *et al.*, 1985; Rockwell and Ehrlich, 1989; Coran *et al.*, 1992). Treatment of septic shock by NSAIDs results in inhibition of arachidonic acid metabolism via the cyclooxygenase pathway and diminished metabolite release, resulting in maintenance of cardiac output and mean arterial pressure. In addition, superoxide radicals are reduced, conferring further protection against the development of shock in treated animals (Rockwell and Ehrlich, 1989; Young and Passmore, 1990). However, long-term survival of septic animals treated with NSAIDs remains unchanged. Concomitant inhibition of arachidonic acid metabolism via the lipoxygenase pathway by leukotriene receptor antagonists results in leukotriene inhibition and improvement in mean arterial pressure and cardiac output, but confers no improvement in long-term survival of septic animals (Young and Passmore, 1990).

Since long-term survival of septic animals is not improved following inhibition of arachidonic acid metabolites by NSAIDs and since the use of cyclooxygenase inhibitors increases endotoxin-induced synthesis of IL-1 and TNF, attention has been focused on the role of cytokines, mainly TNF, IL-1 and IL-6, in septic shock (Dinarello, 1991a). TNF was originally defined as a factor produced in BCG-primed animals in response to endotoxin, and has been referred to by the following names: cachectin, macrophage cytotoxin, necrosin, cytotoxin, TNF-α, haemorrhagic factor, macrophage cytotoxic factor and differentiation-inducing factor. The biological activities of TNF include induction of cachexia and inflammation, mediation of endotoxin-induced shock, antibacterial and antitumour effects, enhancement of wound healing, autoimmunity, increase of angiogenesis and causation of bone resorption (Aggarwal, 1992). Interleukin-1 refers to a group of three polypeptide hormones, IL-1α, IL-1β and IL-1 receptor antagonist, which have a role in the regulation of immune and inflammatory responses. Interleukin-1α and IL-1β are also referred to as osteoclast-activating factor, catabolin, endogenous pyrogen, epidermal cell-derived thymocyte-activating factor, lymphocyte-activating factor, serum amyloid A inducer, hepatocyte-stimulating factor, fibroblast-activating factor, leukocyte endogenous mediator, proteolysis-inducing factor, β-cell-activating factor and haemopoietin-1. The biological activities of IL-1 include the following: induces muscle proteolysis, causes fever, induces inflammation, enhances wound healing, increases bone resorption and causes haematopoiesis (Dower, 1992). Interleukin-6 is a potent cytokine which regulates immune response and acute phase reactions. IL-6 has also been termed the following: IFN-β2, B-cell-stimulating factor, 26-kDa protein, hepatocyte-stimulating factor, thrombopoietin, hybridoma/plasmacytoma growth factor, macrophage granulocyte-inducing protein2 and CTL differentiation factor. The biological activities of IL-6 include: induction of acute phase response, promotion of growth, inhibition of growth, differentiation or regulation of specific gene expression of a variety of cells, increase in levels of ACTH, induction of megakarocyte maturation and enhancement of antibody production against sheep red blood cells (Taga and Kishimoto, 1992).

There is a growing volume of evidence to support the fact that cytokines, predominantly TNF, IL-1 and IL-6, are central mediators of septic shock. A component of Gram-negative bacterial membranes, LPS or endotoxin, binds to lipopolysaccharide binding protein (LBP) to produce an LPS–LBP complex which activates monocyte/macrophage cell lineages, including Kupffer cells and peritoneal, pulmonary and other macrophages, to induce TNF, IL-1 and IL-6 secretion. Once TNF is secreted, a cascade of inflammatory mediators is initiated, with secondary secretion of other mediators, activation of neutrophils and endothelial cells, synthesis of acute phase reactants, onset of coagulopathy, and finally, shock and death (Dinarello, 1991b; Giroir, 1993). Numerous studies have indicated TNF as the causative agent in the septic shock syndrome (Calandra and Glauser, 1990; Giroir, 1993; Molloy *et al.*, 1993). TNF's role in septic shock satisfies Koch's postulates in the following manner:

1. TNF is produced in the septic shock syndrome in both patients and experimental animal models of sepsis.
2. Transfer of TNF to uninfected animals or humans induces the septic shock syndrome.
3. Blocking TNF activity in animals with the septic shock syndrome prevents death and tissue injury commonly seen in sepsis (Tracey and Cerami, 1993).

Mounting evidence has indicated that IL-1 acts synergistically with TNF to stimulate its own release as well as that of TNF, thus amplifying the cascade of inflammatory mediators and further exacerbates the shock state. However, IL-1 itself cannot induce lethal shock (Dinarello, 1991a; Cerami, 1992; Molloy *et al.*, 1993). IL-6, similar to TNF and IL-1, induces fever and increases acute-phase protein synthesis, but does not induce inflammation or shock (Dinarello, 1991a). However, some studies have demonstrated a positive correlation between IL-6 levels and the severity of shock (Dinarello, 1991a), while other studies have shown that serum levels of TNF can prognosticate survival (Fiers, 1991; Tracey and Cerami, 1993; Martin *et al.*, 1994).

Therapeutic options currently being directed at cytokine manipulation in the treatment of septic shock include the neutralization of TNF with anti-TNF antibodies, soluble TNF receptors and soluble TNF receptor fusion proteins; suppression of TNF biosynthesis with glucocorticoids, IL-10, leukaemia inhibiting factor, G-CSF, pentoxifylline, amrinone, PGE2 and cyclosporine A; reduced responsiveness to TNF by infusing low-dose TNF, LPS and hydrazine; and inhibition of secondary mediators (IL-1, IL-6, IFN-γ, platelet-activating factor, eicosanoids and nitric oxide; Tracey and Cerami, 1993).

OWN EXPERIMENTAL STUDIES

The following series of studies were carried out in our laboratory using ten mongrel dogs. The dogs were divided into control or ibuprofen-treated groups based upon treatment regimen. All animals received an LD_{100} i.v. infusion of 10^9 live *E. coli* bacteria. Five animals received an i.v. dose of ibuprofen at 10 mg kg^{-1} of body weight 10 min prior to the *E. coli* infusion. After baseline samples were obtained, the bacteria were infused and haemodynamic monitoring and blood samples were obtained at selected time intervals. Mean cardiac output and blood pressure are presented in Fig. 16.1. In the ibuprofen group, mean cardiac output decreased from

1.1 l min^{-1} to 0.55 l min^{-1} following the bacterial infusion, then gradually increased to 0.99 l min^{-1}. In the control group, mean cardiac output decreased throughout the study period. Mean blood pressure in the control animal group mimicked cardiac output. However, the mean blood pressures

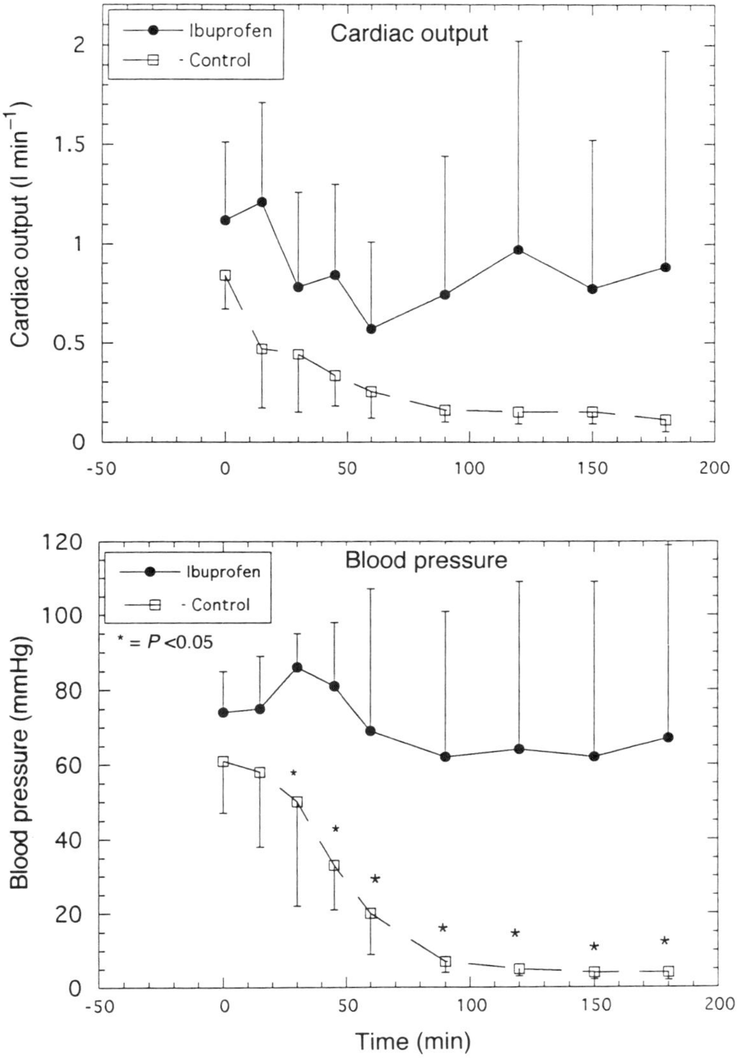

Fig. 16.1. Mean cardiac output and blood pressure following *E. coli* administration in ibuprofen-treated (closed circles) and control dogs (open squares).

in the ibuprofen-treated animals were significantly increased compared to the control animals from 30 min until the conclusion of the observation period.

Mean levels of PGE2 remained stable in the ibuprofen-treated group during the course of the study period. In contrast, PGE2 levels in the control group increased from baseline levels and were significantly increased

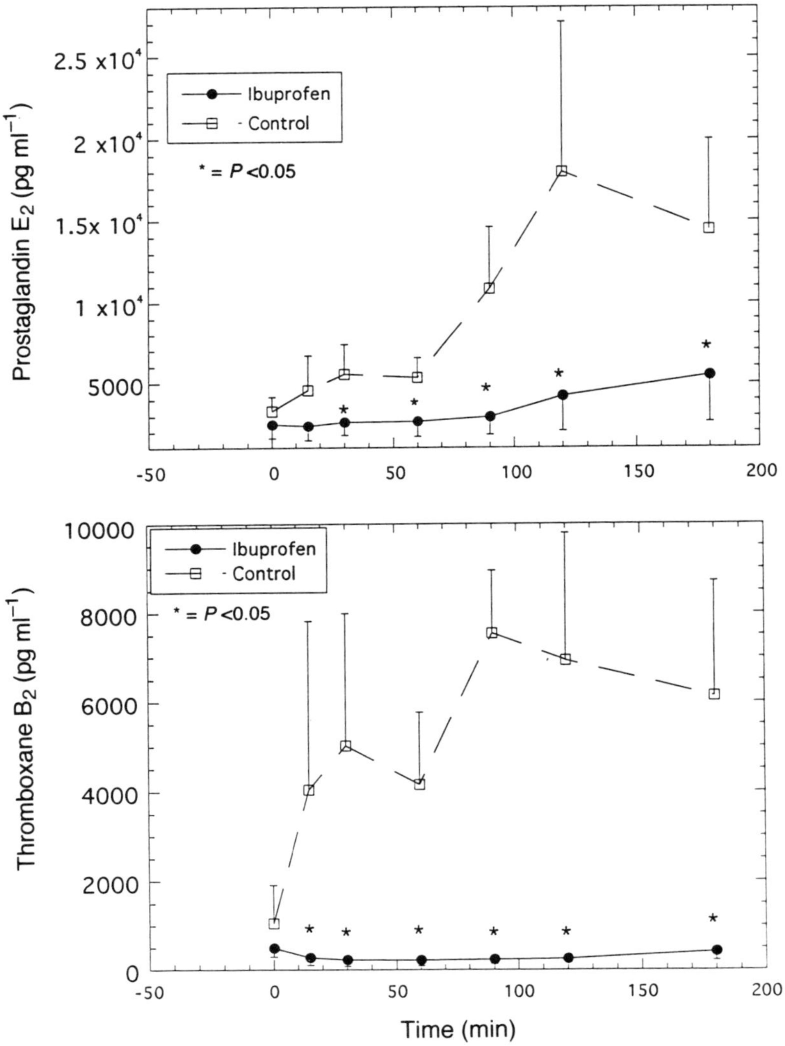

Fig. 16.2. Mean eicosanoid levels following *E. coli* administration in ibuprofen-treated (closed circles) and control dogs (open squares).

compared to the ibuprofen group at 90, 120 and 180 minutes following the *E. coli* infusion. A similar pattern was observed for thromboxane B2 levels (Fig. 16.2).

In contrast to eicosanoid secretion which was inhibited by ibuprofen therapy, mean levels of blood cytokine levels (measured as IL-6 and TNF)

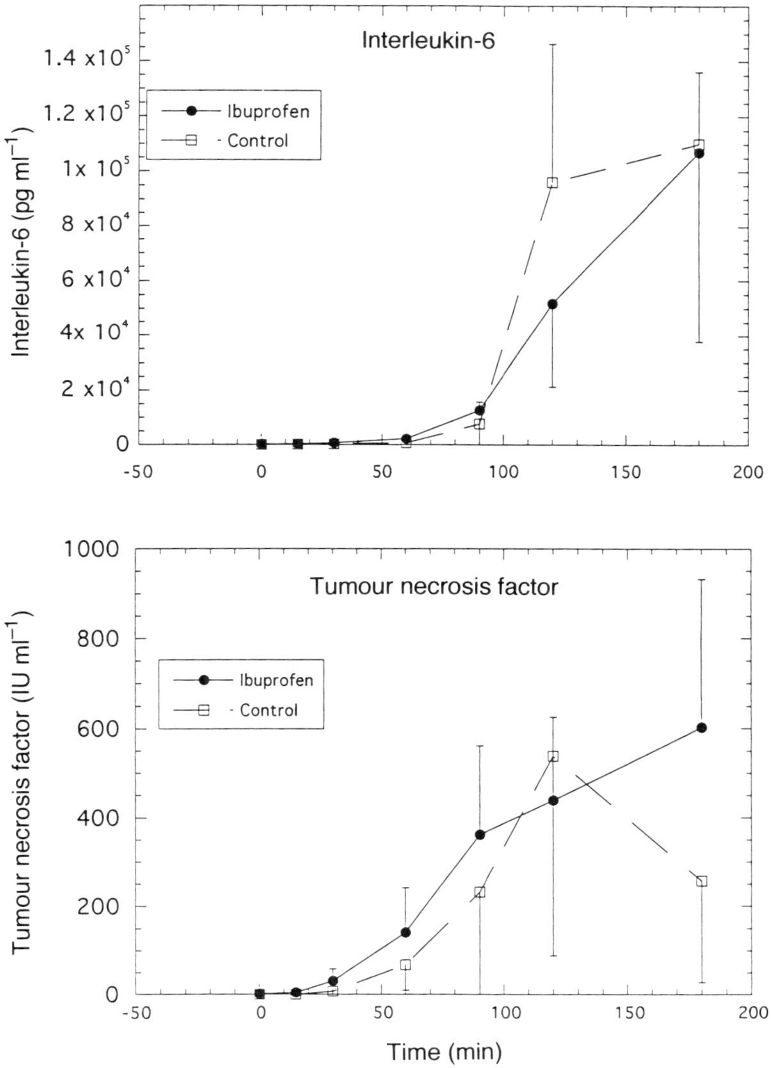

Fig. 16.3. Mean cytokine levels following *E. coli* administration in ibuprofen-treated (closed circles) and control dogs (open squares).

increased significantly from baseline in both the ibuprofen and control animal groups (Fig. 16.3). There were no significant differences between the mean levels of IL-6 or TNF in ibuprofen and control animals.

This study demonstrates the effects of cyclooxygenase inhibition by ibuprofen in a live *E. coli* model of septic shock. Haemodynamic function is restored, eicosanoid production is inhibited, but cytokine levels are significantly increased. Death eventually occurs in these animals in spite of significant haemodynamic improvement in the ibuprofen-treated animals over a short observation period.

In conclusion, this study illustrates the limited effects of cyclooxygenase inhibition in a live *E. coli* model of septic shock in dogs. Although an increase in eicosanoid levels is prevented, cytokine expression continues unabated, which eventually leads to the animals' demise. Concomitant treatment of sepsis employing both NSAIDs as well as cytokine neutralization, suppression, or inhibition may well be the key to the successful treatment of septic shock.

References for Part V

Abranches, P., Santos-Gomes, G., Rachamim, N., Campino, L., Schnur, L.F. and Jaffe, C.L. (1991) An experimental model for canine visceral leishmaniasis. *Parasite Immunology* 13, 537–550.

Aggarwal, B.B. (1992) Tumor necrosis factor. In: Aggarwal, B.B. and Gutterman, J.U. (eds), *Human Cytokines. Handbook for Basic and Clinical Research.* Blackwell Scientific Publications, Boston.

Arana, B.A., Navin, T.R., Arana, F.E., Berman, J.D. and Rosenkaimer, F. (1994) Efficacy of a short course (10 days) of high dose meglumine antimonate with or without IFN-γ in treating cutaneous leishmaniasis in Guatemala. *Clinics of Infectious Disease* 18, 381–384.

Ashford, R.W., Desjeux, P. and DeRaadt, P. (1992) Estimation of population at risk of infection and number of cases of leishmaniasis. *Parasitology Today*, 8, 104–105.

Asthana, D., Yang, W.-C., Fuller, L., Zucker, K., Lu, P., Zheng, S., Esquenazi, V., Carreno, M., Roth, D., Burke, G.W., Nery, J. and Miller, J. (1993) Differential effect of IFN-γ on kidney cell expression of MHC class II molecules, kidney cell associated molecules and their stimulatory capacity in mixed lymphocyte kidney cell culture. *Transplantation Immunology* 1, 282–293.

Beck, R.R. and Abel, F.L. (1987) Effect of ibuprofen on the course of canine endotoxin shock. *Circulatory Shock* 23, 59–70.

Belosevic, M., Davis, C.E., Meltzer, M.S. and Nacy, C.A. (1988) Regulation of activated macrophages antimicrobial activities. Identification of lymphokines that co-operate with IFN-γ for induction of resistance to infection. *Journal of Immunology* 141, 890–896.

Bogdan, C., Gessner, A. and Rollinghoff, M. (1993) Cytokines in leishmaniasis, a complex network of stimulatory and inhibitory interactions. *Immunobiology* 189, 356–396.

Bone, R.C. (1991) Sepsis, the sepsis syndrome, multi-organ failure: a plea for comparable definitions. *Annals of Internal Medicine* 114, 332–333.

Bottazzo, G.F., Dean, B.M., McNally, J.M., McKay, E.H., Swift, P.G. and Gamble, D.R. (1985) *In situ* characterization of autoimmune phenomena and expression of HLA molecules in the pancreas in diabetic insulitis. *New England Journal of Medicine* 313, 353–360.

Bradonisio, O., Altamura, M., Carelli, G., Ceci, L., Antonaci, S. and Jirillo, E. (1988) Lymphocyte functions in *Leishmania infantum* infected dogs. EOS. *Journal of Immunology and Immunopharmacology* 9, 37–40.

Butler, R.R., Wise, W.C., Halushka, P.V. and Cook, J.A. (1983) Effects of gentamycin and indomethacin in the treatment of septic shock: Effects of prostacyclin and thromboxane A2 production. *Journal of Pharmacology and Experimental Therapeutics* 225, 94–101.

Cabral, M., O'Grady, J. and Alexander, J. (1992) Demonstration of leishmania specific cell mediated and humoral immunity in asymptomatic dogs. *Parasite Immunology* 14, 531–539.

Calandra, T. and Glauser, M.P. (1990) Cytokines and septic shock. *Diagnostic Microbiology and Infectious Disease* 13, 377–381.

Carlin, G., Djursater, R., Smedegard, G. and Gerdin, B. (1985) Effect of anti-inflammatory drugs on xanthine oxidase and xanthine oxidase induced depolymerization of hyaluronic acid. *Agents and Actions* 16, 377–384.

Carvalho, E.M., Teixeira, R.S. and Johnson, W.D., Jr (1981) Cell mediated immunity in American visceral leishmaniasis. *Infection and Immunity* 48, 409–414.

Carvalho, E.M., Lata, L.S., Shrivastrarca, S.N., Blackwell, J. and Neva, F.A. (1989) Antigen specific immunosuppression in visceral leishmaniasis is cell mediated. *Journal of Clinical Investigation* 83, 860–864.

Cerami, A. (1992) Inflammatory cytokines. *Clinical Immunology and Immunopathology* 62, S3-S10.

Chan, M.M. (1993) T cell response in murine *Leishmania mexicana* amazonensis infection: production of interferon-γ by CD8$^+$ cells. *European Journal of Immunology* 23, 1181–1184.

Cobbold, S., Holmes, M. and Willett, B. (1994) The immunology of companion animals: reagents and therapeutics strategies with potential veterinary and human clinical applications. *Immunology Today* 15, 347–353.

Cook, J.A., Wise, W.C. and Halushka, P.V. (1980) Elevated thromboxane levels in the rat during endotoxic shock: Protective effects of imidazole, 13-azaprostanoic acid, or essential fatty acid deficiency. *Journal of Clinical Investigation* 65, 227–230.

Cook, J.A., Wise, W.C., Butler, R.R., Reines, H.D., Rambo, W. and Halushka, P.V. (1982) The potential role of thromboxane and prostacyclin in endotoxic and septic shock. *American Journal of Emergency Medicine* 2, 28–37.

Cook, J.A., Olanoff, L.S., Wise, W.C., Tempel, G.E., Reines, H.D. and Halushka, P.V. (1985) Role of eicanosoids in endotoxic and septic shock. In: Janssen, H.F. and Barnes, C.B. (eds), *Circulatory Shock: Basic and Clinical Implications*. Academic Press, Orlando, pp. 102–107.

Coran, A.G., Drongowski, R.A., Paik, J.J. and Remick, D.G. (1992) Ibuprofen intervention in canine septic shock: Reduction of pathophysiology without decreased cytokines. *Journal of Surgical Research* 53, 272–279.

Cox, F.E.G. and Liew, F.Y. (1992) T cell subsets and cytokines in parasitic infections. *Immunology Today* 13, 445–448.

Da Conceicão-Silva, F., Perlaza, B.L., Louis, J.A. and Romero, P. (1994) *Leishmania major* infection in mice primes for specific major histocompatibility complex class I-restricted $CD8^+$ cytotoxic T cell responses. *European Journal of Immunology* 24, 2813–2817.

Dinarello, C.A. (1991a) The proinflammatory cytokines interleukin-1 and tumor necrosis factor and treatment of the septic shock syndrome. *Journal of Infectious Diseases* 163, 1177–1184.

Dinarello, C.A. (1991b) Inflammatory cytokines: Interleukin-1 and tumor necrosis factor as effector molecules in autoimmune diseases. *Current Opinion in Immunology* 3, 941–948.

Dower, S.K. (1992) Tumor necrosis factor. In: Aggarwal, B.B. and Gutterman, J.U. (eds), *Human Cytokines. Handbook for Basic and Clinical Research*. Blackwell Scientific Publications, Boston.

Esquenazi, V., Fuller, L., Pardo, V., Roth, D., Milgrom, M. and Miller, J. (1987) *In vivo* and *in vitro* induction of class II molecules on canine renal cells and their effect on the mixed lymphocyte kidney cell culture. *Transplantation* 44, 680–692.

Ferrick, D.A., Schrenzel, M.D., Mulvania, T., Hsieh, B., Ferlin, W.G. and Lepper, H. (1995) Differential production of interferon-g and interleukin-4 in response to Th1- and Th2-stimulating pathogens by γ/δ T cells *in vivo*. *Nature* 373, 255–257.

Fiers, W. (1991) Tumor necrosis factor. Characterization at the molecular, cellular and *in vivo* level. *Federation of European Biochemical Societies* 285, 199–212.

Freeman, B.A. and Crapo, J.D. (1982) Biology of disease: Free radicals and tissue injury. *Laboratory Investigation* 47, 412–426.

Fuller, L., Fernandez, J., Zheng, S., Carreno, M., Esquenazi, V., Yang, W.-C. and Miller, J. (1992) Immuno- and biochemical characterization of purified canine interferon-γ: Purification of a monoclonal antibody, affinity, purification, and its effect on MLC, and MLKC reactions. *Transplantation* 53, 195–202.

Fuller, L., Carreno, M., Esquenazi, V., Zucker, K., Zheng, S., Roth, D., Burke, G., Nery, J., Asthana, D., Olson, L. and Miller, J. (1994) Characterization of anti-canine cytokine monoclonal antibodies specific for IFN-γ: Effect of anti-IFN-γ on renal transplant rejection. *Tissue Antigens* 43, 163–169.

Galvão-Castro, B., Sá Ferreira, J.A., Marzochi, K.F., Marzochi, M.C., Coutinho, S.G. and Lambert, P.-H. (1984) Polyclonal B cell activation, circulating immune complexes, and autoimmunity in human American visceral leishmaniasis. *Clinical and Experimental Immunology* 56, 58–66.

Giroir, B.P. (1993) Mediators of septic shock: New approaches for interrupting the endogenous inflammatory cascade. *Critical Care Medicine* 21, 780–789.

Glauser, M.P., Zanetti, G., Baumgartner, J.-D. and Cohen, J. (1991) Septic shock: pathogenesis. *Lancet* 338, 732–736.

Haldar, J.P., Ghose, S., Saha, K.C. and Ghose, A.C. (1983) Cell mediated immune responses in Indian kala-azar and post kala-azar dermal leishmaniais. *Infection and Immunity* 42, 702–707.

Hanafusa, T., Pujol-Borell, R., Chiovato, R., Russell, R.C.G., Doniach, D. and Bottazzo, G.F. (1983) Aberrant expression of HLA-DR antigen on thyrocytes in graves disease: Relevance of autoimmunity. *Lancet* 12, 1111–1115.

Hsieh, C.S., Macatonia, S.E., Tripp, C.S., Wolf, S.F., O'Garra, A. and Murphy, K.M. (1993) Development of Th1 $CD4^+$ T cells through IL-12 produced by *Listeria*-induced macrophages. *Science* 260, 247.

Kelker, H.C., Le, J., Rubin, B.Y., Yip, K.Y., Nagler, C. and Vilcek, J. (1984) Three molecular forms of natural human interferon-γ revealed by immunoprecipitation with monoclonal antibody. *Journal of Biological Chemistry* 259, 4301–4304.

Killick-Kendrick, R., Killick Kendrick, M., Pinelli, E., Del Real, G., Molina, R., Vitutia, M.M., Cañavate, M.C. and Nieto, X. (1994) A laboratory model of canine leishmaniasis: the inoculation of dogs with *Leishmania infantum* promastigotes from midguts of experimentally infected phlebotomine sandflies. *Parasite* 1, 311–318.

Klebanoff, S.J., Olszowski, S., VanVoorhis, W.C., Ledbetter, J.A., Waltersdorph, A.M. and Schlecgte, K.G. (1992) Effect of g-interferon on human neutrophils: protection from deterioration on storage. *Blood* 80, 225–234.

Kovacsovics-Bankowsk, M. and Rock, K.L. (1994) Presentation of exogenous antigens by macrophages: analysis of major histocompatibility complex class I and II presentation and regulation by cytokines. *European Journal of Immunology* 24, 2421–2428.

Laison, R. and Shaw, J.J. (1987) Evolution, classification and geographical distribution. In: Peters, W. and Killick-Kendrick, R. (eds), *The Leishmaniasis in Biology and Medicine*, Volume 1. Academic Press, London, pp. 2–104.

Lee, W.M., Reines, D., Watt, G.H., Cook, J.A., Wise, W.C., Halushka, P.V. and Galbraith, R.M. (1989) Alterations in Gc levels and complexing in septic shock. *Circulatory Shock* 28, 249–255.

Liew, F.Y. (1989) Functional heterogeneity of $CD4^+$ T cells in leishmaniasis. *Immunology Today* 10, 40–45.

Liew, F.Y. (1990) Regulation of cell-mediated immunity in leishmaniasis. *Current Topics in Microbiology and Immunology* 155, 53–64.

Liew, F.Y. and O'Donell, C.A. (1993) Immunology of leishmaniasis. *Advances in Parasitology* 32, 161–259.

Lindahl, P., Gresser, I. and Leary, P. (1974) Enhancement of the expression of histocompatibility antigen of mouse lymphoid cells by interferon *in vitro*. *European Journal of Immunology* 4, 779–784.

Martin, C., Saux, P., Mege, J.-L., Perrin, G., Papazian, L. and Gouin, F. (1994) Prognostic values of serum cytokines in septic shock. *Intensive Care Medicine* 20, 272–277.

McCord, J.M. (1985) Oxygen-derived free radicals in postischemic tissue injury. *New England Journal of Medicine* 312, 159–163.

Moll, H., Scollay, R.G. and Mitchell G.F. (1988) Resistance to cutaneous leishmaniasis in nude mice injected with $L3T4^+$ T cells but not with $Lyt2^+$ T cells. *Immunology and Cellular Biology* 66, 57–63.

Molloy, R.G., Mannick, J.A. and Rodrick, M.I. (1993) Cytokines, sepsis and immunomodulation. *British Journal of Surgery* 80, 289–297.

Mosmann, T.R., Cherwinski, H., Bond, M.W., Giedlin, M.A. and Coffmann, R.L. (1986) Two types of murine helper T-cell clones. I: Definition according to profiles of lymphokine activities and secreted proteins. *Journal of Immunology* 136, 2348–2357.

Müller, I., Pedrazzini, T., Farrel, J.P. and Louis, J. (1989) T-cell responses and immunity to experimental infection with *Leishmania major*. *Annual Review of Immunology* 7, 561–578.

Müller, I., Kropf, P., Etges, R.J. and Louis, J. (1993) γ-Interferon response in secondary *Leishmania major* infection: role of $CD8^+$ T cells. *Infection and Immunity* 61, 3730–3738.

Murray, H.W. (1994) Interferon-γ and host antimicrobial defense: current and clinical applications. *American Journal of Medicine* 97, 459–467.

Nieto, J., Cañavate, M.C., De Fuentes, I., Rodriguez, E., Aguiar, E.C., Garcia, E. and Del Real, G. (1991) Ensayo de vacunacion frente a leishmaniasis canina. Analisis de la respuesta a antigenos purificados. Valencia, Spain. *Parasitologia en el sur-oeste de Europa*. p. 186. Abstr. ICASEPI.

Oettinger, W., Peskar, B.A. and Berger, H.G. (1987) Profiles of endogenous prostaglandin F2α, thromboxane A2 and prostacyclin with regard to cardiovascular and organ functions in early septic shock in man. *European Surgical Research* 19, 65–77.

Pfeifer, J.D., Wick, M.J., Roberts, R.L., Findlay, K., Normark, S.J. and Harding, C.V. (1993) Phagocytic processing of bacterial antigens for class I MHC presentation to T cells. *Nature* 361, 359–362.

Pham, T. and Mauël, J. (1987) Studies on intracelllar killing of *Leishmania major* and lysis of host macrophages by immune lymphoid cells *in vitro*. *Parasite Immunology* 9, 721–736.

Pinelli, E., Killick-Kendrick, R., Wagenaar, J., Bernardina, W., Del Real, G. and Ruitenberg, E.J. (1994) Cellular and humoral immune responses in dogs experimentally and naturally infected with *Leishmania infantum*. *Infection and Immunity* 62, 229–235.

Pinelli, E., Gonzalo, R.M., Boog, C.J.P., Rutten, V.P.M.G., Gebhard, D., del Real, G. and Ruitenberg, E.J. (1995) *Leishmania infantum*-specific T cell lines derived from asymptomatic dogs that lyse infected macrophages in a major histocompatibility complex-restricted manner. *European Journal of Immunology* 25, 1594–1600.

Ranjan, D., Roth, D., Esquenazi, V., Carreno, M., Fuller, L., Leif, R., Burke, G. and Miller, J. (1991) The effects of tissue associated and MHC class II antigen presentation on *in vitro* lymphoproliferative responses against canine liver and kidney cell subpopulations. *Transplantation* 51, 475–480.

Rock, K.L. (1996) A new foreign policy: MHC class I molecules monitor the outside world. *Immunology Today* 17, 131–137.

Rock, K.L., Gamble, S. and Rothstein, L. (1990) Presentation of exogenous antigen with class I major histocompatibility complex molecules. *Science* 249, 918–922.

Rockwell, W.B. and Ehrlich, H.P. (1989) Ibuprofen in acute-care therapy. *Annals of Surgery* 211, 78–83.

Roth, D., Fuller, L., Esquenazi, V., Kyriakides, G., Pardo, V. and Miller, J. (1985) The biologic significance of the mixed lymphocyte kidney culture in humans. *Transplantation* 40, 376–383.

Sacks, D.L., Lal, S.L., Shrivastava, S.N., Blackwell, J. and Neva, F.A. (1987) An analysis of T cell responsiveness in Indian kala-azar. *Journal of Immunology* 138, 908–913.

Scharton, T.M. and Scott, P. (1993) Natural killer cells are a source of IFN-γ that drives differentiation of $CD4^+$ T cell subsets and induces early resistance to *L. major* in mice. *Journal of Experimental Medicine* 178, 567–577.

Scott, P. (1991) IFN-γ modulates the early development of Th1 and Th2 responses in a murine model of cutaneous leishmaniasis. *Journal of Immunology* 147, 3149–3155.

Scott, P. (1993) IL-12: Initiation cytokine for cell-mediated immunity. *Science* 260, 496–497.

Scott, P.A., Pearce, E., Cheever, A.W., Coffmann, R.L. and Sher, A. (1989) Role of cytokines and $CD4^+$ T cell subsets in the regulation of parasite immunity and disease. *Immunological Reviews* 112, 161–182.

Simon, L.S. and Mills, J.A. (1980) Nonsteroidal antiinflammatory drugs. *New England Journal of Medicine* 302, 1179–1185.

Simpson, P.J. and Lucchesi, B.R. (1987) Free radicals and myocardial ischemia and reperfusion injury. *Journal of Laboratory and Clinical Medicine* 110, 13–30.

Slappendel, R.J. (1988) Canine leishmaniasis. Review based on 95 cases in The Netherlands. *Veterinary Quarterly* 10, 1–16.

Slotman, G.J., Quinn, J.V., Burcghard, K.W. and Gann, D.S. (1985) Thromboxane, prostacyclin, and the hemodynamic effects of graded bacteremic shock. *Circulatory Shock* 16, 395–404.

Smith, L.E., Rodrigues, M. and Russell, D.G. (1991) The interaction between $CD8^+$ cytotoxic T cells and *Leishmania*-infected macrophages. *Journal of Experimental Medicine* 174, 499–505.

Taga, T. and Kishimoto, T. (1992) Tumor necrosis factor. In: Aggarwal, B.B. and Gutterman, J.U. (eds), *Human Cytokines. Handbook for Basic and Clinical Research.* Blackwell Scientific Publications, Boston.

Titus, R.G. and Ribeiro, J.M. (1988) Salivary gland lysates from the sand fly *Lutzomyia longipalpis* enhance *Leishmania* infectivity. *Science* 239, 1306–1308.

Tracey, K.J. and Cerami, A. (1993) Tumor necrosis factor: An updated review of its biology. *Critical Care Medicine* 21, S415–S422.

Wynn, T.A., Jankovic, D., Hieny, S., Zioncheck, K., Jardieu, P., Cheever, A.W. and Sher, A. (1995) IL-12 exacerbates rather than suppresses T helper 2-dependent pathology in the absence of endogenous IFN-γ. *Journal of Immunology* 154, 3999–4009.

Yellin, S.A., Nguyen, D., Quinn, J.V., Burchard, K.W., Crowley, J.P. and Slotman, G.J. (1986) Prostacyclin and thromboxane A2 in septic shock: Species difference. *Circulatory Shock* 20, 291–297.

Young, J.S. and Passmore, J.C. (1990) Hemodynamic and renal advantages of dual cyclooxygenase and leukotriene blockade during canine endotoxic shock. *Circulatory Shock* 32, 243–255.

Zucker, K.E., Lu, P., Esquenazi, V. and Miller, J. (1992) Cloning of the cDNA for canine interferon γ. *Journal of Interferon Research* 12, 191–214.
Zucker, K.E., Lu, P., Asthana, D., Carreno, M., Yang, W.-C., Esquenazi, V., Fuller, L. and Miller, J. (1993) Production and characterization of recombinant canine interferon-γ from *Escherichia coli*. *Journal of Interferon Research* 13, 91–97.

VI

Feline Cytokines

Edited by V.E.C.J. Schijns

Feline Cytokines

17

V.E.C.J. Schijns[1] and W.G. Bradley[2]

[1] *Veterinary Faculty, Utrecht University, Yalelaan 1, de Uithof, 3584 CL Utrecht, The Netherlands;* [2] *Eckerd College, 4200 54th Avenue South, St Petersburg, Florida 33711, USA*

INTRODUCTION

Cytokines are regulatory polypeptides that function as intercellular messengers to coordinate cellular responses (such as an immune response). They can be produced by lymphocytes, monocytes, fibroblasts and many other of the body's nucleated cells. The messages cytokines carry are typically sent to other cells (paracrine) but can also be used to stimulate the cell that produced them (autocrine). Like hormones, cytokines are very potent and active at picogram to nanogram per millilitre levels and are therefore usually secreted in extremely small amounts to be active in the immediate microenvironment. This low level production makes detection of cytokines in the serum very difficult.

Within the past decade, it has been recognized that subsets of $CD4^+$ T lymphocytes, specifically Th lymphocytes, transcribe and secrete recognizable subsets of cytokines (Mosmann and Sad, 1996). For instance, the Th1 lymphocyte, which participates in cellular immunity, produces IL-2, IFN-γ and lymphotoxin. The Th-2 lymphocyte, which participates in humoral immunity, produces IL-4, IL-5, IL-6, IL-9, IL-10 and IL-13. Interestingly, the characteristic cytokines produced by each subset of Th lymphocytes are mutually inhibitory for the development and activation of the reciprocal subset. The production of IFN-γ inhibits Th2 activation and IL-10 inhibits Th1 cytokine production (Fiorentino *et al.*, 1989). Both subsets of T lymphocytes have been implicated in the recovery from or exacerbation of a variety of diseases, including AIDS (for review see Mosmann and Sad, 1996).

The accumulation of evidence supporting a switch from a Th1 cytokine response to a Th2 cytokine response during human AIDS (Clerici and Shearer, 1993; Mosmann, 1994a) attracted the interest of researchers studying feline AIDS (FAIDS). However, in order to study the patterns of cytokine expression throughout development of FAIDS, investigators first needed to molecularly clone, sequence and characterize the feline cytokine genes.

At the Second International Symposium on Feline Retrovirus Research (6–9 October 1993, Research Triangle Park, North Carolina; hosted by the North Carolina State University College of Veterinary Medicine), investigators presented details of the expression, molecular cloning, nucleotide sequencing and characterization of several feline cytokine genes for the first time (Bradley *et al.*, 1993a; Gebhard *et al.*, 1993; Rottman and Tompkins, 1993). Before this time the only feline cytokine gene that had been cloned was TNF-α (McGraw *et al.*, 1990).

METHODS USED TO CLONE FELINE CYTOKINE GENES

The ability to detect feline cytokine gene expression and to molecularly clone complementary DNAs (cDNAs) derived from these genes relied on the availability of cytokine gene sequences from other mammals (human, mouse, pig, etc.). By aligning the nucleotide sequences of cytokine genes obtained from other mammals, regions of strong sequence conservation were identified. The assumption was made that these regions had also been conserved in the feline genome. Using the nucleotide sequences in the highly conserved regions as a guide, sets of oligonucleotide primers containing these sequences were synthesized. The primers were then used in PCR (Saiki *et al.*, 1988) where cDNAs produced by reverse transcription of total cell RNA of mitogen-stimulated feline PBMC were used as the template. Under appropriate conditions, the PCRs successfully amplified the corresponding sequences of feline cytokine genes. Initial attempts to use RT-PCR to detect transcription from IL-2, IL-4, IL-6, IL-10, IL-12, TNF-α, IFN-γ and β-actin were successful (Bradley *et al.*, 1993a; Gebhard *et al.*, 1993; Rottman and Tompkins, 1993; Rottman *et al.*, 1995), as were initial attempts to molecularly clone several of these feline cytokine genes (Bradley *et al.*, 1993b). More recently a competitive RT-PCR (Babu *et al.*, 1993) has been described which quantitatively detects cytokine gene transcription in the cat, cow, dog, horse and pig (Rottman *et al.*, 1996a). Since the symposium, several laboratories have successfully cloned feline cytokine cDNAs. The nucleotide sequences of these feline genes can be found in GenBank's repository under the accession numbers listed in Table 17.1.

Table 17.1. GenBank accession numbers for feline cytokine and related genes.

Gene	Accession number
IL-2	L19402
IL-2rα	D16143
IL-4	X87408
IL-6	D13227, L16914
IL-10	U39569
IL-12 (p35)	Y07761
IL-12 (p40)	Y07762
IFN-γ	X86972, D30619
TNF-α	M92061

MOLECULAR CLONING OF FELINE CYTOKINES

Three types of feline IFN including IFN-γ have been characterized which resemble those of other mammals in their biological properties (Yamamoto *et al.*, 1986; Weiss and Toivio-Kinnucan, 1988). A feline IFN-α gene, 60% homologous in its nucleotide sequence to human IFN-α1, has been cloned and expressed in a baculovirus system (Nakamura *et al.*, 1992). Feline cells are responsive to human IFN-α (Weiss and Toivio-Kinnucan, 1988; Weiss and Oostrom-Ram, 1990). Feline IFN-γ has been cloned (Argyle *et al.*, 1995; Schijns *et al.*, 1995a) and expressed in a baculovirus system (Schijns *et al.*, unpublished).

Feline Interleukin-1

Feline IL-1 activity was first described in 1987 (Goitsuka *et al.*, 1987) and can be partly blocked by antibody to human IL-1. Cats are responsive to human IL-1 (Sirko *et al.*, 1989). Feline IL-1β has been cloned by Daniel *et al.* (1992, unpublished).

Feline Interleukin-2

Feline IL-2 closely resembles rat and human IL-2 (Goitsuka *et al.*, 1986). Its activity can be partly blocked by mAb to human IL-2 (Lin and Bowman, 1991). Feline lymphocytes are responsive to human IL-2 (Fenwick *et al.*, 1988). Upon sequencing the feline IL-2 cDNA, Bradley and co-workers found that the predicted aa sequence exhibited 81%, 69%, 64% and 60% homology with human, bovine, rat and murine IL-2, respectively. These results were similar to those reported by Cozzi *et al.* (1993). Unexpectedly, they found that when the nucleotide sequences were compared, feline IL-2 exhibited 90%, 83%, 76% and 73% homology with human, bovine, rat and murine IL-2. Due to the degeneracy of the genetic code (wobble at the third position of the triplet codon), less homology at the nucleotide level than at the aa level was expected. Just the opposite was observed. The strongest homology existed between the nucleotide sequences and not the aa sequences. The observation of strong conservation at the nucleotide level was further confirmed after comparison with the nucleotide and predicted aa sequences of killer whale (*Orcinus orca*), manatee (*Trichechus manatus latirostris*), dolphin (*Tursiops truncatus*), polar bear (*Ursa maritimus*), and grey wolf (*Canis lupus*) IL-2 (Ness, 1995; Barber, 1996; Cashman *et al.*, 1996; McMahon, 1996).

The IL-2 receptor α chain has been cloned by Y. Matsumoto *et al.* (1993, unpublished results).

Feline Interleukin-4

The full-length cDNA encoding feline IL-4 has been cloned by Schijns *et al.* (1995b). A cDNA sequence corresponding to feline IL-4 was also reported by Ward (1995). The 380 bp fragment that was cloned contained the start codon and continued to within 25 bp of the predicted stop codon. This feline IL-4 nucleotide sequence was found to be 83%, 82%, 80%, 79% and 61% homologous to IL-4 from killer whale, pig, human, manatee and rat, respectively. The predicted peptide sequence of feline IL-4 exhibited 72%, 71%, 64%, 55% and 38% homology with the killer whale, pig, manatee, human and rat IL-4 peptide, respectively.

Feline Interleukin-6

Feline IL-6 is similar to human and murine IL-6 in its biological activities (Ohashi *et al.*, 1989; Bradley *et al.*, 1993b). In 1993, the molecular cloning, nucleotide sequence and predicted aa sequence of feline IL-6 were reported (Bradley *et al.*, 1993b; Ohashi *et al.*, 1993). The predicted feline IL-6 protein was found to contain at its amino terminus a hydrophobic secretory signal sequence, which if used would result in production of a mature peptide of approximately 20 kDa. Observations by Goitsuka *et al.* (1990) suggest that under non-denaturing conditions, feline IL-6 migrates through a gel filtration column with an apparent molecular mass of 30–40 kDa. This is approximately twice its predicted size. While human IL-6 has been reported to be approximately 21 to 28 kDa, it is possible that the differences observed with feline IL-6 may be associated with post-translational modifications or that feline IL-6 is capable of forming homodimers.

The predicted aa sequence of feline IL-6 was also found to contain a characteristic signature shared by IL-6, G-CSF (Nagata *et al.*, 1986; Tsuchiya *et al.*, 1986) and myelomonocytic growth factor (MGF; Leutz *et al.*, 1989; Bradley *et al.*, 1993b). The conservation of four cysteine residues in feline IL-6 (two in the signal sequence), even in regions where the surrounding residues exhibited little homology, suggest that these cysteines may be critical for establishment of proper secondary and tertiary structures or for formation of homodimers. The expression of feline IL-6 cDNA in Crandell feline kidney (CrFK) cells (Ohashi *et al.*, 1993) should help to decipher the observed differences between the predicted and apparent molecular mass of feline IL-6.

Feline Interleukin-10

Feline IL-10 has been cloned by Scott and O'Reilly (1995, unpublished results). The sequence has been deposited in GenBank under accession number U39569. The cDNA fragment is 737 nucleotides long and encodes a protein of 178 aa.

Feline Interleukin-12

Upon cloning feline IL-12p35, Bush and co-workers observed that like human and murine IL-12p35, the predicted protein exhibited conservation of three leucine zipper motifs (Landschulz *et al.*, 1988; Bush *et al.*, 1994). Since IL-12p35 is one of two peptide chains (the other peptide chain is IL-12p40) that come together to form the biologically active IL-12, they hypothesized that the interaction between IL-12p35 and IL-12p40 may be through the leucine zipper motifs. It is also possible that IL-12p35 may interact with other leucine-zipper-containing proteins to create new complexes with novel biological activities (Bush *et al.*, 1994). Recently, the cDNAs encoding the entire feline p35 and p40 peptides have been cloned (Schijns *et al.*, 1997). The sequences have been deposited in GenBank under accession numbers Y07761 and Y07762.

Feline Tumour Necrosis Factor-α

Tumour necrosis factor exhibits low (if any) species restriction. Lehmann *et al.* (1992) described feline TNF activity in cats experimentally infected with feline immunodeficiency virus. Cats are responsive to human (Aoki *et al.*, 1989) and murine TNF (B. Haagmans, Utrecht, 1996, personal communication). The report by McGraw *et al.* (1990) was the first to describe the molecular cloning of a feline cytokine gene. This gene was TNF-α. Using the sequence published by McGraw *et al.* several groups of investigators have since designed synthetic oligonucleotides to use in the RT-PCR either to monitor gene expression of TNF-α during a variety of infections or to clone the feline TNF-α cDNA (Rimstad *et al.*, 1995).

Interleukin-16, Interleukin-18, Tumour Necrosis Factor Receptors, Stem Cell Factor

At a recent Concerted Action Workshop on Cytokines in FIV and HIV infection held in Pisa, Italy (November 1996, organized in the framework of the European Community), data were presented on the successful cloning of several feline cytokines that have only recently been recognized in mice and man. Leutenegger and co-workers from the University of Zurich, Switzerland, reported the cloning and expression of lymphocyte chemoattractant factor, feline IL-16. Three communications were from the University of Glasgow: Duthie *et al.* have cloned the cDNAs encoding feline TNF receptor-1 and -2; Dunham and Onions reported the cloning, expression and bioactivity of rFeSCF with potential clinical application in anaemia; and Argyle and co-workers presented the cloning of feline IL-18, also termed IFN-γ-inducing factor (IGIF).

IMPORTANCE OF CLONING FELINE CYTOKINE GENES

As can be deduced from the aforementioned descriptions of feline cytokine gene cloning, the power of the PCR has allowed investigators interested in feline immunology to rapidly accumulate a base of knowledge that would have taken many years just a decade and a half ago. Spurred on by the study of FAIDS, continuing efforts are directed at isolating, cloning, sequencing and expressing feline cytokine genes. These efforts are already providing researchers with tools which allow the investigation of the intricate cellular interactions occurring during the feline immune response to FIV and other pathogens. While many important feline cytokine genes have already been cloned, there is still a large number of the genes corresponding to those known to exist in the human or murine systems that need to be cloned. It is likely that there is an even larger number of cytokine genes to be discovered.

Since the transcription of mRNA from a cytokine gene does not necessarily parallel cytokine protein production (James-Yarish *et al.*, 1994), studies describing cytokine gene transcription should include measurements of cytokine protein production. Unfortunately, relatively few assays exist to quantitate feline cytokine protein production. Therefore, the production of recombinant feline (rFe) cytokines and the generation of mAbs to these recombinant molecules is desperately needed. Production of these reagents will greatly aid in the development of feline cytokine-specific assays. Until then we must rely on what is currently available – the ability to quantitate feline cytokine mRNA production.

In Chapters 18 and 19 two virus diseases of cats are discussed, FAIDS and feline infectious peritonitis (FIP), which recently received much attention regarding the role of cytokines in protective immunity and pathogenesis.

Cytokine Expression During Feline Immunodeficiency Virus Infection

18

W.G. Bradley

Eckerd College, 4200 54th Avenue South, St Petersburg, Florida 33711, USA

INTRODUCTION

Prior to discovery of the feline immunodeficiency virus (FIV) by Pedersen *et al.* (1987) comparatively little was known about the feline immune system (Gorham *et al.*, 1971; Tham and Studdert, 1985; Pedersen, 1987; Lin, 1992). The recognition that FIV induces an immunologic disorder that in its natural host is remarkably similar to human AIDS (FAIDS, feline acquired immunodeficiency syndrome), stimulated intense investigations of the feline immune system (for reviews see Pedersen and Barlough, 1991; Egberink and Horzinek, 1992; Elders and Phillips, 1994; Bendinelli *et al.*, 1995).

Cats naturally or experimentally infected with FIV develop many of the signs and symptoms ascribed to human immunodeficiency virus type 1 (HIV-1) infected humans. Like the targets for HIV-1, the *in vivo* targets for FIV include T lymphocytes, B lymphocytes and macrophages (English *et al.*, 1993), as well as astrocytes and microglia cells located within the CNS. Similar to HIV-1, FIV infection of the CNS is associated with development of neurological abnormalities and behaviour changes (Egberink *et al.*, 1989; Dow *et al.*, 1990, 1992; Hurtrel *et al.*, 1992; Gray *et al.*, 1993; Lafrado *et al.*, 1993; Podell *et al.*, 1993, Phillips *et al.*, 1994; Steffan *et al.*, 1994).

Infected cats produce humoral and cellular responses specific for FIV and respond to current FIV vaccines in a manner similar to that observed in primates vaccinated against HIV-1. Interestingly, development of viral-envelope-specific vaccines against FIV are encountering the same difficulties

 Cytokines in Veterinary Medicine (eds V.E.C.J. Schijns and M.C. Horzinek)

in cross-subtype protection as the development of vaccines against HIV-1. Vaccines based upon the envelope of one virus subtype induce protection against homologous virus but fail to protect against a heterologous viral challenge (Hosie *et al.*, 1995). In addition, infected cats respond to antiviral agents currently used to treat HIV-1 infection (Gardner, 1991; Remington *et al.*, 1991; Yates *et al.*, 1992; Meers *et al.*, 1993; Bradley *et al.*, 1995; Vahlenkamp *et al.*, 1995). Taken together these remarkable similarities between FAIDS and human AIDS support the contention that FIV-infected cats are an appropriate small animal model for AIDS.

Similar to the hallmark of AIDS, FAIDS is defined by a progressive immunodeficiency. To be able to treat effectively cats afflicted with FAIDS and to utilize better the FIV-infected cat as a small animal model for AIDS, we need to further our understanding of what constitutes appropriate and inappropriate feline immune function. As we have learned much about human immunology through investigations of HIV-1 and AIDS, the emergence of FIV and FAIDS has been the catalyst for investigations of feline immune function. Currently the feline model is being used to investigate antiviral drug and vaccine development, gene therapy, the impact of cofactors on disease progression, mechanisms associated with vertical transmission, the development of paediatric AIDS and development of AIDS-associated dementia. Another area of intense investigation into feline AIDS is the impact of altered cytokine production on virus replication, immune response and pathogenesis.

CYTOKINE EXPRESSION DURING FELINE IMMUNODEFICIENCY VIRUS INFECTION

Interleukin-2

In a large number of asymptomatic humans infected with HIV-1, there can be detected a marked reduction in the ability of their peripheral blood lymphocytes to produce IL-2 in response to a variety of recall antigens and mitogens (Clerici *et al.*, 1993b). A similar loss of lymphocyte proliferation and IL-2 production in response to cell mitogens has been observed in FIV-infected cats (Barlough *et al.*, 1991; Torten *et al.*, 1991; Lawrence *et al.*, 1995). In experimentally infected cats, a positive correlation between the loss of $CD4^+$ lymphocytes and reduced responsiveness to the cell mitogens, PWM and Con A, has been described (Barlough *et al.*, 1991). In these cats the PWM response tends to decrease prior to loss of the Con A response (Barlough *et al.*, 1991). Similar observations have been made by Lawrence *et al.* (1995). Additionally, it has been reported that augmentation of the lymphocyte cultures with IL-2 could restore the response to Con A but not to PWM (Lawerence *et al.*, 1995).

Interferon-γ

Using a quantitative–competitive RT-PCR for feline IFN-γ, Dean *et al.* (1996) observed a significant elevation of IFN-γ mRNA expression in $CD4^+$ and $CD8^+$ cells found in the thymus, lymph nodes, spleen and peripheral blood during the acute phase of infection (less than 140 days post-infection) occurring in experimentally infected cats. The relative levels of IFN-γ mRNA expression were greatest in the peripheral lymph nodes, followed by mesenteric lymph nodes, spleen, and were lowest in the thymus. After peak production of IFN-γ mRNA, it is thought that expression of IFN-γ returns to baseline. While no measurements of biologically active IFN-γ were made, the temporal measurements of IFN-γ mRNA production suggest that an initial Th1 response to FIV infection occurs but then succumbs to the development of a predominant Th2 response (Dean *et al.*, 1996).

Interleukin-4

In HIV-1-infected individuals where there is a detectable reduction in the production of IL-2 to recall antigens or cell mitogens, there is a concurrent rise in IL-4 production in response to the same molecules (Clerici *et al.*, 1993b). These observations support the hypothesis that progression to AIDS may be intimately related to a switch from a Th1 to a Th2 pattern of cytokine production.

In the study by Dean *et al.* (1996), it was reported that throughout the time course studied (140 days) IL-4 transcription continues to rise in peripheral and mesenteric lymph nodes of acutely infected cats. This increase in IL-4 transcription appeared to be associated with a similar rise in IL-10 transcription, suggestive of a Th2 response. As Th1 and Th2 responses are mutually inhibitory, it will be of interest to determine precisely when the fate of a Th1/Th2 switch is determined in FIV infected cats (i.e. within the first few days following infection or at some later point in time).

Interleukin-6

Several cytokines are thought to be associated with the pathology observed in HIV-1-infected humans. One of these cytokines, IL-6, has been detected at elevated levels in the serum (Breen *et al.*, 1990) and in the cerebrospinal fluid (Laurenzi *et al.*, 1990) of HIV-1-infected humans. Interleukin-6 has also been found to be produced by Kaposi's sarcoma cells and to be required for the optimal growth of these cells (Miles *et al.*, 1990).

In 1992, it was reported that elevations of IL-6-like activity could be detected in the plasma and cultured PBMC of FIV-infected cats (Ohashi *et al.*, 1992). Within 24 h cultures of non-stimulated PBMC, IL-6 activity was detected at significantly higher levels in cultures derived from FIV-infected cats than in cultures derived from non-infected cats

(8.10 ng ml^{-1} ±2.37 vs. 0.72 ng ml^{-1} ±0.41). When the PBMC of FIV-infected cats were fractionated into adherent (predominantly macrophage/monocytes) and non-adherent cells (predominantly lymphocytes) and then cultured for 24 h, a significantly higher level of IL-6 activity was detected in the adherent fraction of PBMC (7.66 ng ml^{-1} ±3.08 vs. 3.07 ng ml^{-1} ±1.82). While the non-adherent fraction of PBMC from infected cats produced less IL-6 activity than the adherent fraction, it still produced more IL-6-like activity than either adherent or non-adherent PBMC from non-infected cats (Ohashi *et al.*, 1992).

When bronchoalveolar lavage (BAL) cells from FIV-infected specified pathogen-free (SPF) cats were compared with similar cells from non-infected cats, there was a detectable but not significant decrease in the production of IL-6. Within the same cells, a significant decrease in TNF-α production was detected (Ma *et al.*, 1995). These results are different from those obtained by Ohashi *et al.* (1992). However, these differences may reside in the differences between the tissues sampled (PBMC vs. BAL), the types of infection (free roaming cats naturally infected vs. SPF cats experimentally infected), the timing of the infections (unknown for the free roaming cats vs. 1–3 years post-infection for the SPF cats) and the presence of potentiating cofactors in the free roaming cats. The differences between the results described above demonstrate one of the advantages of using the feline model to study AIDS. While HIV-1 does not infect pathogen-free humans, the ability to obtain naturally infected free roaming cats or to purposefully infect SPF cats, provides investigators with unique opportunities to examine specific interactions between host and virus (in the absence of other pathogens) and to examine the role of cofactors in the progression of AIDS.

Interleukin-10

Interleukin-10 is a potent inhibitor of Th1 cytokine synthesis (Fiorentino *et al.*, 1989) and is produced by T lymphocytes, macrophages, keratinocytes and as yet unidentified cells in the placenta (Mosmann, 1994b). It has been reported that IL-4 and IL-10 work synergistically to inhibit Th1 cytokine production (Sher *et al.*, 1992). Therefore, the increased levels of IL-4 and IL-10 associated with HIV-1 infections may be sufficient to cause the observed switch to a Th2 pattern of cytokine expression and suppression of the expression of the Th1 cytokines.

In experiments described by Ritchey and Tompkins (1996), levels of alveolar macrophage cytokine mRNA and proteins as well as FIV gag mRNA were measured less than 18 weeks post-FIV infection. Alveolar macrophages were obtained by BAL and cultured in the presence or absence of LPS. When IL-10 mRNA levels were measured in the LPS-stimulated cultures, infected cats were found to produce significantly more IL-10 than non-infected cats. The question remaining is whether the increase in IL-10 transcription is part of the pathology leading to opportunistic infections of the lungs. Since the lungs are major targets for opportunistic infections in HIV-1-infected individuals, further investigations of feline pulmonary immune function during FIV infection may prove to be extremely important. Such investigations may provide insight into the mechanisms associated with these opportunistic infections and may lead to the design of appropriate therapeutics.

When Dean *et al.* (1996) measured IL-10 mRNA production in experimentally infected cats, they found a sustained increase in IL-10 transcription in peripheral and mesenteric lymph nodes, spleen and thymus. The production of IL-10 mRNA in these tissues began shortly after peak viraemia (42–70 days post-infection). It will be most interesting to determine if the increase in IL-4 and IL-10 transcription occurring during the acute phase of FIV infection corresponds with increasing levels of the respective proteins and if these cytokine levels are capable of suppressing a Th1 response.

Interleukin-12

Interleukin-12, also known as NK cell stimulatory factor, is a heterodimeric protein synthesized by macrophages, B cells and other AC; IL-12 is a potent inducer of Th1 cytokine production and suppresses the development of IL-4-producing cells (Manetti *et al.*, 1993; Trinchieri and Scott, 1994). Contrastingly, IL-10 inhibits IL-12 synthesis, thereby inhibiting the development of IFN-γ-producing Th1 cells (D'Andrea *et al.*, 1993). Using the murine AIDS (MAIDS) model to test the effects of *in vivo* treatement with IL-12, Gazzinelli *et al.* (1994) found that infected mice were significantly protected from the severe immune dysfunction typically associated with MAIDS. This effect appeared to be associated with IL-12's ability to induce IFN-γ production. As IL-12 appears to play a major role in regulating the equilibrium between Th1 and Th2, it will be important to follow the expression of IL-12 during the development of FAIDS.

Rottman *et al.* (1996b) monitored the changes in IL-10 and IL-12 mRNA production in SPF cats challenged with a non-pathogenic strain of *T. gondii*. These cats had been infected with FIV 18 weeks before the challenge with *T. gondii*. At 0, 3 and 7 days post-challenge the levels of IL-10 and IL-12 mRNA in bronchial lymph nodes were measured. When compared with FIV-negative cats, FIV-infected cats exhibited less IL-10 and IL-12 mRNA production before challenge with *T. gondii* (day 0). Following the challenge, FIV-negative cats responded by reducing the levels of IL-10 and IL-12 transcription. This action should favour a Th1 response. Contrastingly, FIV-infected cats failed to reduce IL-10 and IL-12 transcription, a condition favouring a Th2 response. Additionally, when compared with the FIV-negative cats, FIV-infected cats produced lower levels of IL-2 and IFN-γ mRNA. By day 14 post-challenge, 75% of the FIV-infected cats had developed complications requiring euthanasia, whereas the FIV-negative cats only developed a mild disease and then recovered by day 14 post-challenge.

These results are similar to those describing the murine immune response to *T. gondii*. In investigations using the murine system, production of a Th1 response (upregulation of IL-2 and IFN-γ) corresponds to protection against *T. gondii* challenge, whereas suppression of the Th1 response (or induction of a Th2 response) corresponds to the inability to control the infection (Sher *et al.*, 1992).

The results presented by Rottman *et al.* (1996b) suggest that FIV-associated alterations in cytokine gene expression significantly influence the ability to mount appropriate responses to invasion by potential pathogens. Results such as these also demonstrate the utility of the feline model to investigate the immunopathogenesis of AIDS.

Tumour Necrosis Factor-α

Increased TNF-α expression has been detected in PBMC and monocytes (Hober *et al.*, 1989; Roux-Lombard *et al.*, 1989; Voth *et al.*, 1990; Wright *et al.*, 1988) and within the serum and cerebrospinal fluid of HIV-1-infected individuals (Gallo *et al.*, 1989; Mintz *et al.*, 1989). While the precise role that TNF-α and IL-6 play in the progression of AIDS is still being debated, some investigators believe that along with IL-6, TNF-α is partially responsible for some of the pathology associated with HIV-1 infection (Odeh, 1990). TNF-α has been shown to augment HIV-1 expression, to upregulate transcription from the viral LTR (Folks *et al.*, 1989; Michihiko *et al.*, 1989; Okamoto *et al.*, 1989; Osborn *et al.*, 1989; Locardi *et al.*, 1990; Odeh, 1990; Poli *et al.*, 1990; Vyakarnam *et al.*, 1990) and to induce apoptosis in HIV-1-infected T lymphocytes (Matsuyama *et al.*, 1989), FIV-infected T lymphocytes (Ohno *et al.*, 1994) and CrFK cells (Ohno *et al.*, 1993).

To investigate the role of TNF-α during the acute phase of FIV infection we assayed TNF-α, FIV p26 and FIV proviral DNA in sequential biopsies of spleen and lymph node sections, PBMC and serum of FIV-infected SPF cats (Kraus *et al.*, 1996). We found that expression of virus and TNF-α were intimately linked. Prior to detection of viraemia in these cats, serum TNF-α levels rose significantly ($P = 0.04$) and remained elevated throughout the viraemic phase. When lymph node sections were subjected to immunohistochemical analysis for the detection of TNF-α and FIV p26 protein expression, immunoprecipitates for the respective proteins were found localized in common foci. Following seroconversion, the levels of serum and lymph node TNF-α dropped and serum FIV p26 became undetectable, but remained inducible *in vitro*. At this time, proviral DNA was significantly reduced in the PBMC ($P = 0.041$) but remained prevalent in the lymph nodes. These results suggest that in the absence of potentiating cofactors, there exists a temporal relationship between TNF-α and virus expression during acute infections.

In the investigation performed by Ma *et al.* (1995), it was found that when BAL cells and non-adherent PBMC from FIV-infected SPF cats (3 years post-infection) were cultured in the presence of LPS or Con A/phorbol myristate acetate, respectively, they produced significantly less TNF-α than did similar cells from uninfected cats. These results differ from results obtained from acutely infected cats within the same lab (Ritchey and Tompkins, 1996). When the BAL cells from acutely infected SPF cats (less than 18 weeks) were tested for production of TNF-α mRNA and protein, it was found that these cells produced significantly more TNF-α than did similar cells from non-infected cats. These results are consistent with those we obtained from acutely infected SPF cats (Kraus *et al.*, 1996).

Therefore it appears that conflicting data found in the literature describing TNF-α production in HIV-1-infected individuals may be due to factors not controllable in that system (i.e. timing of infection, presence of potentiating cofactors and age of the individual). This illustrates another advantage of using the FIV model to study the immunopathogenesis associated with AIDS.

CONCLUSIONS

Driven by the study of feline AIDS, continuing efforts directed at isolating, cloning, sequencing and expressing feline cytokine genes are equipping researchers with tools to investigate the intricate cellular interactions which occur during the immune response to FIV. The investigations of cytokine expression during acute and chronic phases of FIV infection are beginning to provide a window into the complex interactions between host and virus. Changes in cytokine expression are detectable within days after FIV infection. These changes are severe enough that infected cats become susceptible to what would normally be harmless challenges of intracellular pathogens. Then as the disease progresses to FAIDS, the cat loses the ability to produce Th1 responses. The inability to mount a Th1 response sets the cat up for a host of viral, fungal and parasitic opportunistic infections. These are characteristics of human HIV-1 infections as well. For reasons discussed throughout this chapter, it is becoming increasingly clear that in addition to benefiting veterinary medicine, contributions to the understanding of human AIDS will also come from investigations utilizing the feline model of AIDS.

ACKNOWLEDGEMENTS

I thank Drs Niels Pedersen, Gregg Dean, Wayne Tompkins, Mary Tompkins and their colleagues for sharing results before publication. I thank Ted Pierson and Jennifer Nelson for their invaluable assistance with the manuscript. The author's work was supported by grants from Eckerd College's Howard Hughes Summer Research Grant and Faculty Development Grants, the Ford Foundation, grants from the Eleanor Dana Foundation, the Ronald McDonald Charities, and the National Institutes of Health (RO1 CA40931-07).

Cytokines in Feline Infectious Peritonitis

19

V.E.C.J. Schijns, B.L. Haagmans and M.C. Horzinek

Veterinary Faculty, Utrecht University, Yalelaan 1, de Uithof, 3584 CL Utrecht, The Netherlands

PATHOGENESIS

Feline infectious peritonitis (FIP) is a disease in domestic cats and some wild felids that is caused by a virus of the *Coronaviridae* family. Feline infectious peritonitis virus (FIPV) is composed of three structural proteins: a 180- to 210-kDa peplomer protein (S), a 45-kDa nucleocapsid protein (N) and a 25- to 30-kDa matrix or membrane protein (M) (de Groot *et al.*, 1987). Although the pathogenesis of FIP is poorly understood, infection of cells of the monocyte–macrophage lineage appears to be the key event. After an initial enteric (perhaps also respiratory) infection the virus rapidly disseminates to different target organs via blood-borne monocytes.

There is ample evidence for an involvement of the immune system in pathogenesis. The presence of antibody directed against FIPV proteins often leads to an accelerated, fulminating course of the disease: clinical signs and lesions develop earlier, and the mean survival time is dramatically reduced (25–30 days vs. < 10 days for seronegative and seropositive cats, respectively; Pedersen and Black, 1983). This so-called 'early death' phenomenon can be explained by antibody-dependent enhancement (ADE) of virus infection. Antibody binding to the virus results in the formation of infectious virus–antibody complexes, which attach to the surface of the macrophage (via Fc or complement receptors) with higher efficiency than virus without antibody. Antibody-dependent enhancement has been demonstrated as an infection mechanism for a number of human and animal viruses including dengue and related flaviviruses, respiratory syncytial virus, influenza virus type A, rabies virus and HIV (Halstead, 1988; Takeda *et al.*, 1988).

FIPV induces the secretion of feline IL-1 in alveolar and peritoneal macrophages; it is suggested to be involved in vascular lesion formation (Goitsuka *et al.*, 1987, 1988). Remarkably, abnormally high levels of IL-6 activity have been noted in the sera and ascites fluids from cats suffering from FIP hypergammaglobulinaemia, suggesting a relationship with polyclonal B cell activation (Ohashi *et al.*, 1989). Both cytokines, together with TNF, may contribute to the FIP-associated vascular lesions and granuloma formation.

At present, there is no treatment for FIP nor are there effective ways to prevent coronavirus infection. Vaccination seems the most logical approach, but attempts to vaccinate cats have been mostly unsuccessful. We have shown that immunization of kittens with FIPV spike (S) protein expressed by a live recombinant vaccinia virus led to significantly increased sensitivity to challenge infection and ADE, in contrast to immunization with other FIPV proteins (Vennema *et al.*, 1990). Immunization with the FIPV N or M protein expressed by vaccinia virus had no effect on the time course and outcome of a lethal challenge infection (Vennema *et al.*, 1991). However, i.n. instillation of a temperature-sensitive FIPV protected cats against challenge with virulent FIPV (Gerber *et al.*, 1990). Recently, Gonin and co-workers reported 50% protection in cats immunized with replication-defective adenovirus expressing the membrane glycoprotein of FIPV (Gonin *et al.*, 1995).

SIGNIFICANCE OF CELL-MEDIATED IMMUNITY IN PROTECTION

In any particular immune response there is a tendency for either the CMI or the antibody response to predominate (Parish, 1972). Cell-mediated immunity is considered essential for FIPV resistance for various reasons. Firstly, FIP symptoms are greatly increased in the presence of a concurrent infection with feline leukaemia virus (FeLV), a strong suppressant of CMI (Rojko and Olsen, 1984). Secondly, the spleen and lymph nodes of cats suffering from FIP exhibit severe necrosis in T-dependent areas, but contain normal to hyperplastic B cell regions (Weiss and Scott, 1981). Thirdly, enhanced dissemination of FIPV through macrophages and more severe disease is seen in thymectomized kittens (Hayashi *et al.*, 1983). Following a challenge infection with FIPV, the Con A whole blood blastogenic responses, an indicator for non-specific cellular activity, are profoundly reduced in non-protected cats. In immune cats that survive the challenge infection this response recovers and reaches pre-infection levels within 1 week (Stoddart *et al.*, 1988).

Feline infectious peritonitis virus resistant cats develop strong DTH skin responses, show virus-specific lymphocyte blast transformation and remain seronegative. Conversely, FIPV-susceptible cats seroconvert and fail to generate a DTH response (Weiss and Cox, 1989). Finally, cats protected against virulent FIPV challenge by i.n. vaccination with a temperature-sensitive FIPV strain exhibit virus-specific *in vitro* proliferation of peripheral blood lymphocytes in contrast to diseased cats. Protected cats develop low virus neutralizing antibody titres and reduced IgG and IgA responses after challenge infection (Gerber *et al.*, 1990).

Thus humoral immunity is non-protective and can even enhance pathogenesis while CMI is considered to confer protection. Immune therapy or immunization must therefore be directed at stimulating CMI over humoral immunity. These two types of immune reaction are often mutually exclusive (Parish, 1972) and tightly regulated by different physiological cytokine responses (Daynes *et al.*, 1990; Scott, 1991). In general, establishment of a stable protective CMI without induction of antibody can be obtained after repeated administration of low antigen doses (Parish, 1972; Bretscher *et al.*, 1992). This immunization strategy, referred to as 'low zone immune deviation', elicits strong DTH responses and is associated with B cell unresponsiveness to subimmunogenic antigen concentrations. It is explained by different thresholds for induction of CMI versus antibody reactions. Similar findings have been obtained in a SIV/macaque model (Clerici *et al.*, 1993a). In FIPV infection repeated low dose immunizations can elicit protective immunity without signs of illness and induce resistance to secondary challenge infection (Pedersen and Black, 1983).

SIGNIFICANCE OF THE CYTOKINE BALANCE

We anticipate that the FIPV vaccine-induced protective and non-protective immune reactions in the cat are under the control of distinct cytokine cascades. Murine (Heinzel *et al.*, 1991), rat (Papp *et al.*, 1992; Noble *et al.*, 1993) and human models (Romagnani, 1991) have shown that the major mechanism for immune regulation involves the development of functionally different helper T cell subsets secreting distinct patterns of lymphokines (Mosmann and Coffman, 1989). Progression of and recovery from infectious diseases are causally linked to dissimilar expansion of these Th subsets.

A prototypic example is *Leishmania major* infection in mice, in which IFN-γ production correlates with recovery and protection, and IL-4 secretion with exacerbation and disease (Heinzel *et al.*, 1991). Similar associations between Th1 type responses and protection, and between Th2 type responses and susceptibility have been observed in other parasitic and infectious diseases including AIDS (Scott and Kaufman, 1991; Clerici and Shearer, 1993; Mosmann and Sad, 1996). The factors influencing the different helper cell lymphokine profiles are less well understood, but appear to correspond to the type and dose of antigen (Parish, 1972), the lymphoid tissue microenvironment (mucosal versus systemic) where helper cell activation occurs (Daynes *et al.*, 1990) and, most importantly, the initial cytokine profile (Scott, 1991, 1993). Although intermediate patterns do exist, $CD4^+$ Th lymphocytes can be divided in Th1 and Th2 subtypes. In mice, Th1 cells (but not Th2) produce IL-2, IFN-γ and lymphotoxin, and promote CMI, whereas Th2 (and not Th1) cells release IL-4, IL-5, IL-6 and IL-10 which influence B cell development and augment humoral responses. Each subset may produce cytokines that stimulate the expansion of that subset and inhibition of the other. Resting Th cell-precursors produce IL-2 as the principal growth factor, IFN-γ and little or no IL-4. Thus, activation of Th1 and Th2 subsets is often mutually exclusive and appears to correlate with cellular or humoral immune reactions, respectively. Given the dichotomy of Th cells, the balance of existing lymphokines has a profound influence on the type of cell that is clonally expanded. Preferential development of one Th cell subset is often apparent at early stages after immunization or infection (Scott, 1991, 1993; Afonso *et al.*, 1994). *In vivo* cytokine expression patterns are not necessarily linked to Th cells. For example IFN-γ can be produced by γδ T cells, NK cells and $CD8^+$ T cells, while IL-10 is also produced by B cells and macrophages. Cytokine patterns are therefore often referred to as Th1/2-like.

SIGNIFICANCE OF THE CYTOKINE BALANCE IN FIPV INFECTION

Information about a dichotomy in Th1 and Th2 type cytokine profiles and associated immune responses has become available for an increasing number of infectious diseases, allergies and autoimmune diseases (for review see Mosmann and Sad, 1996). In view of published evidence it would appear likely that an intracellular pathogen like FIPV is effectively eliminated by a Th1 type immune effector mechanism but escapes or exacerbates during Th2 type immune responses. FIPV induces the secretion of feline IL-1 in alveolar and peritoneal macrophages; it is suggested to be involved in vascular lesion formation (Goitsuka *et al.*, 1987, 1988). Abnormally high levels of IL-6 activity have been noted in sera and ascites from cats suffering from FIP. The observed reciprocity of FIPV-induced immune responses with on the one hand a protective CMI response, associated with minimal antiviral antibody titres, and on the other hand a non-protective humoral immune response, suggests a dichotomy of Th1 and Th2 type cytokine responses, respectively. Th1-associated CMI is considered to play a major protective role in virtually all viral infections; CTL and complement-fixing antibodies are considered essential to eliminate intracellular viruses. Consistently, abnormally high levels of IL-6 activity noted in sera and ascites from cats suffering from FIP, suggest the prevalence of a Th2 type response in diseased cats. However, the development of immune complexes formed by complement-fixing antibodies, the observed Arthus-like reaction and the high influx of neutrophils into lesions of FIP-diseased cats may also be explained by an inadequate Th1 cytokine response. Neutrophils present in FIPV lesions or stimulated with immune complexes *in vitro* contain high levels of immunosuppressive TGF-β (Haagmans *et al.*, unpublished observations).

Recently, we have observed that soluble mediators released during FIPV infection cause apoptosis and T cell depletion in lymph nodes and spleens of diseased cats (Haagmans *et al.*, 1996). We demonstrated that the immune-mediated disease is characterized by immune complexes in lesions, serum and renal glomeruli and by the depletion of the C3 complement factor (Jacobse-Geels *et al.*, 1982).

The characterization of cytokine responses in resistant cats and the identification of disease-associated responses will lead to a better understanding of the immune dysregulation during FIP. Examination of this issue has direct implications for vaccination and therapy. In this context, the monitoring of cytokine expression in peripheral blood may provide information of potential immune responses, whereas sampling of draining lymphoid tissue may more accurately reflect ongoing responses. Using quantitative RT-PCR, Dean *et al.* observed a marked increase in IL-10 mRNA and only little increase in IL-12p40 expression in FIPV-positive tissues (spleen, thymus, mesenteric, retropharyngeal and popliteal lymph nodes and bone marrow). In general, mRNA levels of IL-4, IFN-γ and TNF-α were increased to variable degrees. This upregulated cytokine transcription occurred early, starting 1 day p.i. and remained high until death. In comparison, cats inoculated with *Listeria monocytogenes* generated an early increase in IL-12p40 mRNA, followed by a peak of IFN-γ mRNA production, and – after bacterial clearance – a peak of IL-10 mRNA. Healthy FIPV-infected cats had a 60-fold lower IL-10 to IL-12p40 ratio when compared to diseased cats (Dean *et al.*, personal communication). If shortage of IL-12 or IFN-γ production is indeed causing the defective outgrowth of FIPV-specific Th1 cells, therapeutic reconstitution of these cytokines or prophylactic use during immunization might be considered.

We have demonstrated the feasibility of this approach in mouse models. Manipulation of the cytokine balance using recombinant cytokines, their antagonists, their inducers or cytokine (receptor) gene-disrupted mouse mutants proved a powerful method to influence the degree of protective antiviral immunity, the magnitude and the quality of antibody responses as well as the priming of virus-specific splenocyte responses in mice (Schijns *et al.*, 1994a, b, 1995c; unpublished observations).

Th1 CYTOKINES AS IMMUNOLOGICAL ADJUVANTS FOR SELECTIVE INDUCTION OF PROTECTIVE IMMUNE RESPONSES

Attempts to vaccinate cats against FIP have been unsuccessful in most cases. Antibodies thus induced did not protect, they rather enhanced infection of macrophages, leading to increased virus titres and accelerated disease. In contrast, a CMI response appears to protect cats. Such responses have been obtained experimentally by Pedersen's group, although not reproducibly, using sublethal inocula in an increasing dose regimen. Induction and/or enhancement of functional CMI specific for FIPV may be of benefit both prophylactically and therapeutically. However, we should not exclude the possibility that certain components of the humoral immune response, especially those evoked by immunization, may exert a protective role.

How can a strong and stable cell-mediated Th1 guided response be induced? One possibility is the application of small doses of antigen which are subimmunogenic for induction of B cell responses but induce solely CMI. It has recently been shown that viable *Leishmania* parasites induce a Th1 response in susceptible mice, when injected at low doses (Bretscher *et al.*, 1992). This reaction is accompanied by a state of long-lasting immunologic memory resulting in protective Th1 type immunity against a high dose of *Leishmania* parasites, that would normally lead to a Th2 response and fatal disease. Similar findings have recently been obtained in a macaque–simian immunodeficiency virus (SIV) model (Clerici *et al.*, 1993a).

Another or additional possibility implies the use of cytokines as immunological adjuvants in conjunction with immunization to 'lock in' a stable Th1-regulated CMI response. Observations from *in vitro* and *in vivo* models support the notion that the cytokine milieu present during primary antigen stimulation are a major factor in determining the differentiation pathway of naive T cells. Th1 non-responsiveness will preferentially cause deficiency of cellular immunity. This will render the infected host susceptible to infection and disease induced by various pathogens.

Thus, particular cytokines administered exogenously or induced endogenously, may selectively stimulate or inhibit certain immune reactions. Cytokines which might be used to induce or strengthen a CMI response include IFN-γ, IFN-α and IL-12; they have been reported to stimulate Th1-type responses (Afonso *et al.*, 1994; Schijns *et al.*, 1995c). These cytokines can either be co-administered with the antigen or co-expressed in a replicating live vector. The latter approach has been shown to result in reduced vector virulence. In particular, infection with recombinant vaccinia viruses encoding IL-2, IFN-γ and TNF results in a self-limiting infection (Ruby *et al.*, 1992). Recently, it has been shown that neutralization of IL-4 at the time of immunization with inactivated respiratory syncytial virus improves Th1 immune responses and reduces illness against challenge infection (Tang and Graham, 1994). A similar inhibition of early IL-4 expression during immunization of cats against FIPV, using IL-4 mutants (Kruse *et al.*, 1992; Carballido *et al.*, 1995) or antagonists may evoke the proper population of effector cells.

In the near future we will attempt to evoke protective immunity by vaccination with low doses of FIPV in the presence of recombinant cytokines which preferentially induce CMI. By this immunological manipulation, using Th1 favouring cofactors of the vaccine like IL-12 and IFN-γ or IL-4 antagonists, we shall try to tip the early cytokine balance in the desired Th1 > Th2 (CMI) direction which may result in a new immunization strategy (even after exposure) against FIPV infection.

MOLECULAR CLONING OF FELINE IFN-γ, IL-4 AND IL-12

Feline Interferon-γ

Until recently, little has been known about the cytokine expression patterns in cats. Since clinical observations suggest that humoral responses are associated with the fatal course of FIP, while CMI is assumed to be protective, and likely afforded by a type-1 cytokine response, we focused on feline IFN-γ and IL-12. However, nothing is known about the physiological role of both cytokines in cats. Because of its species specificity we decided to clone feline IFN-γ (Schijns *et al.*, 1995a). We have cloned a cDNA containing the coding sequence of feline IFN-γ by RT-PCR on mRNA of Con A stimulated feline mononuclear lymphocytes, using primers based on conserved regions of the canine IFN-γ gene (Devos *et al.*, 1992). The 543 bp feline IFN-γ cDNA, which was identical in two independently isolated clones, contains an open reading frame encoding for a 167 aa protein with a predicted molecular mass of 17 kDa. Hydrophobicity analysis suggests that the first 23 aa form the signal peptide that is cleaved from the precursor polypeptide behind the cysteine at position 23. The remaining 144 aa would constitute the mature protein. Alignment of the nucleotide sequence of feline and canine IFN-γ revealed a similarity between the coding regions of 89.1%. At the aa level there is similarity between 145/167 aa (86.8% identity). Transient eukaryotic expression of the cloned cDNA using the vaccinia virus/phage T7 RNA polymerase system (Fuerst *et al.*, 1986) yielded antiviral activity in a VSV inhibition assay on CrFK cells and *Felis catus* whole fetus (fcwf) cells. This activity was not observed using control plasmids or canine (MDCK) cells. Recently, we have expressed feline IFN-γ in *Spodoptera fugiperda* (Sf21) cells using the baculovirus expression system. The recombinant product proved biologically active; it exerted antiviral activity against a variety of viruses (Schijns *et al.*, unpublished). The availability of large amounts of rFeIFN-γ allows us to determine the Th1 favouring activity when administered as adjuvant in a FIPV vaccine. Specific antibodies raised against it will enable studies on the (patho)physiological role of this lymphokine in the cat.

Feline Interleukin-12

The recently discovered IL-12 is a heterodimeric 70-kDa cytokine composed of a disulphide-bonded 35-kDa (p35) subunit and a 40-kDa (p40) subunit. Produced mainly by activated monocytes/macrophages and B cells, it enhances NK/LAK cell cytolytic activity and stimulates CTL responses and IFN-γ production (for review see: Trinchieri, 1995). It is therefore a key regulatory molecule in Th1-driven immune responses that are considered necessary for the elimination of intracellular pathogens (Scott and Kaufman, 1991). *In vivo* administered IL-12 has powerful antitumour and antimetastatic activity (Brunda *et al.*, 1993) and is a promising vaccine adjuvant (Afonso *et al.*, 1994; Schijns *et al.*, 1995c). Due to its inhibitory effects on IL-4 and IgE synthesis, IL-12 may be considered in the prevention of IgE-mediated allergic conditions (Kiniwa *et al.*, 1992).

Since it is likely species specific we decided to clone feline IL-12 (Schijns *et al.*, 1997). We have cloned a cDNA containing the coding sequence of feline IL-12 by the 'rapid amplification of cDNA ends' (RACE) protocol (Frohman *et al.*, 1988). The mRNA of feline PBMC stimulated for 4 to 24 h with either 0.0075% (w/v) of *S. aureus* (Pansorbin Calbiochem-Behring Co., La Jolla, California), or 1–10 μg LPS (Difco Laboratories, Detroit, USA) was reverse transcribed into cDNA, using a 35-base oligonucleotide primer containing 17 dT residues and an adaptor sequence. The cat IL-12 p35 and p40 cDNA were amplified by a first PCR using a sense primer based on regions of the canine, bovine and human IL-12 p35 cDNA (5′-CAGTGCCGGCTCAGCATGTG-3′) and p40 cDNA (5′-ATGCATCCTCAGCAGTTGG-3′) and the antisense adaptor primer (5′-GACTCGAGTCGACATCG-3′). Conditions for PCR have been described (Frohman *et al.*, 1988). In a subsequent semi-nested PCR the feline p35 and p40 cDNAs were amplified using the same sense primers and antisense primers that were predominantly based on sequences of canine IL-12 p35 or p40 cDNA (Belke and Buettner, Genebank accession numbers U49085 and U49100, respectively); for p35 (5′-CTAGGAAGCATTCAGATAGCTC-3′) and for p40 (5′-ATCCTGGGGGTGGAACCTAAC-3′). The PCR products were cloned into plasmid pGEM-T (Promega, Madison, WI) or pNoTa/T7 (5 Prime → 3 Prime, Inc. Boulder, CO, USA) and sequenced. The 685 bp feline IL-12 p35 cDNA sequence, which was identical in at least two independently isolated clones, contains an open reading frame encoding a 222 aa protein. The 1006 bp feline IL-12 p40 cDNA sequence, which was identical in three independently isolated clones, contains an open reading frame encoding a 329 aa protein. Comparison of the predicted aa sequence revealed that the feline IL-12 p35 protein shares 90.5, 81.5, 85.1 and 55.4% identity with the canine, bovine, human and murine IL-12 p35 chain, respectively, while the predicted feline IL-12 p40 protein shares 92.1, 84.8, 84.2 and 68.1% identity with the canine, bovine, human and murine IL-12 p40 chain, respectively. The availability of large amounts of rFeIL-12, and specific antibodies raised against it will enable studies on the (patho)physiological role of this cytokine in the cat. It would be of interest to determine IL-12's immunomodulatory role in tumour bearing and allergic cats as well as in cats infected with feline infectious peritonitis virus (FIPV), feline immunodeficiency virus (FIV) and feline leukaemia virus (FeLV) and to study its adjuvant activity in vaccines against these viruses.

Feline Interleukin-4

Interleukin-4, a lymphokine produced mainly by activated T cells and mast cells, is a key regulatory molecule in Th type-2 driven immune responses that are considered necessary for the elimination of extracellular pathogens (Scott and Kaufman, 1991). Mice treated with IL-4 neutralizing antibodies (Finkelman *et al.*, 1990), as well as mice bearing a defective IL-4 gene (Kühn *et al.*, 1991), fail to produce IgE and show reduced serum levels of IgG1. This indicates a pathophysiological role for IL-4 in the generation of certain disease states, like hyper-IgE syndrome or IgE-mediated allergic conditions (Finkelman *et al.*, 1990; Tepper *et al.*, 1990). IL-4-deficient mice have strongly reduced Th2 cytokine responses (Kopf *et al.*, 1993). Due to its inhibitory effects on IFN-γ, IL-4 is considered to downregulate Th1-associated immune responses that are necessary for the control of many intracellular pathogens.

Nothing is known about the physiological role of IL-4 in cats. Since it is likely species specific we decided to clone feline IL-4 (Schijns *et al.*, 1995b). We have cloned a cDNA containing the coding sequence of feline IL-4 by the 'rapid amplification of cDNA ends' (RACE) protocol (Frohman *et al.*, 1988). The mRNA of Con-A-stimulated feline mononuclear lymphocytes was reverse transcribed into cDNA, using a 35-base oligonucleotide primer containing 17 dT residues and an adaptor sequence. The cat IL-4 cDNA was amplified by PCR using a sense primer based on conserved regions of the human, porcine and bovine IL-4 gene (5′-TATTAATGGGTCTCACC-TACCA-3′) and the antisense adaptor primer (5′-GACTCGAGTCGA-CATCG-3′). Conditions for PCR have been described (Frohman *et al.*, 1988). The PCR products were cloned into plasmid pGEM-T and sequenced. The 522 bp feline IL-4 cDNA sequence, which was identical in three independently isolated clones, contains an open reading frame encoding a 133 aa protein. By analogy with human IL-4, cleavage of the precursor peptide probably occurs after Gly24. Comparison of the predicted aa sequence revealed that feline IL-4 shares 72, 67, 66 and 50% identity with porcine, bovine, cervine and human IL-4, respectively. Eventually the availability of large amounts of rFeIL-4 and specific antibodies raised against it will enable studies on the (patho)physiological role of this lymphokine in coronavirus-infected cats.

ACKNOWLEDGEMENTS

The authors wish to thank Dr Gregg Dean and his colleagues for sharing their unpublished data. This work was supported by a grant from Virbac (Carros, France).

References for Part VI

Afonso, L.C.C., Scharton, T.M., Vieira, L.Q., Wysocka, M., Trinchieri, G. and Scott, P. (1994) The adjuvant effect of interleukin-12 in a vaccine against *Leishmania major*. *Science* 263, 235–237.

Aoki, N., Siegfried, M. and Lefer, A.M. (1989) Anti-EDRF effect of tumor necrosis factor in isolated, perfused cat carotid arteries. *American Journal of Physiology* 256, H1509–1512.

Argyle, D.J., Smith, K., McBride, K., Fulton, R. and Onions, D.E. (1995) Nucleotide and predicted peptide sequence of feline interferon-γ (IFN-γ). *DNA Sequences* 5, 169–171.

Babu, J.S., Kanangat, S. and Rouse, B.T. (1993) Limitations and modifications of quantitative polymerase chain reaction. Application to measurement of multiple mRNAs present in small amounts of sample RNA. *Journal of Immunological Methods* 165, 207–216.

Barber, S.M. (1996) Isolation, Characterization and Expression of Polar Bear *(Ursa maritimus)* and Grey Wolf (*Canis lupus*) Interleukin-2. BS. Thesis, Eckerd College, St Petersburg, Florida, USA.

Barlough, J.E., Ackley, C.D., George, J.W., Levy, N., Acevedo, R., Moore, P.F., Rideout, B.A., Cooper, M.D. and Pedersen, N.C. (1991) Acquired immune dysfunction in cats with experimentally induced feline immunodeficiency virus infection: comparison of short-term and long-term infections. *Journal of Acquired Immune Deficiency Syndromes* 4, 219–227.

Bendinelli, M., Pistello, M., Matteucci, D., Lombardi, S., Baldinotti, F., Bandecchi, P., Ghilarducci, R., Ceccherini-Nelli, L., Garzelli, C., Poli, A., Esposito, F., Malvaldi, G. and Tozzini, F. (1993) Small animal model of AIDS and the feline immunodeficiency virus. In: Friedman, H. *et al.* (eds), *Drugs of Abuse, Immunity, and AIDS*. Plenum Press, New York, pp. 189–202.

Bradley, W.G., Gibbs, C., Kraus, L.A., Good, R.A. and Day, N.K. (1993a) Molecular cloning and characterization of cDNAs encoding feline interleukin-2, -6, and -10. *Proceedings of the Second International Symposium on Feline Retrovirus Research*. Raleigh, North Carolina, 6–9 October, p. 34.

Bradley, W.G., Gibbs, C., Kraus, L., Good, R.A. and Day, N.K. (1993b) Molecular cloning and characterization of a cdna encoding feline interleukin-6. *Proceedings of the Society for Experimental Biology and Medicine* 204, 301–305.

Bradley, W.G., Kraus, L.A., Good, R.A. and Day, N.K. (1995) Dehydroepiandrosterone inhibits replication of feline immunodeficiency virus in chronically infected cells. *Veterinary Immunology and Immunopathology* 46, 159–168.

Breen, E.C., Rezai, A.R., Nakajima, K., Beall, G.N., Mitsuyasu, R.T., Hirano, T., Kishimoto, T. and Martinez-Maza, O. (1990) Infection with HIV is associated with elevated IL-6 levels. *Journal of Immunology* 144, 480–484.

Bretscher, P.A., Wei, G., Menon, J.N. and Bielefeldt-Ohmann, H. (1992) Establishment of stable, cell-mediated immunity that makes 'susceptible' mice resistant to *Leishmania major*. *Science* 257, 539–542.

Brunda, M.J., Luistro, L., Warrier, R.R., Wright, R.B., Hubbard, B.R., Murphy, M., Wolf, S.F. and Gately, M.R. (1993) Antitumor and antimetastatic activity of interleukin-12 against murine tumors. *Journal of Experimental Medicine* 178, 1223–1230.

Bush, K., Day, N.K., Kraus, L.A., Good, R.A. and Bradley, W.G. (1994) Molecular cloning of feline interleukin 12 p35 reveals the conservation of leucine-zipper motifs present in human and murine IL-12 p35. *Molecular Immunology* 31, 1373–1374.

Carballido, J.M., Schols, D., Namikawa, R., Zurawski, S., Zurawski, G., Roncarolo, M.-G. and de Vries, J.E. (1995) IL-4 induces human B cell maturation and IgE synthesis in SCID-hu mice. Inhibition of ongoing IgE production by *in vivo* treatment with an IL-4/IL-13 receptor antagonist. *Journal of Immunology* 155, 4162–4170.

Cashman, M.E., Ness, T.L., Roess, W.B., Bradley, W.G. and J.E. Reynolds III (1996) Isolation and characterization of a cDNA encoding interleukin 2 from the Florida manatee, *Trichechus manatus latirostris*. *Marine Mammal Science* 12, 89–98.

Clerici, M. and Shearer, G.M. (1993) A Th-1 > Th-2 switch is a critical step in the etiology of HIV infection. *Immunology Today* 14, 107–111.

Clerici, M., Axberg, I., Polacino-Firpo, P., Kuller, L., Morton, W., Benviste, R. *et al.* (1993a) SIV specific cellular immunity in seronegative macaques exposed to SIV/Mne. *IXth International Conference on AIDS*, Berlin (Abstract WS-A15–3), p. 31.

Clerici, M., Hakim, F.T., Venzon, D.J., Blatt, S., Hendrix, C.W., Wynn, T.A. and Shearer, G.M. (1993b) Changes in interleukin-2 and interleukin-4 production in asymptomatic, human immunodeficiency virus-seropositive individuals. *Journal of Clinical Investigation* 91, 759–765.

Cozzi, P.J., Padrid, P.A., Takeda, J., Alegre, N.L., Yuhki, N. and Leff, A.R. (1993) Sequence and functional characterization of feline interleukin 2. *Biochemical & Biophysical Research Communications* 194, 1038–1043.

D'Andrea, A., Aste-Amezaga, M., Valiante, N.M., Ma, X., Kubin, M. and Trinchieri, G. (1993) Interleukin 10 (IL-10) inhibits human lymphocyte interferon γ-production by suppressing natural killer cell stimulatory factor/IL-12 synthesis in accessory cells. *Journal of Experimental Medicine* 178, 1041–1048.

Daynes, R.A., Araneo, B.A., Dowell, T.A., Huang, K. and Dudley, D. (1990) Regulation of murine lymphokine production *in vivo*. III. The lymphoid tissue microenvironment exerts regulatory influences over T helper cell function. *Journal of Experimental Medicine* 171, 979–996.

Dean, G., Lavoy, A. and Pedersen, N. (1996) FIV-associated changes in cytokine mRNA *in vivo*. *Proceedings of the Third International Feline Retrovirus Research Symposium*, 6–9 March, Fort Collins, Colorado. p. 16.

Devos, K., Duerinck, F., Van Audenhove, K. and Fiers, W. (1992) Cloning and expression of the canine interferon γ gene. *Journal of Interferon Research* 12, 95–102.

Dow, S.W., Poss, M.L. and Hoover, E.A. (1990) Feline immunodeficiency virus: a neurotropic lentivirus. *Journal of Acquired Immune Deficiency Syndromes* 3, 658–668.

Dow, S.W., Dreitz, M.J. and Hoover, E.A. (1992) Feline immunodeficiency virus neurotropism: evidence that astrocytes and microglia are the primary target cells. *Veterinary Immunology and Immunopathology* 3, 23–35.

Egberink, H. and Horzinek, M.C. (1992) Animal immunodeficiency viruses. *Veterinary Microbiology* 33, 311–331.

Egberink, H.F., Berrocal, A., Bax, H.A., van den Ingh, T.S., Walter, J.H. and Horzinek, M.C. (1989) Feline immunodeficiency virus infection. *Veterinary Immunology and Immunopathology* 21, 111–129.

Elder, J.H. and Phillips, T.R. (1994) Molecular properties of feline immunodeficiency virus (FIV). *Infectious Agents and Disease* 2, 361–374.

English, R.V., Johnson, C.M., Gebhard, D.H. and Tompkins, M.B. (1993) *In vivo* lymphocyte tropism of feline immunodeficiency virus. *Journal of Virology* 67, 175–186.

Fenwick, B.W., Schore, C.E. and Osburn, B.I. (1988) Human recombinant interleukin-2 induces *in vitro* proliferation of equine, caprine, ovine, canine, and feline peripheral blood lymphocytes. *Comparative Immunology, Microbiology and Infectious Diseases* 11, 51–60.

Finkelman, F.D., Holmes, J., Katona, I.M., Urban, J.F. Jr, Beckmann, M.P., Park, L.S., Schooley, K.A., Coffman, R.L., Mosmann, T.R. and Paul, W.E. (1990) Lymphokine control of *in vivo* immunoglobulin isotype selection. *Annual Reviews in Immunology* 8, 303–333.

Fiorentino, D.F., Bond, M.W. and Mosmann, T.R. (1989) Two types of mouse T helper cell. IV. Th2 clones secrete a factor that inhibits cytokine production by Th1 clones. *Journal of Experimental Medicine* 170, 2081–2095.

Folks, T.M., Clouse, K.A., Justement, J., Rabson, A.R., Duh, E.J., Kehrl, J.H. and Fauci, A.S. (1989) Tumor necrosis factor alfa induces expression of human immunodeficiency virus in a chronically infected T-cell clone. *Proceedings of the National Academy of Sciences USA* 86, 2365–2368.

Frohman, M.A., Dush, M.K. and Martin, G.R. (1988) Rapid production of full-length cDNAs from rare transcripts: amplification using a single gene-specific oligonucleotide primer. *Proceedings of the National Academy of Sciences USA* 85, 8998–9002.

Fuerst, T.R., Niles, E.G., Studier, F.W. and Moss, B. (1986) Eukaryotic transient expression system based on recombinant vaccia virus that synthesizes bacteriophage T7 TNA polymerase. *Proceedings of the National Academy of Sciences USA* 83, 8122–8126.

Gallo, P., Frei, K., Rordorf, C., Lazdins, J., Tavolato, B. and Fontana, A. (1989) Human immunodeficiency virus type 1 (HIV-1) infection of the central nervous system: an elevation of cytokines in the cerebrospinal fluid. *Journal of Neuroimmunology* 23, 109–116.

Gardner, M.B. (1991) Simian and feline immunodeficiency viruses: animal lentivirus models for evaluation of AIDS vaccines and antiviral agents. *Antiviral Research* 15, 267–286.

Gazzinelli, R.T., Giese, N.A. and Morse, H.C. (1994) *In vivo* treatment with interleukin 12 protects mice from immune abnormalities observed during murine acquired immunodeficiency syndrome (MAIDS). *Journal of Experimental Medicine* 180, 2199–2208.

Gebhard, D., Rottman, J., Russlander, D., Gager, M., Levy, M. and Tompkins, W.A.F. (1993) Phenotype and cytokine profile of an established feline myelomonocytic-like cell line. *Proceedings of the Second International Symposium on Feline Retrovirus Research*. Raleigh, North Carolina, 6–9 October, p. 38.

Gerber, J.D., Ingersoll, J.D., Gast, A.M., Christianson, K.K., Selzer, N.L., Landon, R.M., Pfeiffer, N.E., Sharpee, R.L. and Beckenhauer, W.H. (1990) Protection against feline infectious peritonitis virus by intranasal inoculation of a temperature-sensitive FIPV vaccine. *Vaccine* 8, 536–542.

Goitsuka, R., Hirota, Y., Hasegawa, A. and Tomoda, I. (1986) Feline interleukin 2 activity. *Japanese Journal of Veterinary Sciences* 48, 529–537.

Goitsuka, R., Hirota, Y., Hasegawa, S. and Tomoda, I. (1987) Release of interleukin 1 from peritoneal exudate cells of cats with feline infectious peritonitis. *Japanese Journal of Veterinary Sciences* 49, 811–818.

Goitsuka, R., Onda, C., Hirota, Y., Hasegawa, A. and Tomoda, I. (1988) Feline interleukin 1 production induced by feline infectious peritonitis virus. *Japanese Journal of Veterinary Sciences* 50, 209–214.

Goitsuka, R., Ohashi, T., Ono, K., Yasukawa, K., Koishibara, Y., Fukui, H., Ohsugi,Y. and Hasegawa, A. (1990) IL-6 activity in feline infectious peritonitis. *Journal of Immunology* 144, 2599–2603.

Gonin, P., Oualikene, W., Fournier, A., Soulier, M., Audonnet, J.-C., Riviere, M. and Eloit, M. (1995) Evaluation of a replication-defective adenovirus expressing the feline infectious peritonitis membrane protein as a vaccine in cats. *Vaccine Research* 4, 217–227.

Gorham, J.R., Henson, J.B. and Dodgen, C.J. (1971) Basic principles of immunity in cats. *Journal of the American Veterinary Medicine Association* 158, 846–853.

Gray, F., Hurtrel, M. and Hurtrel, B. (1993) Early central nervous system changes in human immunodeficiency virus (HIV)-infection. *Neuropathology and Applied Neurobiology* 19, 3–9.

Groot, R.J. de, Terhaar, R.J., Horzinek, M.C. and Zeist, B.A.M. van der (1987) Intracellular RNAs of the feline infectious peritonitis coronavirus strain 79–1146. *Journal of General Virology* 86, 995–1002.

Haagmans, B.L., Egberink, H. and Horzinek, M.C. (1996). Apoptosis and T-cell depletion during feline infectious peritonitis. *Journal of Virology*, 70, 8977–8983.

Halstead, S.B. (1988) Pathogenesis of dengue: challenges to molecular biology. *Science* 239, 476–481.

Hayashi, T., Sasaki, N., Ami, Y. and Fujiwara, K. (1983) Role of the thymus-dependent lymphocytes and antibodies in feline infectious peritonitis after oral infection. *Japanese Journal of Veterinary Sciences* 45, 759–766.

Heinzel, F.P., Sadick, M.S., Mutha, S. and Locksley, R.M. (1991) Production of interferon-γ, interleukin 2, interleukin 4 and interleukin 10 by $CD4^+$ lymphocytes *in vivo* during healing and progressive murine leishmaniasis. *Proceedings of the National Academy of Sciences USA* 88, 7011–7015.

Hober, D., Haque, A., Wattre, P., Beaucaire, G., Mounton, Y. and Capron, A. (1989) Production of tumor necrosis factor-alpha (TNF-α) and interleukin-1 (IL-1) in patients with AIDS. Enhanced level of TNF-α is related to a higher cytotoxic activity. *Clinical and Experimental Immunology* 78, 329–333.

Hosie, M.J., Osborne, R., Yamamoto, J.K., Neil, J.C. and Jarrett, O. (1995) Protection against homologous but not heterologous challenge induced by inactivated feline immunodeficiency virus vaccines. *Journal of Virology* 69, 1253–1255.

Hurtrel, M., Ganiere, J.P., Guelfi, J.F., Chakrabarti, L., Maire, M.A., Gray, F., Montagnier, L. and Hurtrel, B. (1992) Comparison of early and late feline immunodeficiency virus encephalopathies. *AIDS* 6, 399–406.

Jacobse-Geels, H.E.L., Daha, M.R. and Horzinek, M.C. (1982) Antibody, immune complexes, and complement activity fluctuations in kittens with experimentally induced feline infectious peritonitis. *American Journal of Veterinary Research* 43, 666–670.

James-Yarish, M., Bradley, W.G., Emmanuel, P.J., Good, R.A. and Day, N.K. (1994) Detection of cell specific cluster determinant expression by reverse transcription polymerase chain reaction. *Journal of Immunological Methods* 169, 73–82.

Kiniwa, M., Gately, M., Gubler, U., Gizzonite, R., Fargeas, C. and Delespesse, G. (1992) Recombinant interleukin-12 suppresses the synthesis of immunoglobulin E by interleukin-4 stimulated human lymphocytes. *Journal of Clinical Investigation* 90, 262–266.

Kopf, M., LeGros, G., Bachmann, M., Lamers, M.C., Bluethmann, H. and Kohler, G. (1993) Disruption of the IL-4 gene blocks Th-2 cytokine responses. *Nature* 362, 245–247.

Kraus, L.A., Bradley, W.G., Engelman, R.W., Brown, K.M., Good, R.A. and Day, N.K. (1996) Relationship between tumor necrosis factor A and feline immunodeficiency virus expressions. *Journal of Virology* 70, 566–569.

Kruse, N., Tony, H.-P. and Sebald, W. (1992) Conversion of human interleukin-4 into a high affinity antagonist by a single amino acid replacement. *The European Molecular Biology Organization Journal* 11, 3237–3244.

Kühn, R., Rajewsky, K. and Müller, W. (1991) Generation and analysis of interleukin-4 deficient mice. *Science* 254, 707–710.

Lafrado, L., Podell, M., Krakowka, S., Hayes, K., Hanlon, M. and Mathes, L. (1993) FIV: a model for retrovirus-induced pathogenesis. *AIDS Research Reviews* 3, 115–150.

Landschulz, W.H., Johnson, P.F. and McKnight, S.L. (1988) The leucine zipper: a hypothetical structure common to a new class of DNA binding proteins. *Science* 240, 1759–1764.

Laurenzi, M.A., Siden, A., Persson, M.A., Norkrans, G., Hagberg, L. and Chiodi, F. (1990) Cerebrospinal fluid interleukin-6 activity in HIV infection and inflammatory and noninflammatory diseases of the nervous system. *Clinical Immunology and Immunopathology* 57, 233–241.

Lawrence, C.E., Callanan, J.J., Willett, B.J. and Jarrett, O. (1995) Cytokine production by cats infected with feline immunodeficiency virus; a longitudinal study. *Immunology* 85, 568–574.

Lehmann, R., Joller, H., Haagmans, B.L. and Lutz, H. (1992) Tumor necrosis factor α levels in cats experimentally infected with feline immunodeficiency virus: effects of immunization and feline leukemia virus infection. *Veterinary Immunology and Immunopathology* 35, 61–69.

Leutz, A., Damm, K., Sterneck, E., Kowenz, E., Ness, S., Frank, R., Gausepohl, H., Pan, Y., Smart, J., Hayman, M. and Graft, T. (1989) Molecular cloning of the chicken myelomonocytic growth factor (cMGF) reveals relationship to interleukin 6 and granulocyte colony stimulating factor. *European Molecular Biology Organization Journal* 8, 175–181.

Lin, D.S. (1992) Feline immune system. *Comparative Immunology, Microbiology and Infectious Disease* 15, 1–11.

Lin, D.S. and Bowman, D.D. (1991) Cellular responses of cats with primary toxoplasmosis. *Journal of Parasitology* 77, 272–279.

Locardi, C., Petrini, C., Boccoli, G., Testa, U., Dieffenbach, C., Butto, S. and Belardelli, F. (1990) Increased human immunodeficiency virus (HIV) expression in chronically infected U937 cells upon *in vitro* differentiation by hydroxyvitamin D3: roles of interferon and tumor necrosis factor in regulation of HIV production. *Journal of Virology* 64, 5874–5882.

Ma, J., Kennedy-Stoskopf, S., Sellon, R., Tonkonogy, S., Hawkins, E.C., Tompkins, M.B. and Tompkins, W.A. (1995) Tumor necrosis factor-α responses are depressed and interleukin-6 responses unaltered in feline immunodeficiency virus infected cats. *Veterinary Immunology and Immunopathology* 46, 35–50.

Manetti, B., Parronchi, P., Giudizi, M.G., Piccinni, M-P., Maggi, E., Trinchieri, G. and Romagnani, S. (1993) Natural killer cell stimulatory factor (interleukin 12 [IL-12]) induces T helper type 1 (Th-1)-specific immune responses and inhibits the development of IL-4-producing Th cells. *Journal of Experimental Medicine* 177, 1199–1204.

Matsuyama, T., Hamamoto, Y., Soma, G.I., Mizuno, D., Yamamoto, N. and Kobayashi, N. (1989) Cytocidal effect of tumor necrosis factor on cells chronically infected with human immunodeficiency virus (HIV): enhancement of HIV replication. *Journal of Virology* 63, 2504–2509.

McGraw, R.A., Coffee, B.W., Otto, C.M., Dres, R.T. and Rawlings, C.A. (1990) Gene sequence of feline tumor necrosis factor α. *Nucleic Acids Research* 18, 5563.

McMahon, E.J. (1996) Characterization and Expression of a cDNA Encoding IL-2 of the Bottlenose Dolphin, *Tursiops truncatus*. BS Thesis, Eckerd College, St Petersburg, Florida, USA.

Meers, J., del Fierro, G.M., Cope, R.B., Park, H.S., Greene, W.K. and Robinson, W.F. (1993) Feline immunodeficiency virus infection: plasma, but not peripheral blood mononuclear cell virus titer is influenced by zidovudine and cyclosporine. *Archives of Virology* 132, 67–81.

Michihiko, S., Yamamoto, N., Shinozaki, F., Shimada, K., Soma, G.I. and Kobayashi, N. (1989) Augmentation of *in-vitro* HIV replication in peripheral blood mononuclear cells of AIDS and ARC patients by tumor necrosis factor. *Lancet* i, 1206–1207.

Miles, S.A., Rezai, A.R., Salazar-Gonzalez, J.F., Vander Meyden, M., Stevens, R.H., Logan, D.M., Mitsuyasu, R.T., Taga, T., Hirano, T., Kishimoto, T. and Martinez-Maza, O. (1990) AIDS Kaposi sarcoma-derived cells produce and respond to interleukin 6. *Proceedings of the National Academy of Sciences USA* 87, 4068–4072.

Mintz, M., Rapaport, R., Oleske, J.M., Connor, E.M., Koenigsberger, M.R., Denny, T. and Epstein, L.G. (1989) Elevated serum levels of tumor necrosis factor α are associated with progressive encephalopathy in children with acquired immunodeficiency syndrome. *American Journal of Diseases of Children* 143, 771–774.

Mosmann, T.R. (1994a) Cytokine patterns during the progression to AIDS. *Science* 265, 193–194.

Mosmann, T.R. (1994b) Properties and functions of interleukin-10. *Advances in Immunology* 56, 1–26.

Mosmann, T.R. and Coffman, R.L. (1989) Th-1 and Th-2 cells: different patterns of lymphokine secretion lead to different functional properties. *Annual Reviews in Immunology* 7, 145–173.

Mosmann, T.R. and Sad, S. (1996) The expanding universe of T-cell subsets: Th-1, Th-2 and more. *Immunology Today* 17, 138–146.

Nagata, S., Tsuchiya, M., Asano, S., Kaziro, Y., Yamazaki, T., Yamamoto, O., Hirata, Y., Kubota, N., Oheda, M., Nomura, H. and Ono, M. (1986) Molecular cloning and expression of cDNA for human granulocyte colony-stimulating factor. *Nature* 319, 415–418.

Nakamura, N., Sudo, T., Matsuda, S. and Yanai, A. (1992) Molecular cloning of feline interferon cDNA by direct expression. *Bioscience, Biotechnology and Biochemistry* 56, 211–214.

Ness, T.L. (1995) Isolation, Characterization, and Expression of Cytokines from the Florida Manatee, *Trichechus manatua latirostris*, and the Killer Whale, *Orcinus orca*. BS Thesis, Eckerd College, St Petersburg, Florida, USA.

Noble, A., Staynov, D.Z. and Kemeny, D.M. (1993) Generation of rat Th-2-like cells *in vitro* is interleukin-4-dependent and inhibited by interferon-γ. *Immunology* 79, 562–567.

Odeh, M. (1990) The role of tumor necrosis factor-α in acquired immunodeficiency syndrome. *Journal of Internal Medicine* 228, 549–556.

Ohashi, T., Goitsuka, R., Ono, K. and Hasegawa, A. (1989) Feline hybridoma growth factor/interleukin-6 activity. *Journal of Leucocyte Biology* 46, 501–507.

Ohashi, T., Goitsuka, R., Watari, T., Tsujimoto, H. and Hasegawa, A. (1992) Elevation of feline interleukin 6-like activity in feline immunodeficiency virus infection. *Clinical Immunology and Immunopathology* 65, 207–211.

Ohashi, T., Matsumoto, Y., Watari, T., Goitsuka, R., Tsujimoto, H. and Hasegawa, A. (1993) Molecular cloning of feline interleukin-6 cDNA. *Journal of Veterinary Medical Science* 55, 941–944.

Ohno, K., Nakano, T., Matsumoto,Y., Watari, R., Goitsuka, R., Nakayama, H., Tsujimoto, H. and Hasegawa, A. (1993) Apoptosis induced by tumor necrosis factor in cells chronically infected with feline immunodeficiency virus. *Journal of Virology* 67, 24229–24233.

Ohno, K., Okamoto, Y., Miyazawa, T., Mikami, T., Watari, T., Goitsuka, R., Tsujimoto, H. and Hasegawa, A. (1994) Induction of apoptosis in a T lymphoblastoid cell line infected with feline immunodeficiency virus. *Archives of Virology* 135, 153–158.

Okamoto, T., Matsuyama, T. and Mori, S. (1989) Augmentation of human immunodeficiency virus type 1 gene expression by tumor necrosis factor-α. *Journal of Immunology* 142, 702–707.

Osborn, L., Kunkel, S. and Nabel, G. (1989) Tumor necrosis factor-α and interleukin 1 stimulate the human immunodeficiency virus enhancers by activation of the nuclear factor kappaB. *Proceedings of the National Academy of Sciences USA* 86, 2336–2340.

Papp, I., Wieder, K.J., Sablinski, T., O'Connell, P.J., Milford, E.L., Strom, T.B. and Kupiec-Weglinski, J.W. (1992) Evidence for functional heterogeneity of rat $CD4^+$ T cells *in vivo*. Differential expression of IL-2 and IL-4 mRNA in recipients of cardiac allografts. *Journal of Immunology* 148, 1308–1314.

Parish, C.R. (1972) The relationship between humoral and cell-mediated immunity. *Transplantation Reviews* 13, 35–66.

Pedersen, N.C. (1987) Basic and clinical immunology. In: Holzworth, J. (ed.) *Diseases of the Cat. Medicine and Surgery*. Volume 1, Saunders, Philadelphia, Pennsylvania, USA.

Pedersen, N.C. and Barlough, J.E. (1991) Clinical overview of feline immunodeficiency virus. *Journal of the American Veterinary Medical Association* 199, 1298–1305.

Pedersen, N.C. and Black, J.W. (1983) Attempted immunization of cats against feline infectious peritonitis using avirulent live virus or sublethal amounts of virulent virus. *American Journal of Veterinary Research* 44, 229–234.

Pedersen, N.C., Ho, E.W., Brown, M.L. and Yamamoto, J.K. (1987) Isolation of a T- lymphotrophic virus domestic cats with an immunodeficiency-like syndrome. *Science* 235, 790–793.

Phillips, T.R., Prospero-Garcia, O., Puaoi, D.L., Lerner, D.L., Fox, H.S., Olmsted, R.A., Bloom, F.E., Henriksen, S.J. and Elder, J.H. (1994) Neurological abnormalities associated with feline immunodeficiency virus infection. *Journal of General Virology* 75, 979–987.

Podell, M., Oglesbee, M., Mathes, L., Krakowka, S., Olmsted, R. and Lafrado, L. (1993) AIDS-associated encephalopathy with experimental feline immunodeficiency virus infection. *Journal of Acquired Immunodeficiency Syndromes* 6, 758–771.

Poli, G., Kinter, A., Justement, J.S., Kehrl, J.H., Bressler, P., Stanley, S. and Fauci, A.S. (1990) Tumor necrosis factor α functions in an autocrine manner in the induction of HIV expression. *Proceedings of the National Academy of Sciences USA* 87, 782–785.

Remington, K.M., Chesebro, B., Wehrly, K., Pedersen, N.C. and North, T.W. (1991) Mutants of feline immunodeficiency virus resistant to 3′-azido-3′-deoxythymidine. *Journal of Virology* 65, 308–312.

Rimstad, E., Reubel, G.H., Dean, G.A., Higgins, J. and Pedersen, N. (1995) Cloning, expression and characterization of biologically active feline tumour nercrosis factor-α. *Veterinary Immunology and Immunopathology* 45, 297–310.

Ritchey, J. and Tompkins, M. (1996) Alveolar macrophage function during early post-FIV infection. *Proceedings of the Third International Feline Retrovirus Research Symposium*, 6–9 March, Fort Collins, Colorado, p. 19.

Rojko, J.L. and Olsen, R.G. (1984) The immunobiology of the feline leukemia virus. *Veterinary Immunology and Immunopathology* 6, 107–165.

Romagnani, S. (1991) Human Th-1 and Th-2 subsets: doubt no more. *Immunology Today* 12, 256–257.

Rottman, J. and Tompkins, M. (1993) Development of a RT-PCR procedure to measure levels of feline cytokine mRNA. *Proceedings of the Second International Symposium on Feline Retrovirus Research*. Raleigh, North Carolina, 6–9 October, p.34.

Rottman, J.B., Freeman, E.B., Tonkonogy, S. and Tompkins, M.B. (1995) A reverse transcription-polymerase chain reaction technique to detect feline cytokine genes. *Veterinary Immunology and Immunopathology* 45, 1–18.

Rottman, J.B., Tompkins, W.A.F. and Tompkins, M.B. (1996a) A reverse-transcription- quantitative competitive polymerase chain reaction (RT-qcPCR) technique to measure cytokine gene expression in domestic mammals. *Veterinary Pathology* 33, 242–248.

Rottman, J., Levy, J., Ritchey, J. and Tompkins, M. (1996b) Cytokine alterations in FIV-infected cats challenged with *Toxoplasma gondii*. *Proceedings of the Third International Symposium on Feline Retrovirus Research*, Raleigh, North Carolina, 6–9 March, p.16.

Roux-Lombard, P., Modoux, C., Cruchaud, A. and Dayer, J.M. (1989) Purified blood monocytes from HIV-1-infected patients produce high levels of TNF-α and IL-1. *Clinical Immunology and Immunopathology* 50, 374–384.

Ruby, J., Ramsay, A., Karupiah, G. and Ramshaw, I. (1992) Recombinant virus vectors that coexpress cytokines – A new vaccine strategy. *Vaccine Research* 1, 347–356.

Saiki, R.K., Gelfand, D.H., Stoffel, S., Scharf, S.J., Higuchi, R., Horn, G.T., Mullis, K.B. and Erlich, H.A. (1988) Primer-directed enzymatic amplification of DNA with a thermostable DNA polymerase. *Science* 230, 1350–1354.

Schijns, V.E.C.J., Claassen, I.J.Th., Vermeulen, A.A., Horzinek, M.C. and Osterhaus A.D.M.E. (1994a) Modulation of the antiviral immune response by exogenous cytokines: Effects of TNF-α, IL-1α, IL-2, and IFN-γ on the immunogenicity of an inactivated rabies vaccine. *Journal of General Virology* 74, 55–63.

Schijns, V.E.C.J., Haagmans, B.L., Rijke, E.O., Huang, S., Aguet, M. and Horzinek, M.C. (1994b) Interferon-γ receptor-deficient mice generate antiviral Th-1-characteristic cytokine profiles but altered antibody responses. *Journal of Immunology* 153, 2029–2037

Schijns, V.E.C.J., Wierda, C.M., Vahlenkamp, T.W., Horzinek, M.C. and de Groot, R.J. (1995a) Molecular cloning and expression of cat interferon-γ. *Immunogenetics* 42, 440–441.

Schijns, V.E.C.J., Wierda, C.M., van Dam, E.J., Vahlenkamp, T.W. and Horzinek, M.C. (1995b) Molecular cloning of cat interleukin-4. *Immunogenetics* 42, 434–435.

Schijns, V.E.C.J., Haagmans, B.L. and Horzinek, M.C. (1995c) IL-12 stimulates an antiviral type 1 cytokine response but lacks adjuvant activity in IFN-γ-receptor deficient mice. *Journal of Immunology* 155, 2525–2532.

Schijns, V.E.C.J., Wierda, C.M., Vahlenkamp, T.W. and Horzinek, M.C. (1997) Molecular cloning of cat interleukin-12. *Immunogenetics* 45, 462–463.

Scott, P. (1991) IFN-γ modulates the early development of Th-1 and Th-2 responses in a murine model of cutaneous leishmaniasis. *Journal of Immunology* 147, 3149–3155.

Scott, P. (1993) Il-12: initiation cytokine for cell-mediated immunity. *Science* 260, 496.

Scott, P. and Kaufman, S.H.E. (1991) The role of T cell subsets and cytokines in the regulation of infection. *Immunology Today* 12, 346–348.

Sher, A., Gazzinelli, R.T., Oswald, I.P., Clerici, M., Kullberg, M., Pearce, E.J., Berzofsky, J.A., Mosmann, T.R., James, S.L., Morse III., H.C. and Shearer, G.M. (1992) Role of T-cell derived cytokines in the downregulation of immune responses in parasitic and retroviral infection. *Immunological Reviews* 127, 183–204.

Sirko, S., Bishai, I. and Coceani, F. (1989) Prostaglandin formation in the hypothalamus *in vivo*: effect of pyrogens. *American Journal of Physiology* 256, 616–624.

Steffan, A., Lafon, M., Gendrault, J., Koehren, F., De Monte, M., Royer, C., Kirn, A. and Gut, J. (1994) Feline immunodeficiency virus can productively infect cultured endothelial cells from cat brain microvessels. *Journal of General Virology* 75, 3647–3653.

Stoddart, M.E., Gaskell, R.M., Harbour, D.A. and Gaskell, C.J. (1988) Virus shedding and immune responses in cats inoculated with cell culture-adapted feline infectious peritonitis virus. *Veterinary Microbiology* 16, 145–158.

Takeda, A., Tuazon, C.U. and Ennis, F.A. (1988) Antibody-enhanced infection by HIV-1 via Fc receptor mediated entry. *Science* 242, 580–583.

Tepper, R.I., Levinson, D.A., Stanger, B.Z., Campos-Torres, J., Abbas, A.K. and Leder, P. (1990) IL-4 induces allergic-like inflammatory disease and alters T cell development in transgenic mice. *Cell* 62, 457–467.

Tham, K.M. and Studdert, M. J. (1985) The feline immune system – a review. *Kajian Veterinarian* 17, 83–96.

Torten, M., Franchini, M., Barlough, J.E., George, J.W., Mozes, E., Lutz, H. and Pedersen, N.C. (1991) Progressive immune dysfunction in cats experimentally infected with feline immunodeficiency virus. *Journal of Virology* 64, 5652–5655.

Trinchieri, G. (1995) Interleukin 12: a proinflammatory cytokine with immunoregulatory functions that bridge innate resistance and antigen-specific immunity. *Annual Review of Immunology* 13, 251–276.

Trinchieri, G. and Scott, P. (1994) The role of interleukin 12 in the immune response, disease and therapy. *Immunology Today* 15, 460–463.

Tsuchiya, M., Asano, S., Kazioir, Y. and Nagata, S. (1986) Isolation and characterization of the cDNA for murine granulocyte colony-stimulating factor. *Proceedings of the National Academy of Sciences USA* 83, 7633–7637.

Vahlenkamp, T.W., De Ronde, A., Balzarini, J., Naesens, L., De Clercq, E., Eijk, M.J., Horzinek, M.C. and Egberink, H.F. (1995) (R)-9-(2-phosphonylmethoxypropyl)-2,6-diaminopurine is a potent inhibitor of feline immunodeficiency virus infection. *Antimicrobial Agents and Chemotherapy* 39, 746–749.

Vennema, H., De Groot, R.J., Harbour, D.A., Dalderup, M., Gruffydd-Jones, T., Horzinek, M.C. and Spaan, W. (1990) Early death after feline infectious peritonitis virus challenge due to recombinant vaccinia virus immunization. *Journal of Virology* 64, 339–346.

Vennema, H., De Groot, R.J., Harbour, D.A., Horzinek, M.C. and Spaan, W. (1991) Primary structure of the membrane and nucleocapsid protein genes of feline infectious peritonitis virus and immunogenicity of recombinant vaccinia virus in kittens. *Virology* 181, 327–335.

Voth, R., Rossol, S., Klein, K., Hess, G., Schutt, K., Schroder, H., Buschendelde, K.H. and Muller, W. (1990) Differential gene expression of IFN-γ and tumor necrosis factor-α in peripheral blood mononuclear cells from patients with AIDS related complex and AIDS. *Journal of Immunology* 144, 970–975.

Vyakarnam, A., McKeating, J., Meager, A. and Beverley, P. (1990) Tumor necrosis factors (α, β) induced by HIV-1 in peripheral blood mononuclear cells potentiate virus replication. *AIDS* 4, 21–27.

Ward, A.M. (1995) Isolation and Characterization of Feline Interleukin-4: Initial Steps in the Study of Feline AIDS, a Small Animal Model for Human AIDS. BS Thesis, Eckerd College, St Petersburg, Florida, USA.

Weiss, R.C. and Cox, N.R. (1989) Evaluation of immunity to feline infectious peritonitis in cats with cutaneous viral-induced delayed hypersensitivity. *Veterinary Immunology and Immunopathology* 21, 293–309.

Weiss, R.C. and Oostrom-Ram, T. (1990) Effect of recombinant human interferon α *in vitro* and *in vivo* on mitogen-induced lymphocyte blastogenesis in cats. *Veterinary Immunology and Immunopathology* 24, 147–157.

Weiss, R.C. and Scott, F.W. (1981) Antibody-mediated enhancement of disease in feline infectious peritonitis: comparisons with dengue hemorrhagic fever. *Comparative Immunology, Microbiology and Infectious Diseases* 4, 175–189.

Weiss, R.C. and Toivio-Kinnucan, M. (1988) Inhibition of feline infectious peritonitis virus replication by recombinant human leucocyte (α) interferon and feline fibroblastic interferon (β) interferon. *American Journal of Veterinary Research* 49, 1329–1335.

Yamamoto, J.K., Ho, E. and Pedersen, N.C. (1986) A feline retrovirus induced T-lymphoblastoid cell-line that produces an atypical α type of interferon. *Veterinary Immunology and Immunopathology* 11, 1–19.

Yates, K.M., Rosenberg, L.J., Harris, C.K., Bronstad, D.C., King, G.K., Biehle, G.A., Walker, B., Ford, C.R., Hall, J.E. and Tizard, I.R. (1992) Pilot study of the effect of acemannan in cats infected with feline immunodeficiency virus. *Veterinary Immunology and Immunopathology* 35, 177–189.

VII

Avian Cytokines

Edited by K.A. Schat

Avian Cytokines

20

K.A. Schat[1] and P. Kaiser[2]

[1]*College of Veterinary Medicine, Cornell University, Ithaca, New York 14853, USA;* [2]*Institute for Animal Health, Compton, Berkshire RG20 7NN, UK*

INTRODUCTION

The production of poultry meat has increased drastically during the last 20 years and is expected to increase even further over the next decade. Improved resistance to disease and increased viability are two of the key factors necessary to sustain the expected increase in production (Hardiman, 1996). Delivery at an early age of cytokines and/or vaccines may increase the viability and resistance to disease. The development of reliable technology (INOVOJECT[R], Embrex, Inc.) to deliver biologically active substances *in ovo* in a cost-effective manner (Sarma *et al.*, 1995) makes this aim easier to achieve. Unfortunately, one of the main obstacles in manipulating the immune responsiveness of chickens is the lack of progress in the characterization of avian cytokines (Kaiser, 1996). The paucity of information is not only an impediment for the improvement of poultry health but it also presents problems for a fundamental understanding of avian immunology. An extensive review of the current knowledge of avian cytokines has recently been published (Kaiser, 1996). In this chapter we will briefly review the cytokines which have been cloned and the application of these cloned products in research on avian immunology and disease. In addition, avian cytokines which have not been cloned but which have been characterized using reagents obtained from cloned mammalian cytokines will be discussed. It is realized that numerous papers have been published describing partially or poorly defined cytokines present in supernatant fluids of mitogen-stimulated lymphocytes, but these have been reviewed elsewhere (Klasing, 1994; Kaiser, 1996).

Impediments to Progress

The identification of avian cytokine genes has been problematical, due to the lack of both functional and immunological cross-reactivity between most mammalian and chicken cytokine reagents, the absence of specific bioassays for cytokines in the chicken, and the low degree of identity between mammalian cytokine genes and their chicken homologues. Thus far, only five homologues of mammalian cytokines have been cloned in the chicken – type I IFN (Sekellick *et al.*, 1994), type II IFN (Digby and Lowenthal, 1995), IL-8 (Bedard *et al.*, 1987; Sugano *et al.*, 1987; Barker *et al.*, 1993), IL-15 (Sundick and Jones, 1996) and the transforming growth factor-β (TGF-β) cytokine family (Jakowlew *et al.*, 1988a,b,c, 1990). One gene for the type I IL-1 receptor (IL-1R) has also been cloned (Guida *et al.*, 1992). In addition, a chicken cytokine myelomonocytic growth factor (mgF) has been cloned (Leutz *et al.*, 1989; Sterneck *et al.*, 1992) for which no direct mammalian homologue has been found, but which has some homology with human G-CSF and IL-6.

Transforming Growth Factor-β Family

In mammals, TGF-β has a variety of functions involved in immunoregulation. These include induction of cytokine release, chemotaxis of neutrophils and monocytes, and stimulation of IgA secretion by B cells. In the chicken, TGF-β functions are mostly involved in wound repair, cellular differentiation and growth. Four of the five known TGF-β isoforms are expressed in the chicken, including TGF-β4, which so far appears to be unique to the chicken. In most *in vitro* assays, the biological activities of these five isoforms are indistinguishable. However, their sites of synthesis and localization *in vivo* are distinct, and their expression is differentially controlled. The first three TGF-β isoforms cloned in the chicken (TGF-β1, TGF-β4 and TGF-β3) were isolated from cultured chick embryo chondrocyte RNA using a human TGF-β1 cDNA probe (Jakowlew *et al.*, 1988a,b,c). The fourth isoform, TGF-β2, was isolated from a chicken embryo chondrocyte cDNA library using a simian TGF-β2 cDNA probe (Jakowlew *et al.*, 1990). The success of this approach is not surprising when the degree of identity between these isoforms across species is considered (see Table 20.1). Such a high degree of identity appears to be the exception rather than the rule and perhaps explains why other chicken cytokines have not been cloned by hybridization techniques.

Table 20.1. Degree of identity at the amino acid level between the various isoforms of chicken and human TGF-β.

	Amino acid identity (%)						
	hTGF-β1	hTGF-β2	hTGF-β3	cTGF-β1	cTGF-β2	cTGF-β3	cTGF-β4
cTGF-β1	100	71	77	—	72	78	80
cTGF-β2	72	99	79	72	—	79	64
cTGF-β3	76	79	99	78	79	—	71
cTGF-β4	80	64	71	80	64	71	—

Abbreviations: c = chicken; h = human.

Applications of TGF

To the best of our knowledge no studies have been reported on the use of recombinant TGF products in studies on the pathogenesis of avian diseases.

Interleukin-8

Interleukin-8 was characterized in mammals as a neutrophil-activating protein with biological activities similar to known chemotactic substances. It can be produced by a wide variety of cell types in response to cytokines and/or cytokine inducers. IL-8 and other structurally related chemotactic proteins are now termed chemokines.

Two clones, 9E3 (Sugano *et al.*, 1987) and CEF4 (Bedard *et al.*, 1987), were isolated from cDNA libraries from chick embryo fibroblasts (CEF) transformed with Rous sarcoma virus by differential screening. These clones were later shown to encode the same predicted 11-kDa protein, which has homology to members of a superfamily of inflammatory and growth-regulatory proteins including IL-8. The protein expressed from these clones is 6 kDa in size (Gonneville *et al.*, 1991). Barker *et al.* (1993) reported that the 6-kDa 9E3/CEF4 protein has 51% identity with mammalian IL-8 at the aa level. It also has 45% identity with another related protein, gro/melanoma growth-stimulatory activity (gro/MGSA) (Stoeckle and Barker, 1990). Both proteins are chemotactic for neutrophils, while gro/MGSA is mitogenic for melanoma cells. When Barker *et al.* (1993) examined the ability of 9E3/CEF4 to stimulate chemotaxis and mitogenicity in chicken PBLs and fibroblasts, its strongest activity was shown to be as a chemoattractant. These results suggest that 9E3/CEF4 may be the chicken homologue of IL-8.

Applications of Interleukin-8

Thus far no studies have been reported on the effect of recombinant proteins of this potential IL-8 homologue on disease interactions in chickens.

Interleukin-15

Interleukin-15, a recently isolated mammalian cytokine, is functionally comparable to IL-2 and stimulates the proliferation of T cells (Grabstein *et al.*, 1994), NK cells (Carson *et al.*, 1994), lymphokine-activated killer cells (Gamero *et al.*, 1995) and other cells. IL-15 uses the same receptor as IL-2, but in contrast to IL-2 is produced by a wide variety of cells but not by activated T cells. Sundick and Jones (1996) reported the isolation of a cDNA clone from mRNA from chicken spleen cells 8 h after activation with Con A. COS-7 cells transfected with the cDNA clone produced an IL which is able to stimulate the proliferation of T lymphoblasts. The cDNA encodes an open reading frame (ORF) of 429 nucleotides. Interestingly, the predicted protein of 143 aa has significant homology with monkey IL-15. Sundick and Jones (1996) speculated that this protein is the primordial T cell stimulation cytokine and that IL-2 evolved relatively late in phylogeny. Recently, Sundick and Gill-Dixon (1997) reported that this chicken gene resembles IL-2 in several aspects. The gene also codes for four cysteines forming two intrachain disulphide bonds, which is a highly conserved feature of IL-15. It emphasizes the importance of using defined reagents before assigning cytokine functions to specific interleukin

INTERFERONS AND INTERFERON-ASSOCIATED PROTEINS

Chicken Type I Interferons

Interferon was first described in the chicken, as an antiviral activity in conditioned media obtained from chorioallantoic membranes which had been exposed to inactivated influenza virus (Isaacs and Lindenmann, 1957). Lampson *et al.* (1963) subsequently purified this antiviral component and identified it as a 20- to 34-kDa protein. Subsequently it was shown that mammalian species have several types of IFN which can be divided into two broad categories: type I and type II. The former group consists of four different species (IFN-α, -β, -ω and -τ) while the latter has only one (IFN-γ). Chickens express at least two different species of type I IFN and IFN-γ.

The chicken homologue of mammalian type I IFN was cloned by Sekellick *et al.* (1994). They used an RT-PCR approach with primers based on mammalian IFN-α/β aa sequences and RNA from virus-induced 'aged' CEF. The resulting 269 bp PCR product was then used as a probe against a cDNA library derived from the induced CEF. A cDNA was thus isolated which encodes the chicken homologue of mammalian type I IFN. At the aa level, chicken type I IFN has 20–24% sequence identity with consensus mammalian type I IFN sequences, and only 3% identity with mammalian type II IFN. The coding region encodes a signal peptide of 31 aa, and a mature protein of 162 aa with a predicted molecular mass of 19 kDa.

Mammalian type I IFNs share a structure of primarily α-helices and both the conserved features and the predicted secondary structure of the predicted chicken IFN protein fit this structural model. Northern blot analysis showed that chicken IFN mRNA production correlated with chicken IFN induction as determined by a bioassay. Also, biologically active chicken IFN synthesized in mouse cells, from mRNA prepared from the cloned sequence, was neutralized by a mAb prepared against purified chicken IFN. It has recently been shown that, like mammals, the chicken has a large number of intron-less type I IFN genes coding for at least two serologically distinct antiviral activities (Sick *et al.*, 1996).

Chicken Type II Interferon

Digby and Lowenthal (1995) described the cloning and expression of chicken type II IFN from a cDNA expression library generated from a T cell line producing high levels of IFN-γ-like activity, using a virus-inhibition assay. The predicted mature protein consists of 145 aa with a molecular mass of 16.8 kDa, with two potential *N*-glycosylation sites located near the *N* terminus. Chicken type II IFN is 32% and 35% identical to human and equine IFN-γ, respectively. There is only 15% identity between chicken type I and type II IFNs, suggesting that the two principal types of IFN emerged before the divergence of avian and mammalian species that occurred 350 million years ago.

A genomic clone of chicken IFN-γ has recently been isolated (Kaiser *et al.*, 1996). The structure of the chicken IFN-γ gene (Fig. 20.1) is remarkably similar to that of mammalian IFN-γ genes; the gene consists of four exons and three introns. The size of the exons and two of the introns are similar to those of the mammalian equivalents, while the third intron is dramatically reduced in the chicken. As in mammals, chicken IFN-γ is a single-copy gene.

Interferons in Other Avian Species

Type I IFN has been cloned from at least two other avian species: ducks and turkeys. Duck type I IFN was cloned by hybridization using the chicken type I IFN cDNA clone as a probe (Schultz *et al.*, 1995b). An intron-less ORF was identified encoding a putative 30 aa signal protein and a mature protein of 161 aa. The mature protein has 50% identity with chicken type I IFN. The four cysteine residues which are conserved in mammalian IFN-α, ω, and τ and chicken type I IFN are also conserved in the duck and turkey (Fig. 20.2).

The turkey type I IFN gene was amplified using primers based on the sequence of chicken type I IFN. The cloned fragment contains an ORF coding for a signal protein of 30 aa and a mature protein of 162 aa with 82% identity with chicken type I IFN at the aa level (Fig. 20.2) (Suresh *et al.*, 1995).

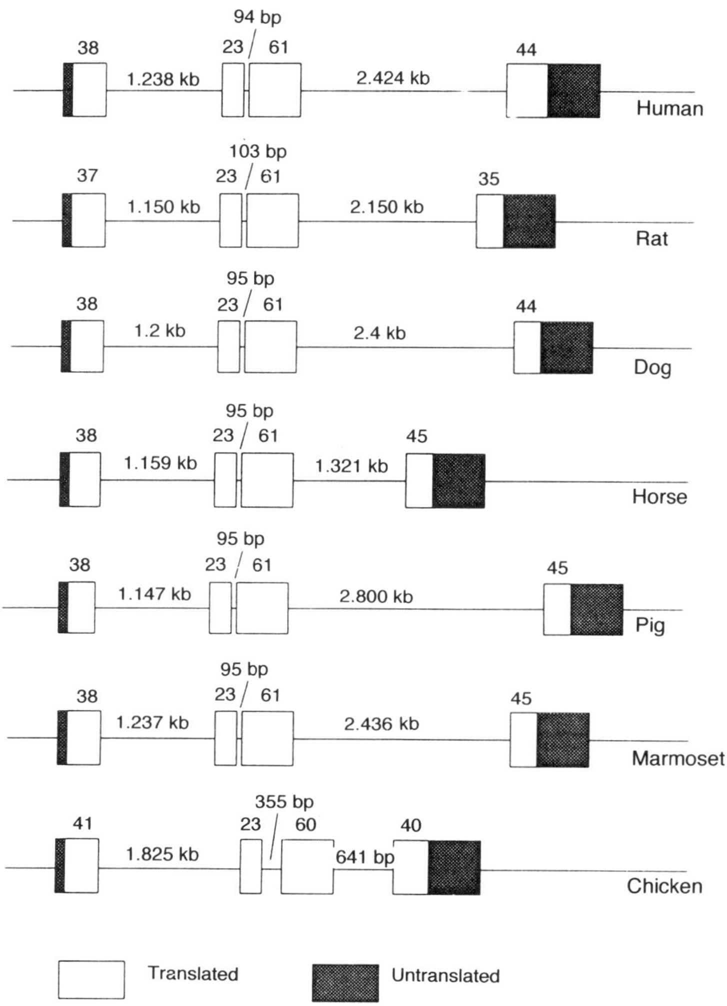

Fig. 20.1. Comparison of the structure of mammalian and chicken interferon-γ genes.

Interferon-activated Proteins

Type I and type II IFNs are able to induce the transcription of several cellular proteins through binding to specific cell surface receptors and activation of a direct signalling pathway to the nucleus (Shuai, 1994). The transcriptional regulation is mediated through enhancer elements which are located 5′ upstream from IFN-inducible genes. Two broad categories of enhancer elements responding to IFN-α/β have been described in mammalian cells:

```
                                        *
Chicken  MAVPASPQHPRGYGILLLTLLLKALATTASACNHLRPQDATFSHDSLQLL  50
         ||||||||||||||||||||| ||||  | |||||||||||||| ||||||
Turkey   MAVPASPQHPRGYGILLLTLLMKALA.AAAACNHLRPQDATFSRDSLQLL  49
         |  |  |  |  |. | | ||    |  |  |  || .|  | .||||||
Duck     MPGPSAPPPPAIYSALALLLLLTPPA.NAFSCSPLRLHDSAFAWDSLQLL  49

                   *      *
Chicken  RDMAPTLPQLCPQHNASCSFNDTILDTSNTRQADKTTHDILQHLFKILSS  100
         |||||  || |||||| ||||||.||| || |||||||.||||||||||.
Turkey   RDMAPSPPQPCPQHNAPCSFNDTVLDTNNTQQADKTTHNILQHLFKILSG  99
         |.||||| |||||. ||||| ||.||||.||||   |    .|||||  ||.
Duck     RNMAPSPTQPCPQQHAPCSFPDTLLDTNDTQQAAHTALHLLQHLFDTLSS  99

                                     *
Chicken  PSTPAHWNDSQRQSLLNRIHRYTQHLEQCLDSSDTRSRTRWPRNLHLTIK  150
         | ||||| ||||||||| |.|| |||||||  |.|||||||| | ||||
Turkey   PTTPAHWIDSQRQSLLNQIQRYAQHLEQCLADSHTRSRTRWPHNPHLTIN  149
         | |||||.   |. ||||.|    ||| |.     |   | | | || ||
Duck     PSTPAHWLHTARHDLLNQLQHHIHHLERCFPADAARLHRRGPRNLHLSIN  149

             *            *
Chicken  KHFSCLHTFLQDNDYSACAWEHVRLQARAWFLHIHNLTGNTRT         193
         ||||||| ||.|||||||||.|||| ||||.||||||||||||
Turkey   KHFSCLHAFLHDNDYSACAWDHVRLRARAWLLHIHNLTGNTRT         192
         | |.|.. ||..  || |||||||| | | .  || ||   |
Duck     KYFGCIQHFLQNHTYSPCAWDHVRLEAHACFQRIHRLTRTMR          191
```

Fig. 20.2. Alignment of chicken, turkey and duck type I IFN aa sequences. The putative signal peptides of the three proteins are underlined. Conserved cysteine residues are indicated by asterisks.

IFN-stimulated response elements or ISRE (Reid *et al.*, 1989) and the interferon consensus sequence (ICS) with a related motif (Friedman and Stark, 1985). Although there is limited homology between avian and mammalian type I IFN, there is a remarkable similarity between the two groups in these promoters suggesting a similar pathway for the IFN-induced transcription of inducible genes. An ISRE-like sequence has been identified in the 5′ upstream region of the IFN-inducible Mx protein in chickens (Schumacher *et al.*, 1994; Schultz *et al.*, 1995a) and is probably present in ducks (Schultz *et al.*, 1995b). Zöller *et al.* (1992) reported that the 5′ upstream region of the chicken class I gene BF-IV contains sequence motifs similar to mammalian ICS motifs. This chicken ICS motif is indeed recognized by chicken type I IFN. Transfection of CEF with a plasmid containing one or multiple copies of the chicken ICS sequence resulted in enhanced CAT activity upon stimulation with IFN, but a decreased CAT activity was noted in the absence of IFN. The ISRE and ICS sequences are recognized by a family of IFN regulatory proteins or factors (IRF) and several mammalian IRFs have been cloned and sequenced. Recently, three proteins of this family have been cloned and sequenced from chickens: cIRF-1, cIRF-3 and cICSBP

(Grant *et al.*, 1995; Jungwirth *et al.*, 1995). The predicted proteins have a remarkable sequence similarity with their mammalian counterparts, suggesting that the IFN-induced transcription pathways may be well conserved between mammalian species and chickens.

Transcriptional activation by IFN-γ is regulated through a recently defined pathway (Schindler *et al.*, 1992; Shuai *et al.*, 1992). Activation of the gamma activation site (GAS) results in the activation of IFN-γ-dependent genes. The first GAS site was identified upstream of the guanylate-binding protein (GBP) (Lew *et al.*, 1991). Thus far there is little information available on this pathway in avian species, although the recent cloning of the chicken GBP gene with remarkable sequence similarity to mammalian GBP suggests that this pathway may also be conserved (Schwemmle *et al.*, 1996). S. Kaplan and K.A. Schat (unpublished results, 1996) have identified several GAS-like sequences upstream of Marek's disease virus (MDV) immediate early genes, which may play a role in MDV gene regulation in lymphoblastoid tumour cell lines (see below).

APPLICATION OF INTERFERONS

Resistance to *Salmonella* Infection

Fulton *et al.* (1995) reported that treatment of QT35 cells with reovirus-induced, natural chicken IFN resulted in an increased resistance to challenge with *S. typhimurium*, suggesting that IFN could play a role in genetic resistance. This is of interest because: (i) chicken lines may differ in their resistance to challenge with *S. typhimurium* and *S. enteritidis* (Barrow and Bumstead, 1996); and (ii) sequence differences for the chicken NramplI genotypes have recently been described (Hu *et al.*, 1996). This gene codes for the natural resistance-associated macrophage protein I (NramplI), which has been associated with resistance to *Salmonella* infection in mice. IFN-γ response elements have been identified upstream of the chicken NramplI gene. Unfortunately, there was no correlation between NramplI genotypes and resistance to *Salmonella* challenge (Barrow and Bumstead, 1996).

Marek's Disease

Conditioned medium from Con-A-stimulated splenocytes contains factors that are able to maintain MDV in its latent form when splenocytes from latently infected chickens are cultured *in vitro* (Buscaglia and Calnek, 1988). Recently, Volpini *et al.* (1995) used rtype I IFN to demonstrate that type I IFN is at least one of the factors responsible for the maintenance of latency in *in vitro* spleen cell cultures. Interestingly, the IFN was more effective in maintaining latency when spleen cells were obtained 14 days p.i. than 7 days p.i. The reasons for this observation are poorly understood. Volpini *et al.*

(1996) used a MD lymphoblastoid cell line, MDCC-CU184, to study further the effect of recombinant type I IFN on the modulation of MDV protein expression. Addition of recombinant type I IFN to this cell line resulted in downregulation of two immediate early or early MDV proteins, pp14 and pp38 as well as the glycoprotein B. Schat (1995, unpublished results) also found that expression of the immediate early gene ICP4 was downregulated. This is of interest because GAS and ISRE sequences have been identified upstream of ICP4 (S. Kaplan and K.A. Schat, 1996, unpublished results). B.W. Calnek *et al.* (New York, 1996, personal communication) demonstrated that treatment of MDCC-CU184 with rIFN-γ also downregulated the expression of MDV proteins, but they did not identify which proteins were involved. These observations clearly indicate that IFNs may play a role in the pathogenesis of MD. However, inoculation of chicks with recombinant type I IFN did not influence the lytic infection phase of MDV infection (K.A. Schat, 1995, unpublished results; B.W. Calnek, New York, 1996, personal communication).

Recombinant Viruses Expressing Interferon Genes

The type I IFN gene has been inserted in fowlpox virus (rFP-IFN)(Karaca *et al.*, 1995). CEF inoculated with rFP-IFN were protected against challenge with herpesvirus of turkeys and VSV. It also enhanced the LPS-induced activation of macrophages *in vitro*. Chickens inoculated with rFP-IFN which also expressed a Newcastle disease virus protein (rFP-IFN-NDV) had similar levels of protection against challenge with NDV as chickens vaccinated with rFP-NDV (Sharma *et al.*, 1996). Chicks vaccinated with rFP-NDV or with rFP-IFN-NDV by *in ovo* inoculation at 18 days of embryonation were equally well protected, but the latter group was protected against weight loss associated with *in ovo* vaccination with rFP-NDV.

Duck Hepatitis B Virus

Chronic infection with human hepatitis B virus is a major problem worldwide. Treatment with IFN-α is the only established therapy but has only a limited success rate (Offensperger *et al.*, 1994). White Pekin ducks carry duck hepatitis B virus, which is similar to the human virus (Mason *et al.*, 1980), and they are one of the few available models to study hepadnavirus infection *in vivo*. The availability of cloned duck IFN will allow critical studies on the *in vivo* effects of IFN treatment on chronic hepatitis B virus infections.

The Use of Cloned Interferons to Examine Interactions with Other Cytokines

Interferons are important components in the regulation of immune responses and can potentiate the actions of many other cytokines (Belardelli,

1995). The availability of cloned avian IFNs will facilitate the search for other specific cytokines. Suresh *et al.* (1995) have used cloned turkey IFN DNA in their search for the identification of turkey IL-8-like and IL-6-like gene activation in turkey cells. It is expected that they will be able to clone these genes in the near future.

INTERLEUKIN-1 TYPE I RECEPTOR

The chicken homologue of the type-I IL-1R (Guida *et al.*, 1992) has 64% and 61% similarity at the aa level with human and murine IL-1R, respectively. The avian IL-1R has a structure typical of IL-1Rs, with four separate domains: a *N*-terminal signal peptide; an extracellular binding region; a hydrophobic transmembrane segment; and a cytoplasmic domain. The latter is responsible for signal transduction, and has 79% similarity with the corresponding region in the human IL-1R. This conservation of the cytoplasmic domain has functional importance, as demonstrated by site-directed mutagenesis experiments (Heguy *et al.*, 1992), which identified essential functional/structural sites in this region. The extracellular binding domain has a typical Ig-like structure, as seen in mammalian IL-1Rs.

MYELOMONOCYTIC GROWTH FACTOR AND STEM CELL FACTORS

First described as a novel haematopoietic growth factor (Leutz *et al.*, 1984), mgF stimulates the growth of retrovirus-transformed chicken myeloblasts and induces macrophage colony formation in uninfected chick bone-marrow cultures. Myelomonocytic growth factor has been shown to be intimately involved in leukaemia formation in the chicken (Graf *et al.*, 1986). The gene encoding mgF was cloned by Leutz *et al.* (1989) from a chicken myeloblast cDNA library, screened using oligonucleotides corresponding to peptide sequences produced from trypsin treatment of purified mgF. The gene encodes a 201 aa secretory protein, which is later modified by signal peptide cleavage and glycosylation (Leutz *et al.*, 1988). Myelomonocytic growth factor is related to G-CSF and IL-6, but hybridization experiments suggest that the three cytokines are encoded by different genes in the chicken. Sterneck *et al.* (1992) showed that the promoter region of the chicken mgF gene is different to that of mammalian G-CSF and IL-6. Reporter constructs were made containing the mgF promoter, and could only be expressed in myelomonocytic cells. So far, mgF is apparently unique to the chicken. Recently, an unrelated haematopoietic growth factor has been described that was derived from Rous sarcoma virus-transformed spleen stromal cells obtained from 16-day-old chicken embryos. Although this factor, SSL-1, induces similar cell populations to recombinant mgF, it is not related to mgF (Siatskas *et al.*, 1996). Zhou *et al.* (1993) isolated a cDNA clone encoding the chicken homologue of SCF, a pleiotropic cytokine that is essential for the development of haematopoietic cells, germ cells and melanoblasts in mammalian species. Recently, Zhou *et al.* (1996) cloned the chicken SCF gene into a baculovirus vector and expressed a *N*-glycosylated protein with a molecular mass of 20 to 25 kDa. Addition of rSCF to cultures of chicken dorsal root ganglia resulted in the enhanced outgrowth of neurites. Thus far no information has been published on the effect of rSCF on haematopoietic cells.

York *et al.* (1996) inserted the gene for mgF into a fowlpox virus vaccine strain (fp/mgF). Inoculation of 1-day-old chicks with fp/mgF resulted in a rapid increase in the number of monocytes in blood between 2 and 16 days p.i. These monocytes produced increased levels of nitrite *in vitro* after stimulation with LPS or chicken IFN-γ and had an increased phagocytic activity compared to monocytes from control chickens. It will be interesting to determine if inoculation of chicks with fp/mgF will lead to increased resistance to bacterial or viral infections.

OTHER CYTOKINES

Although relatively few chicken cytokines have been cloned, other cytokine-like activities have been identified in the chicken. IL-1-like activity can be measured in conditioned media (CM) from stimulated splenocytes (Hayari *et al.*, 1982). This CM is also cross-reactive in mammalian IL-1 assays (Klasing and Peng, 1987, 1990). IL-2-like activity has been observed by many authors (for a review see Kaiser, 1996). Antibodies to at least one component of the IL-2R have also been identified (Hala *et al.*, 1986; Schauenstein *et al.*, 1988; Lee and Tempelis, 1991, 1992). There has been one report of an IL-3-type response in the chicken (Chang *et al.*, 1990). Cross-species reactivity has been described for IL-6 between mammals and chickens (Amrani, 1990; Samad *et al.*, 1993; Lynagh and Kaiser, 1996). TNF-α activity has also been reported in the chicken (Qureshi *et al.*, 1990), as well as cross-species reactivity between chicken TNF-α and murine cells (Byrnes *et al.*, 1993).

PROSPECTS

New molecular biological techniques, such as the PCR, have provided novel approaches to cloning avian cytokines. However, generally the degree of identity between chicken and mammalian genes is very low, and the efficacy of PCR may be more limited than initially promised. The development of sensitive, reproducible bioassays for chicken cytokines, coupled with the use of expression systems (e.g. COS cells), is possibly the most promising approach.

Lymphocyte Transformation and Lymphokine Production During Chicken Anaemia Virus Infection

21

B.M. Adair and M.S. McNulty

Veterinary Sciences Division, Stormont, Belfast BT4 3SD, UK

INTRODUCTION

In 1979, Yuasa *et al.* described the isolation from domestic fowl of an infectious agent which they named chicken anaemia agent (CAA); it caused increased mortality, anaemia, generalized lymphoid atrophy, and subcutaneous and intramuscular haemorrhages in young chicks. The physico-chemical properties of CAA suggested that it was a small, thermostable, non-enveloped virus. As subsequently shown by other workers CAA is a naked, icosahedral virus, 23–25 nm in diameter, containing a single-stranded, circular DNA genome of approximately 2.3 kb; it was renamed chicken anaemia virus (CAV) (reviewed by McNulty, 1991). Disease occurs when breeder flocks are infected with CAV for the first time when they are in lay. No effect is seen in the breeders, but the virus is vertically transmitted to their progeny. Vertically infected chicks hatch normally, and clinical signs do not normally occur until they are 10–14 days of age. The field disease is known by a number of synonyms, including blue wing disease, anaemia dermatitis syndrome, infectious anaemia syndrome, etc.

Fortunately, outbreaks are rare, because most breeder flocks become infected with CAV during the rearing period, and develop serum antibodies which appear to prevent subsequent vertical transmission. Outbreaks of blue wing disease are frequently complicated by secondary bacterial and fungal infections, indicating that affected flocks are immunosuppressed. The nature of this immunosuppression has been investigated in experimentally infected chicks, and is described in this chapter.

EFFECTS ON LYMPHOID ORGANS

The immunosuppressive effects of CAV arise from the destruction of cells, principally in the bone marrow and thymus, which are the first lymphoid organs affected following experimental parenteral inoculation of 1-day-old chicks (Taniguchi *et al.*, 1983). Analysis of the lymphocyte populations in affected tissues reveals cellular changes (Jeurissen *et al.*, 1989; Adair *et al.*, 1993), which correlate with the results of the pathological studies. The lymphocyte populations in the primary and secondary lymphoid organs show an age-related increase in normal uninfected young chickens. However, following CAV infection, lymphocyte populations are severely depressed and the birds show a generalized lymphocytopenia, with marked thymic atrophy (Taniguchi *et al.*, 1983; Goryo *et al.*, 1989). Bursal cells are not susceptible to infection (Goryo *et al.*, 1989; Smyth *et al.*, 1993), and when antibodies to CAV are produced, the infection is halted and rapidly cleared. Granulopoiesis and erythropoiesis are restored at around 16–18 days p.i. (Taniguchi *et al.*, 1983), and the surviving birds rapidly recover.

Changes in the Bone Marrow

The changes in the bone marrow include decreases in the numbers of mature erythrocytes and haematopoietic cells, and the appearance of large blast cells (Bülow *et al.*, 1986). At the same time, CAV antigen is detectable in cells in the bone marrow (Smyth *et al.*, 1993). Virus antigen is seen in haemocytoblasts, which are the precursor cells of the erythroid and myeloid series, both in the sinusoids, and in the extra-sinusoidal spaces. Depletion of haemocytoblasts in the sinusoids was observed by 8 days post-inoculation, and depletion of these cells, which give rise to the erythrocytes, precipitates the anaemia. These progenitor cells also give rise to the thrombocytes, and since these are essential to the blood clotting mechanism, their depletion is probably related to the observed increase in the number of intramuscular and subcutaneous haemorrhages, which is also a characteristic of the experimental disease (McNulty, 1991). Infection and killing of the haemocytoblasts in the extra-sinusoidal spaces has a major effect on the granulocyte series, and is responsible for the fall in the numbers of circulating granulocytes, which follows infection (Taniguchi *et al.*, 1983). This effect is probably related to the observed increase in susceptibility to opportunistic infections (Vielitz and Landgraf, 1988).

The studies outlined above demonstrate that haematopoiesis in the bone marrow of CAV-infected chickens is halted by infection and destruction of the haemocytoblasts in the sinusoids and extra-sinusoidal spaces, from which all the blood-borne erythrocyte and leukocyte elements are derived. Haematopoiesis is controlled by cytokines produced by the developing cells, and the precise effects of the virus on these factors inside and outside the bone marrow is not known.

Changes in the Thymus

Coincident with the appearance of infected cells in the bone marrow, infected cells also appear in the thymus (Smyth *et al.*, 1993; Fig. 21.1). This suggests that infection takes place in both organs independently and simultaneously. In the thymus, infected cells are concentrated very largely in the cortical area (McNeilly *et al.*, 1991). Cortical thymocytes are destroyed in large numbers, whereas non-lymphoid leukocytes and stromal cells are unaffected by the virus (Taniguchi *et al.*, 1983; Jeurissen *et al.*, 1989). The concentration of virus antigen in the thymic cortex (Fig. 21.1) suggests selective infection of immature T cells in the thymus, and this has been confirmed by dual antibody staining of thymocytes to determine the phenotype of the cells containing virus antigen (Adair *et al.*, 1993). There is evidence that the virus has a preference for CTL which are recognized by mAbs specific for the CD8 marker. Thus in thymus and spleen from CAV-infected birds, a higher percentage of infected cells were $CD8^+$ than were $CD4^+$ (Adair *et al.*, 1993), and studies by Cloud *et al.* (1992) have shown marked increases in Th cell-to-CTL ratios in CAV-infected thymus, spleen and bone marrow. Also, transformed T cell lines which are $CD8^+$ are infected *in vitro* with a higher frequency than cells which are CD8– (Lucio

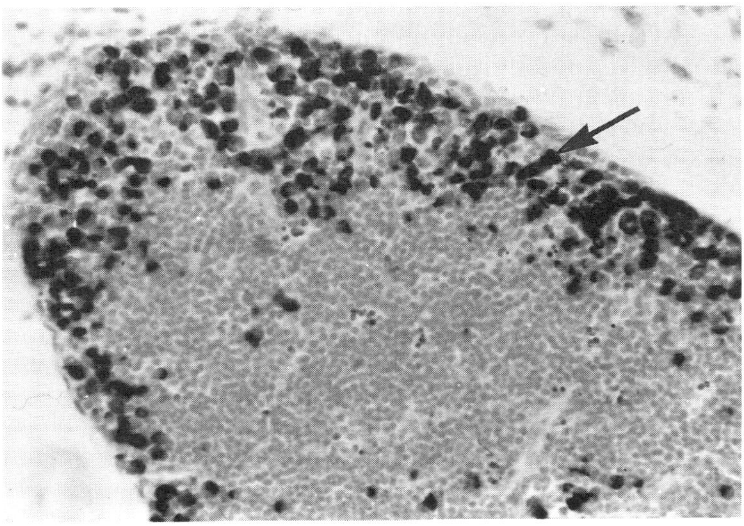

Fig. 21.1. Immunoperoxidase staining of CAV-infected, formalin-fixed thymus tissue, showing concentration of infected cells mainly in the cortical area (arrow).

et al., 1992). However, the preference for $CD8^+$ cells is not exclusive. Cells which bear the $CD4^+$ marker are also destroyed in large numbers, and the MDCC-MSB1 cell line which is used extensively for laboratory cultivation of the virus, is a transformed T cell line, which bears the $CD4^+$ marker (Adair *et al.*, 1993).

EFFECTS ON LYMPHOCYTE TRANSFORMATION

Lymphocytes separated from spleens of CAV-infected chickens at intervals p.i. show decreased transformation responses following stimulation with mitogens (Otaki *et al.*, 1988; Adair *et al.*, 1991). Significant reductions were observed at 8 and 15 days p.i., after which a recovery in activity was evident (Fig. 21.2). By 22 days, no significant difference in transformation response was observed between infected birds and uninfected controls, and by 29 days the activity of the lymphocytes from the infected birds was significantly greater than in the controls, before returning to control levels by 43 days p.i. (Adair *et al.*, 1991).

EFFECTS ON CYTOKINES

Interleukin-2

Interleukin-2 production was measured in a biological assay in which supernatants from Con-A-stimulated splenocytes from infected and uninfected birds were tested for their ability to stimulate Con-A-activated blast cells (derived from adult SPF hens) to multiply and incorporate [^{3}H]thymidine (Adair *et al.*, 1991). The time course of IL-2 suppression by CAV was similar to the effect on the lymphocyte transformation response, with maximal

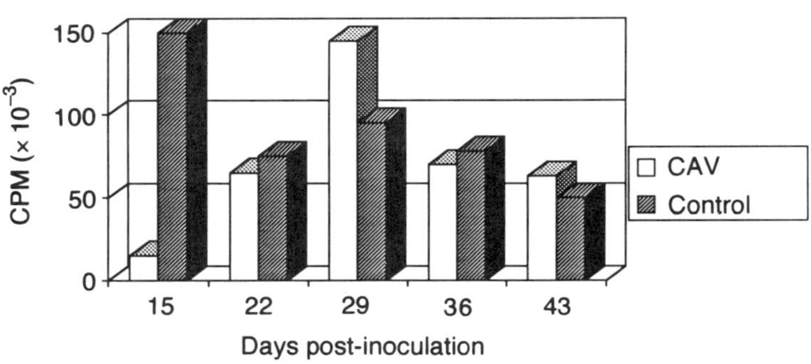

Fig. 21.2. Lymphocyte transformation response of splenocytes from chickens inoculated with CAV at 1 day old.

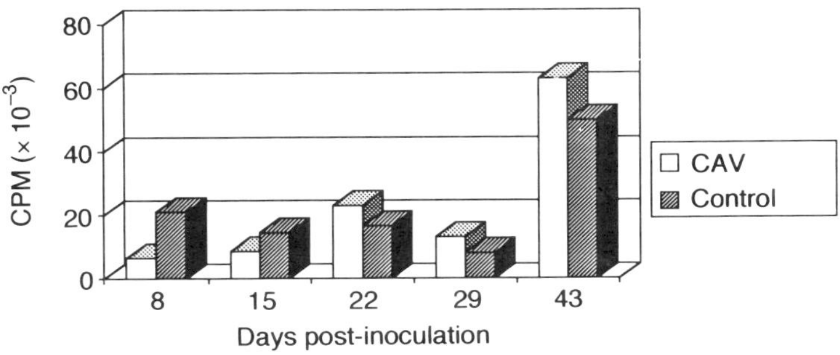

Fig. 21.3. Interleukin-2 production by splenocytes from chickens inoculated with CAV at 1 day old.

suppression recorded at 8 and 15 days post-inoculation (p.i.), compared to controls (Fig. 21.3). Comparison of the effects on the transformation response and IL-2 production indicates a close relationship between these activities. Maximal suppression of the lymphocyte transformation response was observed at 8 and 15 days p.i. At 8 days, the response of the lymphocytes from the infected birds was 65% of the control response, and fell further to 10% of the control by 15 days p.i. However, at 8 days after inoculation, IL-2 production was already at its lowest level, i.e. 30% of the control, and by 15 days (when the transformation response was now lowest) it had recovered to 59% of the control. At 22 days (when the transformation response showed no difference from control), IL-2 production in splenocytes from the infected birds was 40% greater than in the controls. Thus, the effect on the lymphocyte transformation response appeared to follow the effect on IL-2 production. This may suggest that the decline and subsequent recovery in the transformation response is dependent, to an extent at least, on IL-2 production, and therefore IL-2 production may be a more accurate indicator of the effects of the virus on lymphocyte function.

Interferons

The effects of CAV on IFN production have been measured in a biological assay in which dilutions of supernatants from Con-A-stimulated splenocytes recovered from CAV-infected and uninfected chickens were tested for their ability to protect cultures of chicken embryo fibroblasts from Semliki Forest virus-mediated cytolysis (Adair *et al.*, 1991). In samples from CAV-infected birds, IFN titres were initially higher than in controls at 8 days p.i., possibly representing an initial burst of IFN production following infection with the virus (Fig. 21.4). At 15, 22 and 29 days, however, significant reductions in IFN titres were evident in samples from the CAV-infected birds, compared to controls, with the maximum reduction recorded at 22 days p.i. (Fig. 21.4).

Interleukin-1

Substantial reductions in macrophage functions, including IL-1 production have also been demonstrated following CAV infection of 1-day-old chickens (McConnell *et al.*, 1993a). IL-1 production by macrophages from CAV-infected and control chickens was measured in a biological assay, in which macrophages, recovered from spleen suspensions by overnight adherence to plastic dishes, were incubated for 72 h with an endotoxin preparation from *Staphylococcus aureus*. The IL-1-containing supernates were recovered and tested for activity in a chicken thymocyte proliferation assay (McConnell *et al.*, 1993a). Following CAV infection, significant reductions in IL-1 production were recorded. The greatest reduction was observed 14 days p.i. (Fig. 21.5), after which IL-1 production appeared to recover, although levels were

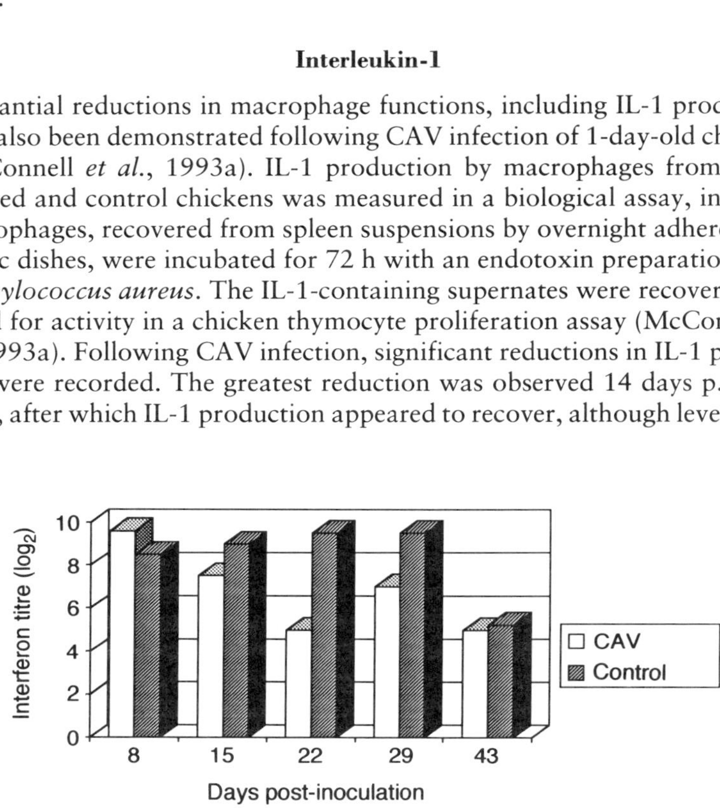

Fig. 21.4. Interferon production by splenocytes from chickens inoculated with CAV at 1 day old.

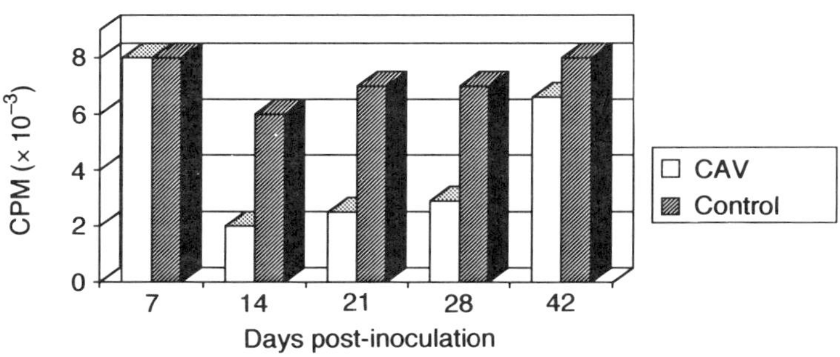

Fig. 21.5. Effect of CAV on IL-1 production by spleen-derived macrophages, following inoculation of chickens at 1 day old (25% IL-1/0.625 μg Con A ml^{-1}).

still significantly lower than controls at the end of the observation period, at 42 days p.i. Significant reductions in macrophage bactericidal activity and macrophage Fc receptor expression were also demonstrated (McConnell *et al.*, 1993a).

Macrophages exposed to CAV *in vitro* show similar functional changes (McConnell *et al.*, 1993a,b). The mechanism by which the virus effects these changes is not clear, since there is no evidence that CAV is capable of infecting macrophages. Thus no virus antigen was detectable in macrophage cultures exposed to the virus *in vitro*, and while virus antigen is infrequently observed within macrophage-like cells following infection of chickens, this is considered to result from ingestion of CAV-infected cells (Smyth *et al.*, 1993). Thus it seems unlikely that the changes are brought about by direct infection of the macrophages with subsequent reduction of functions as normal cell metabolic processes are shut down.

Macrophage Suppressor Factor

Chicken anaemia virus also stimulates production of a macrophage soluble factor which reduces the ability of non-infected chicken spleen mononuclear cells to respond to mitogen (McConnell, 1992). In these experiments, spleen mononuclear cell suspensions from uninfected chickens were incubated in close proximity to, but not in contact with, adherent spleen-derived macrophages from CAV-infected chickens. The two populations of cells were separated by a membrane which allowed passage of soluble factors, but not cells. The responsiveness of the uninfected spleen mononuclear cells was significantly reduced when incubated in the presence of the adherent cells from CAV-infected chickens (McConnell, 1992). The nature of the macrophage suppressor factor was not investigated. The effect may have been induced by an inhibitory cytokine, or by other compounds such as prostaglandins. Whether it is stimulated directly by contact of the macrophages with CAV, virion proteins, or by other products released from CAV-infected cells is not clear. Similar effects have been noted with other avian viruses, e.g. infectious bursal disease virus (IBDV), reticulo-endotheliosis virus and MDV (Carpenter *et al.*, 1978; Lee *et al.*, 1978).

IMMUNOSUPPRESSION IN 3-WEEK-OLD CHICKENS

Similar reductions in lymphocyte and macrophage functions were also demonstrated when chickens were inoculated orally with CAV at 3 weeks of age (McConnell *et al.*, 1993b), although the effects took longer to become apparent after inoculation. In this case, maximal reduction was observed at 14 and 21 days p.i., with recovery taking place after this time. Oral exposure to CAV in chickens of this age produced a subclinical infection. The effects observed on lymphocyte and macrophage functions may explain, in part, the adverse effects on economic performance of subclinically infected broiler flocks in the field. Subclinically infected flocks were the progeny of breeder flocks naturally exposed to CAV during the rearing period, and had developed serum antibodies to CAV by the time they were slaughtered. Such flocks showed a 13% decrease in net income, a 2% increase in feed conversion, and a 2% decrease in average bird weights, compared with similar flocks with no antibodies to CAV (McNulty *et al.*, 1991).

CONSEQUENCES OF CHICKEN ANAEMIA VIRUS INTERACTION WITH LYMPHOCYTES AND MACROPHAGES

The mechanisms by which CAV effects these changes in lymphocytes and macrophages require further investigation. However, it is known that Con A stimulates T cells by binding to carbohydrate moieties present on the CD3 receptor on the T cell membrane (Toivanen and Toivanen, 1973). Signal transduction within the cell is effected through interaction with protein kinase C and a series of calcium-dependent effector molecules which either bind to calcium directly, or associate with specialized calcium binding proteins. The process ultimately leads to induction of mRNA for cytokines, including IL-2, and stimulation of IL-2 receptor production. It has been shown that the mitogen-induced calcium flux in splenocytes from CAV-inoculated chickens was lower at the times when lymphocyte transformation is decreased (Bounous *et al.*, 1995). This suggests interference by the virus with these processes within the cell. However, the main effects of CAV on the lymphocyte transformation response, and on IL-2 production, most likely arise from the destruction of precursor T cells in the thymus, resulting in a fall in the numbers of mature cells in the secondary lymphoid organs and in the blood. The effect on IFN production probably arises by the same means. The mechanism by which the virus affects macrophage functions may be different, since mature macrophages are not susceptible to CAV infection. It is possible that the effects are initiated by binding of viral proteins to receptors on the macrophage membrane, or indeed to the presence of an inhibitory substance, possibly another cytokine, produced by the cells in which the virus was grown, and therefore present in the inoculum. Other viruses have been shown to adversely affect cellular functions without infection of the cells. Thus IBDV infection of chickens is accompanied by a multifocal thymic cortical necrosis and by a decrease in responsiveness of lymphocytes to T cell mitogens, despite the fact that the virus does not grow in T lymphocytes (Hirai and Calnek, 1979; Confer *et al.*, 1981). Whatever the mechanism, the depletion of these essential cytokines, even for the 2 week period following infection, has far reaching consequences for immune function in the infected chickens.

EFFECTS ON IMMUNE COMPETENCE

Observations of the effects of natural and experimental CAV infection indicate a failure of the immune processes responsible for elimination of pathogens, and adverse effects on antibody production. This is seen as an increased susceptibility to infection with opportunistic bacteria and fungi, and lower than expected responses to vaccines (Vielitz and Landgraf, 1988; De Boer *et al.*, 1994). Examination of the normal biological activities of the cytokines known to be affected by the virus indicates how the observed consequences which accompany infection with CAV might arise.

The spectrum of biological activities attributed to IL-2 in mammals includes direct effects on the growth, differentiation and normal functions of a variety of immune and non-immune cells, and the effects in chickens are probably similar. Specifically, the effects include clonal expansion of T cells following activation, growth of B cells, and Ig synthesis in B cells, enhancement of cytolytic capabilities of T cells and NK cells, proliferation and differentiation of macrophage precursors, and maturation of thymocytes (Morsey *et al.*, 1995). Depletion of IL-2 therefore has a direct effect on the host's ability to mount an immune response.

Interleukin-1 is also of crucial importance for normal immune function. In mammals, IL-1 enhances antibacterial resistance to a variety of infectious agents, through its ability to promote production of other cytokines, and also increases production and expression of important receptors. It potentiates T cell activation by inducing IL-2 receptor synthesis by Th cells, and also by increasing IL-2 receptor expression on both CTL and Th cells. Therefore the decrease in IL-1 production in CAV-inoculated chickens is likely to result in a major defect in Th cell activation and ultimately in reduced proliferation of both Th cells and CTLs. The process of activation, division and differentiation of B cells is modulated by IL-1 and B cell growth factors like B cell growth factor-2 (Howard and Paul, 1983; Kishimoto, 1985). These initial signals stimulate the B cell from a quiescent state into the early stage of activation, and increase the expression of lymphokine receptors on the B cell surface (Melchers and Anderson, 1984). The reduction in IL-1 production may therefore be related to the observed adverse effect of CAV on antibody production, in particular the reduction in vaccinal antibody responses (Box *et al.*, 1988; De Boer *et al.*, 1994). However, the dramatic effects of the virus on the production of mature T cells, including Th cells through destruction of precursor cells in the thymus, may be the major reason for the effects on antibody production.

Some reports in the literature have suggested that chickens do not produce a homologue of mammalian IFN-γ (Dijkmans *et al.*, 1990), while others have shown that supernates from mitogen-stimulated chicken splenocytes contain two species of IFN which differ in their acid and heat stability, and therefore probably represent the chicken equivalents of mammalian

α/β (resistant), and γ (labile) IFNs (Bülow *et al.*, 1984; Weiler and Bülow, 1987). Indeed chicken IFN-γ has recently been cloned and expressed (Digby and Lowenthal, 1995). IFN-γ is produced by Th cells of the type recognized in mammals as Th1 cells. They play an important role in recovery from viral infections, not only by inhibiting growth of virus in cells, but also by promoting other activities of T cells, B cells, macrophages, and granulocytes which contribute to the recovery process. The effects of CAV on IFN production probably contribute to the enhancement of virus pathogenicity which has been described when chickens are dually infected with CAV and other viruses such as reoviruses and IBDV (Engstrom *et al.*, 1988; Rosenberger and Cloud, 1989; McNeilly *et al.*, 1995).

As stated above, the major effect of CAV on the cytokines produced by lymphocytes and macrophages probably results from the destruction of the precursors of these cells in the bone marrow and thymus. However, the virus is also capable of infecting and killing mature lymphocytes in the spleen and other secondary lymphoid tissues. While this also undoubtedly results in reduced cytokine production, the total numbers of infected mature cells in these tissues are only a small proportion of the number of immature lymphocytes which are infected and destroyed. Unlike the situation with mature lymphocytes, CAV appears to lack the ability to infect and replicate in macrophages. Despite this, macrophages which come into contact with the virus, or virus products, show reduced functional capability.

As described above, experimental inoculation of CAV into young chickens results in a reproducible series of disease signs, and immunopathological lesions. These features of the experimental disease, plus the ease of accessibility of the cellular elements in the young birds, make the CAV disease model a useful system for studies in many areas of virology and immunology. The experimental system provides the potential for studies of the mechanisms involved in virus-induced immunosuppression, and also for the investigation of how viruses interact with each other and with other pathogens, to overwhelm immune defences, and produce disease. Further investigations with this experimental model should provide useful results, essential for an understanding of the avian immune system where so much basic information is still lacking.

ACKNOWLEDGEMENT

The authors are grateful to the Editors of *Avian Diseases* for permission to use data contained in Figs 21.2–21.5.

References for Part VII

Adair, B.M., McNeilly, F., McConnell, C.D.G., Todd, D., Nelson, R.T. and McNulty, M.S. (1991) Effects of chicken anemia agent on lymphokine production and lymphocyte transformation in experimentally infected chickens. *Avian Diseases* 35, 783–792.

Adair, B.M., McNeilly, F., McConnell, C.D.G. and McNulty, M.S. (1993) Characterisation of surface markers present on cells infected by chicken anemia virus. *Avian Diseases* 37, 943–950.

Amrani, D.L. (1990) Regulation of fibrinogen biosynthesis: glucocorticoid and interleukin-6 control. *Blood Coagulation and Fibrinolysis* 1, 443–446.

Barker, K.A., Hampe, A., Stoeckle, M.Y. and Hanafusa, H. (1993) Transformation-associated cytokine 9E3/CEF4 is chemotactic for chicken peripheral blood mononuclear cells. *Journal of Virology* 67, 3528–3533.

Barrow, P.A. and Bumstead, N. (1996) Salmonellosis – A paradigm of genetic resistance to disease in chickens. *Proceedings of the XXth World Poultry Congress*, World Poultry Science Association (India Branch), New Delhi, India, pp. 609–620.

Bedard, P.-A., Alcorta, D., Simmons, D.L., Luk, K.-C. and Erikson, R.L. (1987) Constitutive expression of a gene encoding a polypeptide homologous to biologically active human platelet protein in Rous sarcoma virus-transformed fibroblasts. *Proceedings of the National Academy of Sciences USA* 84, 6715–6719.

Belardelli, F. (1995) Role of interferons and other cytokines in the regulation of the immune response. *Acta Pathologica et Immunologica Scandinavica* 103, 161–179.

Bounous, D.I., Goodwin, M.A., Brooks, R.L., Jr, Lamichhane, C.M., Campagnoli, R.P., Brown, J. and Snyder, D.B. (1995) Immunosuppression and intracellular calcium signalling in splenocytes from chicks infected with chicken anemia virus, CL-1 isolate. *Avian Diseases* 39, 135–140.

Box, P.G., Holmes, H.C., Bushall, A.C. and Finney, P.M. (1988) Impaired response to killed Newcastle disease vaccine in chickens possessing circulating antibody to chicken anaemia agent. *Avian Pathology* 17, 713–723.

Bülow, V. von, Weiler, V. and Klasen, A. (1984) Activating effects of interferons, lymphokines and viruses on cultured chicken macrophages. *Avian Pathology* 13, 621–637.

Bülow, V. von, Rudolph, R. and Fuchs, B. (1986) Erhöhte Pathogenität des Erregers der aviären infektiösen Anämie bei Hühnerküken (CAA) bei simultaner Infektion mit dem Virus der Marekschen Krankheit (MDV), Bursitisvirus (BDV) oder Reticuloendotheliosevirus (REV). *Journal of Veterinary Medicine* B33, 93–116.

Buscaglia, C. and Calnek, B.W. (1988) Maintenance of Marek's disease herpesvirus latency *in vitro* by a factor found in conditioned medium. *Journal of General Virology* 69, 2809–2818.

Byrnes, S., Eaton, R. and Kogut, M. (1993) *In vitro* interleukin-1 and tumour-necrosis-factor-α production by macrophages from chickens infected with either *Eimeria maxima* or *Eimeria tenella*. *International Journal for Parasitology* 23, 639–645.

Carpenter, C.R., Rubin, A.S. and Bose, H.R. (1978) Suppression of the mitogen-stimulated blastogenic response during reticuloendotheliosis virus-induced tumorigenesis. Investigations into the mechanism of action of the suppressor. *Journal of Immunology* 120, 1313–1320.

Carson, W.E., Giri, J.G., Lindemann, M.J., Linett, M.L., Ahdieh, M., Paxton, R., Anderson, D., Eisenmann, J., Grabstein, K. and Caligiuri, M.A. (1994) Interleukin (IL) 15 is a novel cytokine that activates human natural killer cells via components of the IL-2 receptor. *Journal of Experimental Medicine* 180, 1395–1403.

Chang, E.B., Musch, M.W. and Mayer, L. (1990) Interleukins 1 and 3 stimulate anion secretion in chicken intestine. *Gastroenterology* 98, 1518–1524.

Cloud, S.S., Lillehoj, H.S. and Rosenberger, J.K. (1992) Immune dysfunction following infection with chicken anaemia agent and infectious bursal disease virus I. Kinetic alterations of avian lymphocyte subpopulations. *Veterinary Immunology and Immunopathology* 34, 337–352.

Confer, A.W., Springer, W.T., Shane, S.M. and Donovan, J.F. (1981) Sequential mitogen stimulation of peripheral blood lymphocytes from chickens inoculated with infectious bursal disease virus. *American Journal of Veterinary Research* 45, 2109–2113.

De Boer, G.F., Van Roozelaar, D.J., Moorman, R.J., Jeurissen, S.H.M., Van Der Wijngaard, J.C., Hilbink, F. and Koch, G. (1994) Interaction between chicken anaemia virus and live Newcastle disease vaccine. *Avian Pathology* 23, 263–275.

Digby, M.R. and Lowenthal, J.W. (1995) Cloning and expression of the chicken interferon-γ gene. *Journal of Interferon and Cytokine Research* 15, 939–945.

Dijkmans, R., Creemers, J. and Biliau, A. (1990) Chicken macrophage activation by interferon: do birds lack the molecular homologue of mammalian interferon-γ. *Veterinary Immunology and Immunopathology* 26, 319–332.

Engstrom, B.E., Fossum, O. and Luthman, M. (1988) Blue wing disease of chickens: experimental infection with a Swedish isolate of chicken anaemia agent and avian reovirus. *Avian Pathology* 17, 35–50.

Friedman, R.L. and Stark, G.R. (1985) α-Interferon-induced transcription of HLA and metallothionein genes containing homologous upstream sequences. *Nature* 314, 637–639.

Fulton, R.W., Morton, R.J., Burge, L.J., Short, E.C. and Payton M.E. (1995) Action of quail and chicken interferons on a quail cell line, QT35. *Journal of Interferon and Cytokine Research* 15, 297–300.

Gamero, A.M., Ussery, D., Reintgen, D.S., Puleo, C.A. and Djeu, J.Y. (1995) Interleukin 15 induction of lymphokine-activated killer cell function against autologous tumor cells in melanoma patient lymphocytes by a CD18-dependent perforin-related mechanism. *Cancer Research* 55, 4988–4994.

Gonneville, L., Martins, T.J. and Bedard, P.-A. (1991) Complex expression pattern of the CEF-4 cytokine in transformed and mitogenically stimulated cells. *Oncogene* 6, 1825–1833.

Goryo, M., Suwa, T., Umemura, T., Itakura, C. and Yamashiro, S. (1989) Histopathology of chicks inoculated with chicken anaemia agent (MSB1-TK5803 strain). *Avian Pathology* 18, 73–89.

Grabstein, K.H., Eisenman, J., Shanebeck, K., Rauch, C., Srinivasan, S., Fung, V., Beers, C., Richardson, J., Schoenborn, M.A., Ahdieh, M., Johnson, L., Alderson, M.R., Watson, J.D., Anderson, D.M. and Giri, J.G. (1994) Cloning of a T cell factor that interacts with the β chain of the interleukin 2 receptor. *Science* 264, 965–968.

Graf, T., Weizsäcker, F. von, Grieser, S., Coll, J., Stehelin, D., Patschinsky, T., Bister, K., Bechade, C., Calothy, G. and Leutz, A. (1986) V-mil induces autocrine growth and enhanced tumorigenicity in v-myc transformed avian macrophages. *Cell* 100, 357–364.

Grant, C.E., Vasa, M.Z. and Deeley, R.G. (1995) cIRF-3, a new member of the interferon regulatory factor (IRF) family that is rapidly and transiently induced by dsRNA. *Nucleic Acids Research* 23, 2137–2146.

Guida, S., Heguy, A. and Melli, M. (1992) The chicken IL-1 receptor: differential evolution of the cytoplasmic and extracellular domains. *Gene* 111, 239–243.

Hala, K., Schauenstein, K., Neu, M., Kromer, G., Wolf, H., Bock, G. and Wick, G. (1986) A monoclonal antibody reacting with a membrane determinant expressed on activated chicken T lymphocytes. *European Journal of Immunology* 16, 1331–1336.

Hardiman, J.W. (1996) Broiler breeding by the year 2006. *Proceedings of the XX World Poultry Congress*, World Poultry Science Association (India Branch), New Delhi, India, pp. 461–467.

Hayari, Y., Schauenstein, K. and Globerson, A. (1982) Avian lymphokines, II: interleukin-1 activity in supernatants of stimulated adherent splenocytes of chickens. *Developmental and Comparative Immunology* 6, 785–789.

Heguy, A., Baldari, C.T., Macchia, G., Telford, J.L. and Melli, M. (1992) Amino acids conserved in interleukin-1 receptors (IL-1Rs) and the *Drosophila* Toll protein are essential for IL-1R signal transduction. *Journal of Biological Chemistry* 267, 2605–2609.

Hirai, K. and Calnek, B.W. (1979) *In vitro* replication of infectious bursal disease virus in established lymphoid cell lines and chicken B lympyocytes. *Infection and Immunity* 25, 964–970.
Howard, M. and Paul, W.E. (1983) Regulation of B-cell growth and differentiation by soluble factors. *Annual Review of Immunology* 1, 307–333.
Hu, J., Bumstead, N., Skamene, E., Gros, P. and Malo, D. (1996) Structural organization, sequence, and expression of the chicken NRAMP1 gene encoding the natural resistance-associated macrophage protein I. *DNA and Cell Biology* 15, 113–123.
Isaacs, A. and Lindenmann, J. (1957) Virus interference. I. The interferon. *Proceedings of the Royal Society B* 147, 258–267.
Jakowlew, S.B., Dillard, P.J., Sporn, M.B. and Roberts, A.B. (1988a) Nucleotide sequence of chicken transforming growth factor-β 1 (TGF-b 1). *Nucleic Acids Research* 16, 8730.
Jakowlew, S.B., Dillard, P.J., Kondaiah, P., Sporn, M.B. and Roberts, A.B. (1988b) Complementary deoxyribonucleic acid cloning of a novel transforming growth factor-b messenger ribonucleic acid from chick embryo chondrocytes. *Molecular Endocrinology* 2, 747–755.
Jakowlew, S.B., Dillard, P.J., Sporn, M.B. and Roberts, A.B. (1988c) Complementary deoxyribonucleic acid cloning of a messenger ribonucleic acid encoding transforming growth factor b4 from chicken embryo chondrocytes. *Molecular Endocrinology* 2, 1186–1195.
Jakowlew, S.B., Dillard, P.J., Sporn, M.B. and Roberts, A.B. (1990) Complementary deoxyribonucleic acid cloning of an mRNA encoding transforming growth factor-b2 from chicken embryo chondrocytes. *Growth Factors* 2, 123–133.
Jeurissen, S.H.M., Pol, J.M.A. and De Boer, G.F. (1989) Transient depletion of cortical thymocytes induced by chicken anaemia agent. *Thymus* 14, 115–123.
Jungwirth, C., Rebbert, M., Ozato, K., Degen, H.J., Schultz, U. and Dawid, I.B. (1995) Chicken interferon consensus sequence-binding protein (ICSBP) and interferon regulatory factor (IRF) 1 genes reveal evolutionary conservation in the IRF gene family. *Proceedings of the National Academy of Sciences USA* 92, 3105–3109.
Kaiser, P. (1996) Avian cytokines. In: Davison, T.F., Morris, T.R. and Payne, L.N. (eds), *Poultry Immunology*. Carfax Publishing Company, Abingdon, UK, pp. 83–114.
Kaiser, P., Rothwell, L., Wain, H., Davison, T.F. and Kaufmann, J. (1996) Structure of the chicken interferon-γ gene, and comparison to mammalian homologues. *Journal of Interferon and Cytokine Research* 16, 669.
Karaca, K., Winslow, B., Cochran, M., Reddy, S. and Sharma, J.M. (1995) Immunomodulating effects of avian recombinant cytokines. *Proceedings of the 132nd Annual Meeting of the American Veterinary Medical Association*, Pittsburg, USA, p. 143.
Kishimoto, T. (1985) Factors affecting B cell growth and differentiation. *Annual Review of Immunology* 3, 133–157.
Klasing, K.C. (1994) Avian leukocytic cytokines. *Poultry Science* 73, 1035–1043.
Klasing, K.C. and Peng, R.K. (1987) Influence of cell sources, stimulating agents, and incubation conditions on release of interleukin-1 from chicken macrophages. *Developmental and Comparative Immunology* 11, 385–394.

Klasing, K.C. and Peng, R.K. (1990) Monokine-like activities released from a chicken macrophage line. *Animal Biotechnology* 1, 107–120.

Lampson, G.P., Tytell, A.A., Nemes, M.M. and Hilleman, M.R. (1963) Purification and characterization of chick embryo interferon. *Proceedings of the Society for Experimental Biology and Medicine* 112, 468–478.

Lee, L.F., Sharma, J.M., Nazerian, K. and Witter, R.L. (1978) Suppression of mitogen-induced proliferation of normal spleen cells by macrophages from chickens inoculated with Marek's disease virus. *Journal of Immunology* 120, 1554–1559.

Lee, T. and Tempelis, C.H. (1991) Characterization of a receptor on activated chicken T cells. *Developmental and Comparative Immunology* 15, 329–339.

Lee, T. and Tempelis, C.H. (1992) Possible 110 kDa receptor for interleukin 2 in the chicken. *Developmental and Comparative Immunology* 16, 463–472.

Leutz, A., Beug, H. and Graf, T. (1984) Purification and characterization of cMGF, a novel chicken myelomonocytic growth factor. *European Molecular Biology Organization Journal* 3, 3191–3197.

Leutz, A., Beug, H., Walter, C. and Graf, T. (1988) Hematopoietic growth factor glycosylation: multiple forms of chicken myelomonocytic growth factor. *Journal of Biological Chemistry*, 263, 3905–3911.

Leutz, A., Damm, K., Sterneck, E., Kowenz, E., Ness, S., Frank, R., Gausepohl, H., Pan, Y.-C.E., Smart, J., Hayman, M. and Graf, T. (1989) Molecular cloning of the chicken myelomonocytic growth factor (cMGF) reveals relationship to interleukin 6 and granulocyte colony stimulating factor. *European Molecular Biology Organization Journal* 8, 175–181.

Lew, D.J., Decker, T., Strehlow, I. and Darnell, J.E., Jr (1991) Overlapping elements in the GBP gene promoter mediate transcriptional induction by α and γ interferon. *Molecular and Cellular Biology* 11, 182–191.

Lucio, B., Calnek, B.W., Schat, K.A. and Buscaglia, C. (1992) Comparative susceptibility of avian cell lines to chicken infectious anaemia virus. *Proceedings of the 129th Annual Meeting of the American Veterinary Medical Association*, Boston, 1992.

Lynagh, G. and Kaiser, P. (1996) The role of interleukin-6 in the immune response to coccidial infection. *Proceedings of the International Symposium, Cytokines and the Type I and Type II Paradigm.* Cairns, Australia, p. 29.

Mason, W.S., Seal, G. and Summers, J. (1980) Virus of Pekin ducks with structure and biological relatedness to human hepatitis B virus. *Journal of Virology* 36, 829–836.

McConnell, C.D.G. (1992) Effects of Chicken Anaemia Virus on Cell-mediated Immune Function in Chickens. PhD Thesis, Queen's University, Belfast.

McConnell, C.D.G., Adair, B.M. and McNulty, M.S. (1993a) Effects of chicken anemia virus on macrophage function in chickens. *Avian Diseases* 37, 358–365.

McConnell, C.D.G., Adair, B.M. and McNulty, M.S. (1993b) Effects of chicken anemia virus on cell mediated immune function in chickens exposed to the virus by a natural route. *Avian Diseases* 37, 366–372.

McNeilly, F., Allan, G.M., Moffat, D.A. and McNulty, M.S. (1991) Detection of chicken anaemia agent in chickens by immunofluorescence and immunoperoxidase staining. *Avian Pathology* 20, 125–132.

McNeilly, F., Smyth, J.A., Adair, B.M. and McNulty, M.S. (1995) Synergism between chicken anemia virus (CAV) and avian reovirus following dual infection of 1-day-old chicks by a natural route. *Avian Diseases* 39, 532–537.

McNulty, M.S. (1991) Chicken anaemia agent: a review. *Avian Pathology* 20, 187–203.

McNulty, M.S., McIlroy, S.G., Bruce, D.W. and Todd, D. (1991) Economic effects of subclinical chicken anemia agent in broiler chickens. *Avian Diseases* 35, 263–268.

Melchers, F. and Anderson, J. (1984) B cell activation: three steps and their variations. *Cell* 37, 715–720.

Morsey, M.A., Cox, G.J.M., Van Kessel, A.G., Campos, M. and Babiuk, L.A. (1995) Interleukin-2 In: Myers, M.J. and Murtagh, M.P. (eds), *Cytokines in Animal Health and Disease*. Marcel Dekker, New York, Basel, Hong Kong, pp. 89–120.

Offensperger, W.B., Blum, H.E. and Gerok, W. (1994) Molecular therapeutic strategies in hepatitis B virus infection. *Clinical Investigation* 72, 737–741.

Otaki, Y., Tajima, M., Saito, K. and Nomura, Y. (1988) Immune response of chicks inoculated with chicken anemia agent alone or in combination with Marek's disease virus or turkey herpesvirus. *Japanese Journal of Veterinary Science* 50, 1040–1047.

Qureshi, M.A., Miller, A., Lillehoj, H.S. and Ficken, M.D. (1990) Establishment and characterization of a chicken mononuclear cell line. *Veterinary Immunology and Immunopathology* 26, 237–250.

Reid, L.A., Brasnett, A.H., Gilbert, C.S., Porter, A.C.G., Gewert, D.R., Stark, G.R. and Kerr, I.M. (1989) A single DNA response element can confer inducibility by both α- and γ-interferons. *Proceedings National Academy of Sciences USA* 86, 840–844.

Rosenberger, J.K. and Cloud, S.S. (1989) The effects of age, route of exposure, and coinfection with infectious bursal disease virus on the pathogenicity and transmissibility of chicken anemia agent (CAA). *Avian Diseases* 33, 753–759.

Samad, F., Bergtrom, G., Eissa, H. and Amrani, D.L. (1993) Stimulation of chick hepatocyte fibronectin production by fibroblast-conditioned medium is due to interleukin-6. *Biochimica et Biophysica Acta* 1181, 207–213.

Sarma, G., Greer, W., Gildersleeve, R.P., Murray, D.L. and Miles, A.M. (1995) Field safety and efficacy of *in ovo* administration of HVT + SB-1 bivalent Marek's disease vaccine in commercial broilers. *Avian Diseases* 39, 211–217.

Schauenstein, K., Kromer, G., Hala, K., Bock, G. and Wick, G. (1988) Chicken-activated-T-lymphocyte-antigen (CATLA) recognised by monoclonal antibody INN-CH 16 represents the IL-2 receptor. *Developmental and Comparative Immunology* 11, 69–77.

Schindler, C., Shuai, K., Prezioso, V.R. and Darnell, J.E., Jr (1992) Interferon-dependent tyrosine phosphorylation of a latent cytoplasmic transcription factor. *Science* 257, 809–815.

Schultz, U., Kaspers, B., Rinderle, C., Sekellick, M.J., Marcus, P.I. and Staeheli, P. (1995a) Recombinant chicken interferon: a potent antiviral agent that lacks intrinsic macrophage activating factor activity. *European Journal of Immunology* 25, 847–851.

Schultz, U., Köck, J., Schlicht, H.-J. and Staeheli, P. (1995b) Recombinant duck interferon: a new reagent for studying the mode of interferon action against hepatitis B virus. *Virology* 212, 641–649.

Schumacher, B., Bernasconi, D., Schultz, U. and Staeheli, P. (1994) The chicken Mx promoter contains an ISRE motif and confers interferon inducibility to a reporter gene in chick and monkey cells. *Virology* 203, 144–148.

Schwemmle, M., Kaspers, B., Irion, A., Staeheli, P. and Schultz, U. (1996) Chicken guanylate-binding protein. Conservation of GTPase activity and induction by cytokines. *Journal of Biological Chemistry* 271, 10304–10308.

Sekellick, M.J., Ferrandino, A.F., Hopkins, D.A. and Marcus P.I. (1994) Chicken interferon gene: cloning, expression and analysis. *Journal of Interferon Research* 14, 71–79.

Sharma, J.M., Karaca, K., Erickson, S., Winslow, B., Junker, D.E. and McMillen J.K. (1996) Effect of cytokines on the immune system following *in ovo* administration. *Journal of the American Medical Association* 209, 398.

Shuai, K. (1994) Interferon-activated signal transduction to the nucleus. *Current Opinion in Cell Biology* 6, 253–259.

Shuai, K., Schindler, C., Prezioso, V.R. and Darnell, J.E., Jr (1992) Activation of transcription by IFN-γ. Tyrosine phosphorylation of a 91 kD DNA binding protein. *Science* 259, 1808–1812.

Siatskas, C., McWaters, P.G., Digby, M., Lowenthal, J.W. and Boyd, R.L. (1996) *In vitro* characterization of a novel avian haemopoietic growth factor derived from stromal cells. *Developmental and Comparative Immunology* 20, 139–156.

Sick, C., Schultz, U. and Staeheli, P. (1996) A family of genes coding for two serologically distinct chicken interferons. *The Journal of Biological Chemistry* 271, 7635–7639.

Smyth, J.A., Moffat, D.A., McNulty, M.S., Todd, D. and Mackie, D.A. (1993) A sequential histopathologic and immunocytochemical study of chicken anemia virus infection at one day of age. *Avian Diseases* 37, 324–338.

Sterneck, E., Blattner, C., Graf, T. and Leutz, A. (1992) Structure of the chicken myelomonocytic growth factor gene and specific activation of its promoter in avian myelomonocytic cells by protein kinases. *Molecular and Cellular Biology* 12, 1728–1735.

Stoeckle, M.Y. and Barker, K.A. (1990) Two burgeoning families of platelet factor 4-related proteins: mediators of inflammatory responses. *New Biologist* 2, 313–323.

Sugano, S., Stoeckle, M.Y. and Hanafusa, H. (1987) Transformation by Rous sarcoma virus induces a novel gene with homology to a mitogenic platelet protein. *Cell* 49, 321–328.

Sundick, R.S. and Gill-Dixon, C. (1997) A cloned chicken lymphokine homologous to both mammalian IL-2 and IL-15. *Journal of Immunology* 159, 720–725.

Sundick, R.S. and Jones, L.A. (1996) The sole T cell stimulatory cytokine cloned from a mitogen-activated chicken spleen cDNA library is homologous to mammalian IL-15, and not IL-2. *FASEB Journal* 10, A1057.

Suresh, M., Karaca, K., Foster, D. and Sharma, J.M. (1995) Molecular and functional characterization of turkey interferon. *Journal of Virology* 69, 8159–8163.

Taniguchi, T., Yuasa, N., Maeda, M. and Horiuchi, H. (1983) Chronological observations on hemato-pathological changes in chicks inoculated with chicken anemia agent. *National Institute of Animal Health Quarterly* 23, 1–12.

Toivanen, P. and Toivanen, A. (1973) Selective activation of chicken T lymphocytes by concanavalin A. *Journal of Immunology* 111, 1602–1610.

Vielitz, E. and Landgraf, H. (1988) Anaemia-dermatitis of broilers: field observations on its occurrence, transmission and prevention. *Avian Pathology* 17, 113–120.

Volpini, L.M., Calnek, B.W., Sekellick, M.J. and Marcus, P.I. (1995) Stages of Marek's disease virus latency defined by variable sensitivity to interferon modulation of viral antigen expression. *Veterinary Microbiology* 47, 99–109.

Volpini, L.M., Calnek, B.W., Sneath, B., Sekellick, M.J. and Marcus, P.I. (1996) Interferon modulation of Marek's disease virus genome expression in chicken cell lines. *Avian Diseases* 40, 78–87.

Weiler, H. and Bülow, V. von (1987) Detection of different macrophage activating factor and interferon activities in supernatants of chicken lymphocyte cultures. *Avian Pathology* 16, 439–452.

York, J.J., Strom, D.G., Connick, T.E., McWaters, P.G., Boyle, D.B. and Lowenthal, J.W. (1996) *In vivo* effects of chicken myelomonocytic growth factor. Delivery via a viral vector. *Journal of Immunology* 156, 2991–2997.

Yuasa, N., Taniguchi, T. and Yoshida, I. (1979) Isolation and some characteristics of an agent inducing anemia in chicks. *Avian Diseases* 23, 366–385.

Zhou, J.H., Ohtaki, M. and Sakurai, M. (1993) Sequence of a cDNA encoding chicken stem cell factor. *Gene* 127, 269–270.

Zhou, J.H., Ohtaki, M., Inumaru, S., Kitani, H. and Sakurai, M. (1996) Purification and activity of recombinant chicken stem cell factor produced by using a baculovirus vector. *Journal of Veterinary Medical Science* 58, 173–175.

Zöller, B., Ozato, K., Kroemer, G., Auffray, C. and Jungwirth, C. (1992) Interferon induction of chicken MHC class I gene expression: phylogenetic conservation of the interferon-responsive element. *Virology* 191, 141–149.

Index